Encyclopedia of Scanning Electron Microscopy

Edited by **Lisa Page**

NY RESEARCH
P R E S S

New York

Published by NY Research Press,
23 West, 55th Street, Suite 816,
New York, NY 10019, USA
www.nyresearchpress.com

Encyclopedia of Scanning Electron Microscopy
Edited by Lisa Page

International Standard Book Number: 978-1-63238-166-8 (Hardback)

Printed in the United States of America.

Contents

Preface

This book focuses on various issues concerned with scanning electron microscopy, as well as its theoretical and practical applications. Fine focused electron and ion beams constitute(s) an inevitable part of methods and instruments employed in various science fields. SEMs are well instrumented and supplemented with advanced techniques and methods and thereby present endless possibilities in the areas of quantitative measurement of object topologies, surface imaging, performing elemental analysis and local electrophysical characteristics of semiconductor structures. Creation of micro and nanostructures involves extensive use of fine focused e-beam. Numerous topics are covered under two sections "Instrumentation, Methodology" and "Biology, Medicine" for electronic industry. This book includes contributions by renowned researchers and experts in this field.

Significant researches are present in this book. Intensive efforts have been employed by authors to make this book an outstanding discourse. This book contains the enlightening chapters which have been written on the basis of significant researches done by the experts.

Finally, I would also like to thank all the members involved in this book for being a team and meeting all the deadlines for the submission of their respective works. I would also like to thank my friends and family for being supportive in my efforts.

Editor

Part 1

Instrumentation, Methodology

Interactions, Imaging and Spectra in SEM

Rahul Mehta

University of Central Arkansas,
USA

1. Introduction

In microscopy the question arises- Why employ electron beams instead of light beams to produce magnified images and the answer has to do with resolution. When doing microscopy to produce magnified image of objects, diffraction (bending of waves around narrow openings and obstacles) limits the resolution and hence the quality of image in terms of fine details one can see. The optical wavelengths from deep UV to IR are in range of hundreds of nanometers while electron beam of energy in keV have wavelengths in fractions of nanometers. The dependence of diffraction on the wavelength of the beam makes electron beam more suitable than beams of wavelengths in the optical region. The diffraction also depends on the size of the objects. A Scanning Electron Microscope (SEM) with electron beams in the keV range allows one to produce image (Fig. 1) of objects in the micro to nanometer range with relatively lower diffraction effects. Using a SEM to produce proper image requires a judicious choice of beam energy, intensity, width and proper preparation of the sample being studied. The electron beam in a SEM is nowadays generated using a field emission filament that uses ideas of quantum tunneling. Other methods are also available. The deflection of electron beam of certain energy E is accomplished by means of electromagnetic lenses. Typical E values for conventional SEM can range from as low as 2-5 keV to 20-40 keV.

A basic SEM consists of an electron gun (field emission type or others) that produces the electron beams, electromagnetic optics guide the beam and focus it. The detectors collect the electrons that come from the sample (either direct scattering or emitted from the sample) and the energy of the detected electron together with their intensity (number density) and location of emission is used to put together image. Present day SEM also offer energy dispersive photon detectors that provide analysis of x-rays that are emitted from the specimen due to the interactions of incident electrons with the atoms of the sample.

2. Interaction

Assume that an electron beam of energy E, with a circular cross-section A and a beam current I is incident on a sample with atomic number Z. We will assume that the energy E is typically much less than 100 keV in the following discussions. As the electron beam enters the sample it interacts with the atoms of the samples. This interaction of the electrons is not confided to the surface layers only but also with the atoms and molecules inside. The electron interaction with the atom consists of coulomb attraction with the nuclear positive charge. The interaction of the electron beam with the electrons from the sample is of repulsive nature as the electrons are deflected by the target electrons. The electrons can undergo change in momentum and/or change in energy or both in these interactions. So an entering electron beam can scatter elastically and/or inelastically.

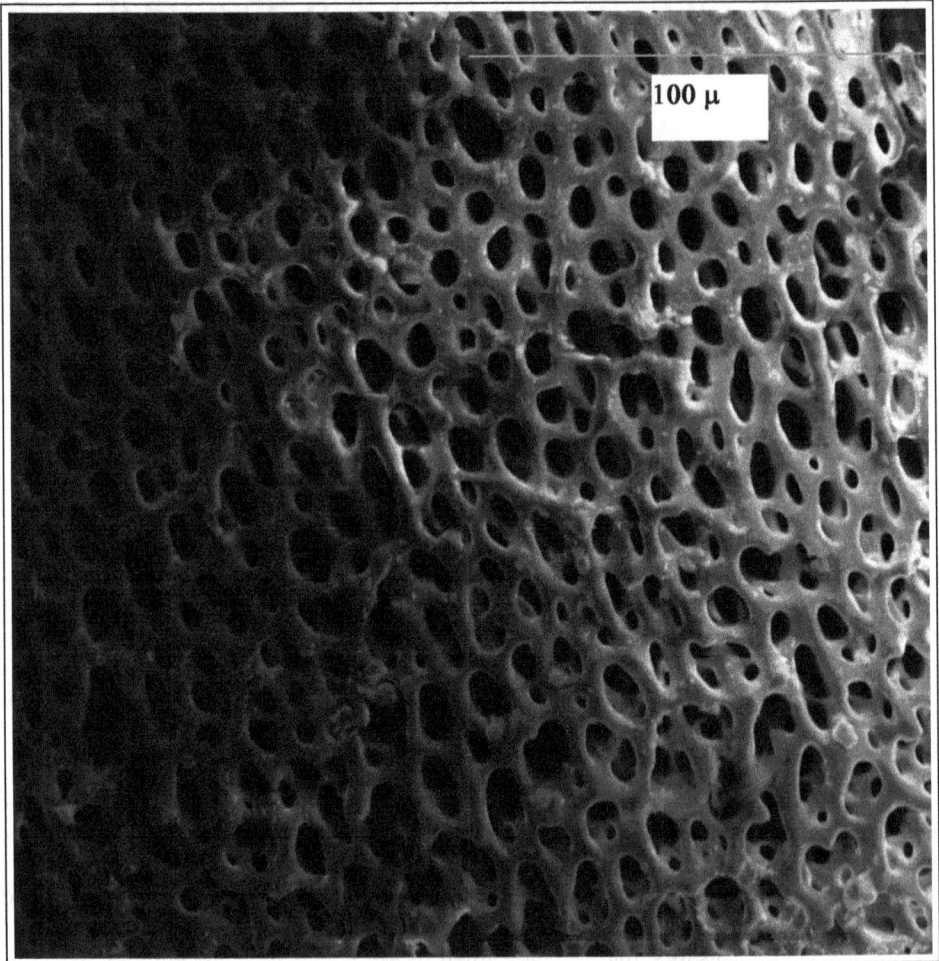

Fig. 1. Biological sample showing kT pores imaged with 20 keV electron beam using a quad backscattered detector. Scale shown by line of 100 μm.

2.1 Elastic scattering

If the scattering involves no loss of energy it is Rutherford scattering (Rutherford, 1911,1914) which is peaked in the forward direction with the probability of scattering decreasing dramatically with increase of angle of scattering and the electron trajectory is modified from some small angle elastic scattering to large angle deviation. Some of the electrons can travel laterally while others can even back scatter. After many of these events it is possible for some of the electrons to leave the sample and these backscattered electrons provide one way of imaging the sample. Probability of elastic scattering depends on inverse square of energy E which means a higher energy beam will start to spread out much later in its path than a smaller energy beam. An electron can transfer energy to the conduction electrons or to a single valence electron – but this will not be important in SEM imaging as the mean free paths for both of these is large, the scattering angles are small and energy loss less than an eV.

2.2 Inelastic scattering

An electron can interact with the solid as a whole generating vibrations (phonon scattering). The energy of the electron goes into overall heating of the solid slightly. The overall energy loss is less than 1 eV and this channel is probably more important near the end of the path of the electron. The scattering results in electron being scattered by larger angle. This effect will be important for image resolution and contrast. The energy loss from inelastic scattering is related inversely with E therefore a higher energy incident electron will keep more of its energy at a depth than a lower energy incident electron at the same depth. If the scattering involves loss of energy then it cannot be described by Rutherford formula. There are many channels by which an electron can lose energy in a sample but here we will look at some that are more pertinent for SEM imaging.

The channels that are useful for imaging are the ones that results in radiative or non-radiative transitions to occur in the sample atom. This is when the electron transfers energy to one of the inner shell electrons and then this result in ionization or electronic rearrangement. The atom that absorbs the energy this way will either give out a photon (radiative process) or eject an electron from same or different shell (Auger process- non-radiative). The radiative photon is generally in the x-ray region of electromagnetic spectrum. The probability of radiative versus non-radiative process taking place defines the fluorescence yield ω. In energy dispersive analysis of a sample using SEM- ω plays an important role in conversion of x-ray intensities (from x-ray spectrum) into absolute numbers. These absolute numbers are related to sample elemental thicknesses and overall compositions.

2.3 Energy loss

The energy loss of the electron in scattering is dependent in a complex way on the atomic number Z of the sample atom, on their mass number A and the density ρ of those atoms. The energy lost by the electron can be transferred to the sample atoms in inelastic scattering. The rate of energy loss with the path length x, dE/dx, was described by Han Bethe (Bethe, 1930) mathematically. Calculations based on this formula suggest that dE/dx increases with Z while increasing E lowers this rate. The dependence on E is much more dramatic than with Z. Monte-Carlo type simulations (Metropolis & Ulam,1949; Newberrry & Myklebust, 1979; Rubinstein & Kroese, 2007) of trajectories of electrons (as they interact with the sample) suggest visualization in terms of an interaction volume. The size and depth of the volume is dependent on energy of the electron beam, their number density and the details about the interacting atoms. volume The probabilities of the electron interactions drops off by a large factor outside this volume.

2.4 Radiative and non-radiative mechanisms

The interaction between the incident electrons and the sample target atoms provides rich information about the chemical environment of the target atoms. This information is in the form of radiative and non-radiative transitions and subsequent emissions that take place in the atoms. The ion-atom collision results in transitions that involve energy transfer through the mechanisms (both radiative and non-radiative type). The radiative transitions in the atoms can lead to emission of photons mainly in the form of x-rays from K, L, M- shells. These x-rays are characteristics of the elements they come from and the x-ray spectra has signature to that effect. Recognizing these x-rays and then measuring them provides relative abundance of elements in the sample. To get an absolute value (e.g. # of atoms of one type as a fraction of all atoms) generally specified as parts per million (ppm)) normalization of the emission yields has to be done. This requires measuring the emission yields from the sample and from a standard sample under identical conditions so that ratios can be formed.. The standard must have been measured independently and sometimes with a different spectroscopic method (e.g. mass spectroscopy or infrared spectroscopy) and for it ppm needs to be available. The non-radiative transitions can result in emission of Auger electrons and Auger spectroscopy can provide information about the intensities there. Normally the standard SEM may not have capability of differentiating and measuring the auger electrons. What is done in that case is to use the value of fluorescence yield ω (which relates the radiative yilds to non-radiative yields) and determine fraction of time an energy transfer to an atom will result in some form of radiative emission. The fluorescence yield then allows one to convert cross section for ionization into cross section for production of x-rays. The fluorescence yield factor F which is related to the ratio of radiative to non-radiative transition has to be carefully used or determined in the normalization procedure and plays a role in correction factors to get the absolute numbers. The correction factors take into account the fact that ratios of intensities are substantially different than the ratio of concentrations of elements in a sample. The atomic number Z and the mass absorption of x-rays in the volume of the sample A are the other two effects that go into the ZAF correction factor and they will be discussed in more detail later on..

2.5 Imaging

In usage the electron beam is incident on a target region from the specimen sample. The energy of the electron E, the mass density of the target, and the atomic number Z of the sample determines the relative intensities of various types of electron scattering. The penetration depth of the electrons, the mean free path and the strengths of different scattering (which are also dependent on both the Z and E) play a role in the information one gets (in the form of images) about the sample. Primarily the back scattering electrons provide an electronic signal that delineates the interaction volume and carries details about the scattering. In addition the information about the specimen is also comes from the production of secondary electrons from the sample.

3. X-ray imaging, analysis and other techniques

3.1 Elemental profile using SEM

Before one can do spectroscopy using a SEM, the sample has to be prepared correctly, mounted on special sample holders and oriented properly. Metallic stubs with sticky carbon surface allows one to present the sample in a particular orientation to the beam. Samples that are placed on a goniometer can even be rotated to image the sample from a different direction. In a typical preparation of samples for SEM analysis: the sample has to be cleaned to remove contamination, dried in most cases and the surface to be analyzed prepared so that the analyzed surface is flat and electrically conducting. The cleaning starts with sample placed in ethanol baths. Part of this fixes the sample and also replaces the water content. For a biological sample -like a bone -first the bone has to be cleaned of most of soft tissues and then the remaining soft tissues are removed by placing the sample with dermestid beetles. The sample is observed under light microscope and if needed other techniques are used to remove any more soft tissue in the area of interest. More ethanol baths for different lengths of time and different concentration of ethanol may have be used. Cleaned samples are sectioned using high speed Dremil and other cutting tools. The surface to be analyzed has to be flat, smooth as possible and without any intruding parts in front of them. The samples are dried using the critical point dryer, if needed, and then sputter coated with Au to make them electrically conductive. For electron beam to be incident on the sample normally, the sample is placed on the mounting stub (with a sticky carbon tape exposed in the normal direction). The prepared flat cross section needs to be positioned correctly on the metal stub. This then ensures the proper orientation of the sample in the beam. The conductive gold layer allows the electrons a path to the local ground – absence of which will result in area of the sample acting as non-conductors (insulator). Electron beam incident, on the non-conductive area, will result in electrical charge getting collected. When seen in the SEM image, the area that is non-conducting will show up as whitish region with very less details to be seen. Over time the whitish area will get brighter losing even more details and also may grow in size (Figure 2). A layer of conducting metal like gold (few atom layer thick) will be sufficient to alleviate this charge clumping and in the SEM image the whitish appearance will disappear. If the image continue to show incomplete charge conduction from an area then a second layer of gold can help to minimize the charge clumping. In extreme cases, one has to use a lower energy and intense electron beam. One of the affect of an extra layer on a sample is to mask some of the features that are being imaged. Other difficulty that arises from a thicker coating of metal is x-ray interference. The metal coating (e.g. gold) emits characteristic x-rays from that metal. These x-rays can overlap partially the x-ray spectra coming from the sample being studied.

Samples that are to be studied in their original conditions have to be handled differently. Some of these are wet samples. Other samples that are not fixed and non- conductive create imaging problems that are tackled differently. These samples generally outgas in vacuum of the SEM chamber and have to be studied in a mode in SEM that allows for differential pumping in different sections of the SEM. For these samples high vacuum (like $\sim 10^{-6}$ Torr) cannot be achieved and so resolution is not as good and images are not as crystallized as a dry sample will do. But the SEM images will still provide details that are useful for the researcher.

Once the sample is placed in the SEM chamber and the detector is chosen (between secondary electron detector and/or backscattered electron detector) image is generated. The image details including the resolution are dependent on the energy of the electron beam type of sample, its geometry and atomic numbers of the atoms present. When the image shows the proper details and is magnified correctly one can open the energy dispersive system to do x-ray spectroscopy. The Energy Dispersive Analysis (EDS) mode of the SEM provides the x-ray spectra for elemental analysis. In order to quantify the elemental yield one needs standard samples. For example in the study of bones, standards representing Calcium Phosphate, are used. Also to get a good calibration of the detector's response in the energy region being studied, other standard elemental samples are employed. For example a pure copper sample has L-shell x-rays around 1 keV and K-shell x-rays around 9 keV. A pure gold or lead sample will give M-shell x-rays in 2-3 keV range and L-shell x-rays around 10 keV. It is essential that the range of x-ray energies being studied be understood in terms of the response of the detector. This response also needs to be established for the range of electron beam energies to be used. The x-ray spectra from standards and from the samples are analyzed using software that is specially developed for analysis needed with corrections built in for various effects that may be important at some energies and not at others. FLAME (fuzzy logic software for spectral analysis and elemental ratio determination) is one of those software. The software, with statistical capabilities provides identification of the elements, atomic and weight percent of elements, intensities of the x-rays and other parameters that are electron beam and elemental atomic number dependent. The software generates a table showing the elemental ratios (weight and atomic) among the elements detected: e.g. oxygen, phosphorus, and calcium in the bone samples.

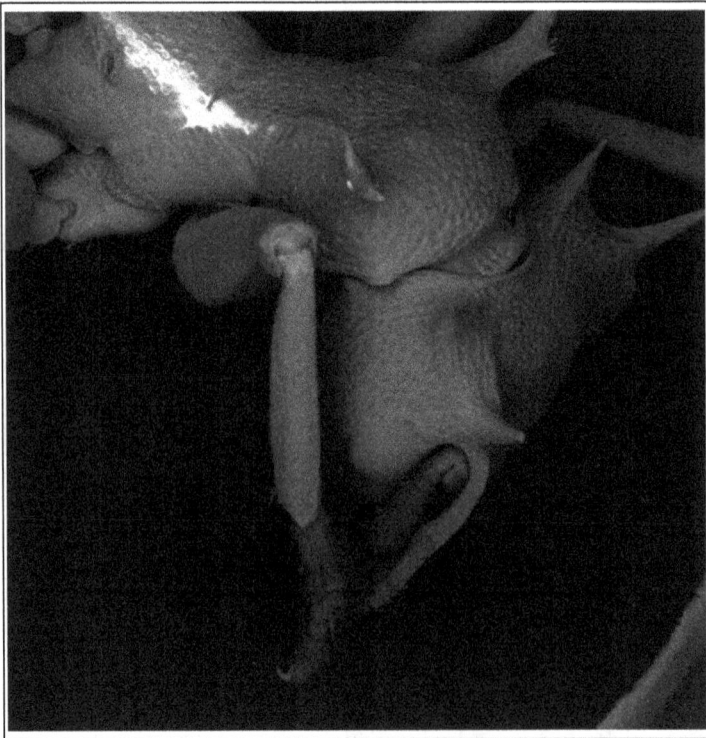

Fig. 2. SEM image of a biological sample(cephalotes) using quad backscattered detector. The sample was not sputter coated resulting in excessive charging(white area) on the sample.

3.2 ZAF correction factors

Castaing (Castaing,1951,1966; Castaing & Henoc, 1966) showed that the k-ratio, which is the ratio of sample x-ray intensity to standard sample x-ray intensities, is proportional to the ratio of the mass fraction of the sample element to that for the standard sample. But experimentation has shown that there are deviation of this k-ratio from the actual concentration ratios. These differences arise from many parameters of the sample but mainly density, electron backscattering, x-ray ionization and production cross section (these are connected by the fluorescene yield) , energy loss of the electron beam and the absorption in the sample matrix. In samples that contain many elements and the mixture is not very homogeneous the measured intensity may vary by a large factor on variation in elastic , inelastic scatterings, and the absorption of the x-rays though the elements of the sample before reaching the detector. In general these various effects coming from the sample matrix on the measured intensity can be lumped into correction due to atomic number (the Z-effect), the absorption of the x-rays in the sample (the A effect) and the F effect due to x-ray fluorescence yield. In total the correction is called ZAF factor and in a simplified equation it is given by eq. (1) as

$$C_i / C_{(i-std)} = \{ZAF\}_i \cdot \left(I_i \middle| I_{(i-std)} \right) \tag{1}$$

where C_i and $C_{(i-std)}$ are the fractional sample weight of element i and for the same element in the standard sample. Here $\left(I_i \middle| I_{(i-std)} \right)$ are the intensities as measured for the same element in sample and in the standard sample. In order to understand the Z,A and F factors, one has to visually assimilate the various processes taking place as an electron beam traverses the sample, loses energy by scattering processes and excitation of the host atoms of the sample takes place.

Z-factor: When an electron beam is backscattered, the backscattering mechanism removes part of beam of electrons which then reduces the number of interactions that can lead to ionization and production of x-rays. In samples with many elements the kinematics of scattering results in greater spread of the beam . The scattering results in greater spread in the energy for the scattered electron. Kinematics suggests that a greater number of electrons backscatter when atomic number Z is greater. The higher Z elements then remove a larger fraction of electron energies. The energy loss from inelastic scattering tends to remove electron energy due to thickness (defined as a product of the thickness as measured along the path and the density). The low atomic number remove this energy at a higher rate than higher atomic number. A Monte Carlo simulation of the trajectory of electron suggests that as the electron traverses a sample it is losing energy. The ionization of an atom and subsequent production of x-rays is critically dependent on if the energy available is above the excitation energy for the particular atom. So the energy may be enough to excite L-shell x-rays but not excite higher K-shell x-rays or in the heavy elements like gold the energy may excite M-shell x-rays but not L-shell x-rays and definitely not K-shell x-rays. During elastic scattering, the kinetic energy conservation tends to scatter electrons at larger angles and hence deviate from its path more. These scattered electron would be less likely to produce ionization and x-rays then if it did not interact elastically. Thus the distribution of the electron in the sample, their energies at a point in the sample and the x-ray production depends strongly on the atomic number of sample atoms. This distribution can be defined in terms of a function φ (ρZ). An area under the plot of this function φ (ρZ). versus ρZ allows one to integrate for the intensities that would be generated. The atomic number effect (the Z-factor) for each element is then the ratio of this function φ (ρZ) for the sample versus for the standard sample.

A-Factor: Inner shell ionization followed by x-ray production takes place over a range of thickness in the sample. The volume from which x-rays come from increases with energy of the incident electrons and scattered electrons can come from deeper region and overall a larger volume. Ionization followed by a radiative transfer of energy leads to the production of x-ray. The x-rays on their way to the detector gets absorbed by the matter they have to pass through. This absorption can be defined in terms of an exponential function. This exponential decrease is given as eq. (2)

$$I = I_0 \, e^{-\mu \rho t} \tag{2}$$

where I and I_0 are the intensity of the x-ray at the detector versus intensity when produced, μ is the mass absorption coefficient, ρ is the density of matter the x-ray passes through and t is the path length of this matter layer and ρt gives the thickness in units of mass per unit area. The exponential term representing the fraction by which incident intensity is reduced is calculated for each of the layers the x-rays have to pass through. The direction in which a generated x-ray has to travel to get to the detector defines the path length. This is related to the takeoff angle, the angle between the incident electron beam and the direction of the x-rays. The incident energy of the electron beam and the takeoff angle can affect the fraction absorption by a large factor. X-ray absorption factor A generally is the largest factor in the ZAF factor. Again the plot of φ(ρt) versus with ρt is used to determine the A-factor from difference in area under the curves of φ for generated x-rays and for emitted one. The emitted x-ray intensity contains the absorption effect using the exponential law.

F-factor: In addition to x-rays being produced following ionization of the atoms by the electron beam, the x-rays themselves can fluoresce more x-rays from the atoms of the sample they pass near. The x-rays fluoresced have energies less than the energy of the x-ray (E_0) that fluoresced them. This has to do with the threshold excitation energy E_c needed for fluorescence. The fluorescing becomes negligible if E_0 is greater than E_c by 5 keV or more.

3.3 Comparative techniques

The x-ray spectra obtained from an SEM is analyzed with special software to determine the yield of x-rays. The spectra is generally shown as intensity versus the energy of the x-rays (Figure 3 and 4). The detector normally ised in a SEM is a (Si(Li) detector with a resolution of about 140 eV at 5.9 keV for $_{54}$Mn x-rays. This resolution is enough to resolve x-rays from adjacent elements and also can differentiate some of the individual transitions within the x-rays from the same element. Si(Li) detectors uses a Silicon crystal which is Lithium doped (has to be cooled below liquid nitrogen temperatures for it to work). The response of the crystal to photons in the 1- 100 keV region is generally depicted with an efficiency curve. This curve shows the percent detection of the photons arriving in the active region of the detector. Other than the geometry of the detection system, a typical efficiency may be 1 out of 10000 (or 1% or less). The physical region between location where x-ray photons are generated and their passage through the in-between matter before reaching the active silicon region of the detector determines the attenuation fraction of the original x-ray signal. In a typical SEM, this attenuation takes place in the layers of air (in the high vacuum chamber), beryllium window layer as the front window of the detector, the gold contact layer and the dead layer of silicon. This absorption and attenuation depends on the energy of the x-ray photon and also the thickness of each layer. For energies above 3-4 keV, the efficiency is smoothly varying (fairly constant in the 5-20 keV range). There are many calibrated photon sources available to measure the efficiency in this region. Experimentally measured efficiencies, together with that predicted and calculated from models are compared. The calculated efficiency includes the attenuation of photon intensities in the layers described above. measured and calculated efficiencies are normalized to each other using the measured energy point (Gallagher & Cipolla,1974; Lennard & Phillips, 1979; Papp, 2005; Maxwell & Campbell, 2005, Mehta etal., 2005)). This procedure results in normalized efficiency curves. The efficiency in the 5-20 keV region can be determined to uncertainties of few percent but for energies of x-rays in the 1-3 keV efficiency is lot more uncertain especially below 1 keV and there lies the problem.

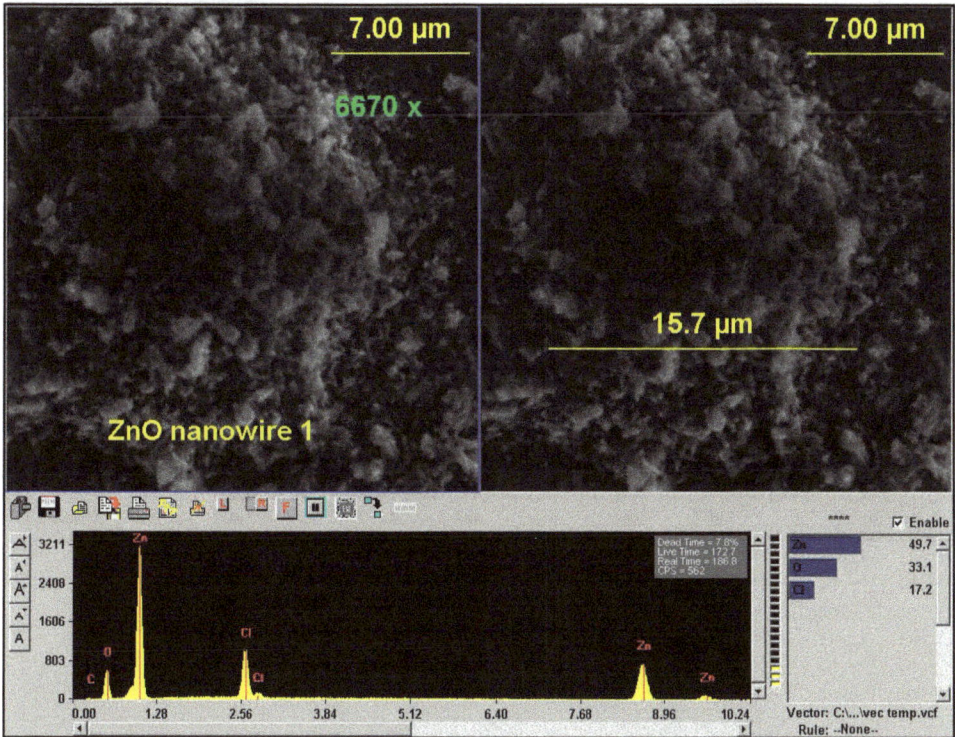

Fig. 3. SEM Image (magnification x6670 and scale as shown) and x-ray spectrum showing L-shell (~1 keV) and K-shell (~ 9 keV) x-rays from zinc in a zinc oxide Nanowire. Also chlorine K_α and K_β can be seen as just resolved. The K-shell x-rays of zinc clearly show separated K_α and K_β peak with a peak intensity ratio of 4:1. Right side table show relative percentages of the elements in the sample (not corrected with k-ratio).

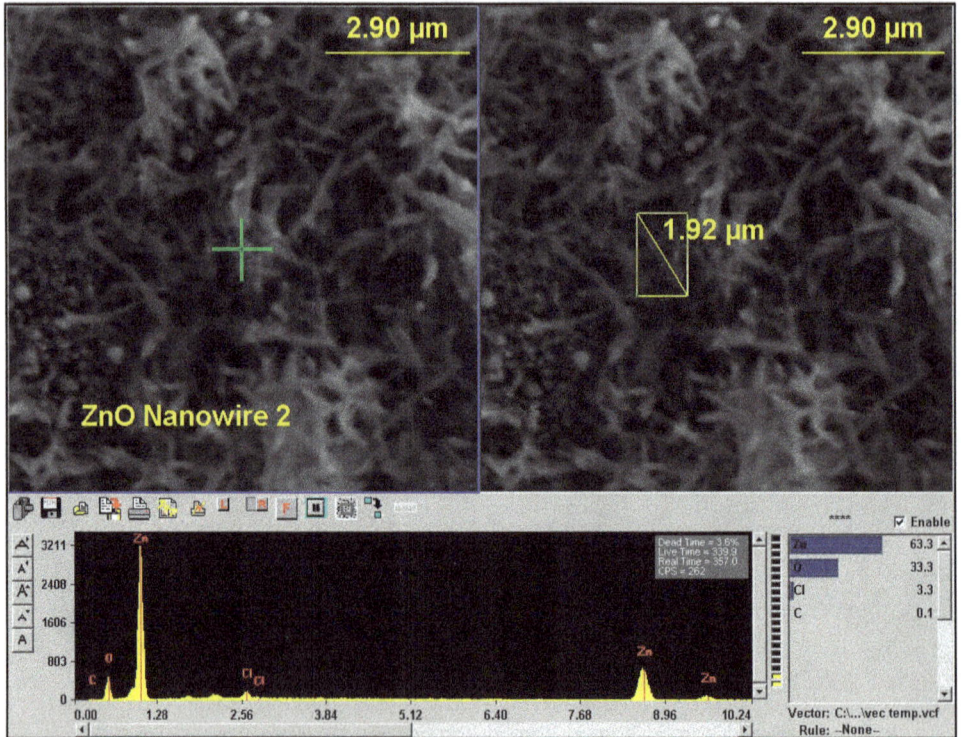

Fig. 4. SEM Image(from a box < 2 μm on the side) and x-ray spectrum showing L-shell (~1 keV) and K-shell (~ 9 keV) x-rays from zinc in a zinc oxide nanowire. The k-shell x-rays clearly show separated K_α and K_β peak with a ratio of 4:1. The image clearly shows the wires of ZnO.

The x-rays from K-shell of carbon, oxygen, up to sodium are all ~1 keV or less. L-shell x-rays below 1 keV come from elements Calcium(Z=20) through Zinc (Z=30) while M_shell x-rays are all less than 3.5 keV (highest M-shell x-rays for Uranium Z=92). For lanthanum (Z=57) the M-shell x-rays are less than 1 keV. The x-rays generated in an SEM are limited by the maximum energy the electrons can have. For a typical SEM that has a maximum voltage available for accelerating of say 20 kV – the electron beam has maximum possible energy of 20 keV. The x-rays generated from samples by such beams can then only be up to 20 keV. So depending upon the elements present in the sample, the x-ray data can give yields that are uncertain by above uncertainties. Yields can be converted to absolute numbers if the number of electrons involved in the generation of x-rays can be determined and standard samples for the elements are available. This leads to the realization that any absolute numbers have to be checked against absolute numbers from other comparable technique. Any normalization procedure among the techniques have to find a unique common point.

3.3.1 X_ray fluorescence (XRF)

For large Z elements (Z> 45) XRF (Bundle et al., 1992) can provide information about x-rays greater than 20 keV that the SEM cannot. XRF is used in that situation and again normalize K-shell x-ray production using XRF with L- or M-shell x-ray production by the electron beam of an SEM. Some of the analyzed samples are fluoresced using radioactive sources of Fe-55, Cm-244 and Am-241 in the XRF. EDS analysis from SEM is energy limited by the electron beam energy used, while XRF is not. XRF spectra is measured to provide x-ray measurements that are outside of the energy range of the SEM measurements. In addition, the lower energy L and M-shell x-rays are measured to provide another set of elemental ratio data. This allows for comparison between elemental ratios determined using SEM and XRF.

3.3.2 Neutron activation analysis

A standard neutron source (Pu-Be in a Howitzer or a neutron generator) can be used to do neutron activation work. The energy of the neutron beam and the flux coming from the source may determine if this technique can allow one to analyze a sample also analyzed with SEM. The incident energy of the neutrons from the source will determine if neutron-atom interaction can lead to compound nucleus formation. In order to see any particular decay mode from this compound nucleus, there has to be appropriate isotopes formed with half lives of transitions in that isotope suitable for decay measurement. Also the yield of these newly made isotopes will depend upon the cross section for absorption of the neutrons in the sample. In order to do neutron activation analysis (NAA), the table of isotopes is used to determine the isotopes that can be produced in activation of the samples. The suitability of the radiation these isotopes produce for analysis has to be established too. Once this is established the uncoated samples are prepared for neutron activation and activated for an optimum length of time. The activated samples are analyzed using gamma ray spectroscopy using a combination of Geiger counter , Sodium iodide detector and/or germanium type high resolution gamma detectors. Intensities of photo peaks can be used to form ratios in a particular photon energy range. This divides out any effect due to efficiency variation. Next taking into account other parameters (like neutron cross section, atomic number, branching ratio etcetera) and comparing the ratios of intensities from a standard sample and from the measured sample, a normalized absolute intensities can be determined. For example standard samples can be used to provide a baseline for radioactivity measurements and dose dependent measurement of other standards to be used. This baseline can provide a scheme for normalizing the intensities from different samples. Comparison among elemental ratios determined using SEM, XRF and NAA is possible then.

3.3.3 Other comparative methods

Another technique that provides absolute weight and atomic percent of the elements in the samples is Particle Induced X-ray Emission (PIXE) (Flewitt & Wild, 2003). This is performed at an accelerator lab facility. PIXE analysis at an accelerator lab can be used to study biological samples using microprobe beam. The samples and standards are mounted as targets on special sample holders. Proton or alpha particle beams interacting with the targets provides an absolute value for weight percent and atomic percent of the elements in the samples. Again an elemental ratio from this technique can be compared to ratios from other techniques described earlier. The goal is to determine a normalization procedure that can be used to efficiently determine a normalized absolute weight or atomic percent of the elements in the sample. The reliability of the results and efficiency of the technique allows researchers to choose one of these techniques to produce reliable results using the normalization procedure established. The goal of any normalization technique is to decrease the uncertainties in the measurements including those done with SEM.

3.3.4 Statistical analysis

A crucial factor in coming to any conclusion in all these techniques is appropriate application of Statistical analysis. It is imperative to the researcher that they analyze the data using statistical packge (e.g. student t-test or ANOVA) after establishing normal distribution of data and homogeneity of variances.

4. Conclusions

SEM is suitable to look at micro- and nano- structural characteristics of solid objects. Visual images obtained from electron detectors combined with characteristic x-rays mapping allow for detailed micro- and nano-compositional analysis. SEM combined with XRF,NAA and PIXE provide a platform to quantify and produce absolute numbers related to compositional elemental and molecular structures.

The sample that is to be investigated has to be specially prepared so as to provide images and spectral information meaningful to the investigation. Many factors play a role here: the type of sample (say biological sample versus a sample for material science study has to be prepared differently at some stage of preparation), the appropriate energy of the beam, angle of incidence, beam intensity (resolution will be affected greatly from this), the counting time and statistics and others. SEM imaging is done differently for a wet cell sample than a critically dried and sputter-coated solar cell slides.

The other crucial factor is the methodology or methodologies adopted for data analysis and the subsequent results determination. Once the images and the spectra have been collected, the data has to be sorted, analyzed and mathematical functionality recognized and established. Statistical analysis then provides the basis for the eventual conclusions and their validity. For example topographical and/or compositional images can be used to generate structural patterns leading to understanding of type of crystalline lattice underlying a bone. This can then provide the basis for determining the strength of a bone or its elasticity or the reason a bone under microgravity conditions leads to Osteoporosis. SEM spectra that can be analyzed to determine the elemental composition of a certain bone have inherent uncertainties. When studying changes in bone composition these uncertainties will affect the determination of the conclusion.

5. Acknowledgment

The author would like to acknowledge the support for his research from the Arkansas Space Grant Symposium, College of Natural Sciences and Mathematics and the department of Physics and Astronomy at University of Central Arkansas. Also the author would like to acknowledge Dr. Jingbio Cui and Alan Thomas, Physics department, University of Arkansas at Little Rock and Devika Mehta (Highschool Senior at Arkansas School for mathematics, Sciences and Arts), for use of the SEM images (Fig 3 and 4) of their ZnO nanowires.

6. References

Bethe H.; *Ann. Phys.* (Leipzig) Vol 5, (1930) 325.

Bundle C. Richard, Evans C.A. Jr, & Wilson S; *Encyclopedia of Material Characterization*, Butterworth-Heinemann, Boston (1992)

Castaing R, *Ph.D. Thesis*, University of Paris (1951)

Castaing R, *Advances in Electronics and Electron Physics*, 13 (1960) 317

Castaing R & Henoc J, *Proc 4th Int. Conf. On X-ray Optics and Microanalysis* (1966) 120

EDS, SEM from RJLee Instruments Limited, Trafford, Pa

Flame, ASPEX corporation of Delmont, Pa. WWW.ASPEXCORP.com

Flewitt P.E.J. and Wild R. K. *Physical methods for Materials Characterization*, Institute of Physics Publishing, Bristol, UK (2003)

Gallagher W.J. and Cipolla S.J., *Nucl. Inst & Meth.* Vol 122 (1974) p.405.

Krane K.S.; *Introductory Nuclear Physics*, John Wiley and Sons (1988) pp.396-405

Lennard W.N. and Phillips D., *Nucl. Inst. & Meth.* vol 166 (1979)p.521

Mehta R.,Puri N.K., Kumar Ajay, Kumar A., Mohanty B.P., Balouria P., Govil I.M., Garg M.L., Nandi T., Ahamad A., and G. Lapicki, *Nucl. Inst. & Meth.* vol B241 (2005)pp.63-68

Metropolis, N. & Ulam S.; "The Monte Carlo Method" *Journal of the American Statistical Association* Vol 44 (1949) pp. 335–341.

Maxwell J.A. and Campbell J.L., X-ray Spctrometry Vol 34 (2005) p. 320

Newberrry D.E. & Myklebust R.L.; *Ultramicrosopy*, vol 3, (1979) 391.

Rubinstein, R. Y. & Kroese, D. P. ; *Simulation and the Mont Carlo Method* (2nd ed.). New York: John Wiley & Sons. (2007)

Rutherford E.;"The scattering of alpha and beta particles by matter and the structure of the atom", *Philosophical Magazine*, , Vol 21 (1911), pp. 669-688.

Rutherford E.,"The structure of the Atom", *Philosophical Magazine*, vol 27 (1914), pp. 488-498.

Papp T., *X-ray Spectrometry* Vol 34 (2005) p.320

2

In Situ Experiments in the Scanning Electron Microscope Chamber

Renaud Podor, Johann Ravaux and Henri-Pierre Brau
Institut de Chimie Séparative de Marcoule, UMR 5257 CEA-CNRS-UM2-ENSCM
Site de Marcoule, Bagnols sur Cèze cedex,
France

1. Introduction

Since the first scanning electron microscope by Knoll (1935) and theoretical developments by von Ardenne (1938a, b) in the 30's, this imaging technique has been widely used by generations of searchers from all the scientific domains to characterize the inner structure of matter. Even if the obtained information is essential for matter description or comprehension of matter transformation, the main constraints associated with classical electron microscopy, i.e. the necessity to work under vacuum and the necessity to prepare the sample before imaging, have always limited the possibilities to "post mortem" characterisation of samples and avoided observation of biological samples.

Electron microscopists early identified the necessity to undergo these limits. The development of a SEM chamber that is capable of maintaining a relatively high pressure and that allows imaging uncoated insulating samples began in the 70's and has been "achieved" in the late 90's – early 00's (Stokes, 2008) with the commercialisation of the low-vacuum and environmental SEM. The availability of new generations of electron guns (and more particularly the field effect electron gun characterized by a very intense brightness), as well as the new generation of electronic columns that are now commonly associated with the environmental scanning electron microscopes opens new possibilities for material characterisation up to the nanometer scale. The development of this generation of microscopes have opened the door for performing real time experiments, using the electron microscope chamber as a microlab allowing direct observation of reactions at the micrometer scale. Many SEM providers or researchers have developed specific stages that can be used for the *in situ* experimentation in the scanning electron microscope chamber. This field is one of the most interesting uses of the ESEM that offers fantastic opportunities for matter properties characterisation. Even if numerous recent articles and reviews are dedicated to *in situ* experimentation in the VP/ESEM (Donald, 2003 ; Mendez-Vilas et al., 2008 ; Stokes, 2008 ; Stabentheiner et al., 2010 ; Gianola et al., 2011 ; Torres & Ramirez, 2011), no one describes all the possibilities of this technique. The present chapter will provide a large – and as exhaustive as possible – overview of the possibilities offered by the new SEM and ESEM generation in terms of "*in situ* experiments" focussing specifically on the more recent results (2000-2011).

This chapter will be split into five parts. We will first discuss the goals of *in situ* experimentation. Then, specific parts will be devoted to *in situ* mechanical tests, experiments under wet conditions, and a forth part dedicated to high temperature experiments in the SEM. Last, a specific part will be devoted to the "future" of in-SEM experiments. In each part, the main limits of the technique as well as the detection modes will be reported. Each part will be focussed on examples of the use of the technique for performing *in situ* experiments.

2. Goals and implementation requirements of *in situ* experimentation

The main goal of *in situ* experimentation in the SEM (or ESEM) chamber is to determine properties of matter through the study of its behaviour under constraint. This requires the combination of data collection over a given duration (on a unique sample) and image treatment for information extraction. The studied properties are generally related to microscopic phenomena and hardly assessable by other techniques. *In situ* experiment in the SEM chamber corresponds to both imaging systems in evolution under a constraint and imaging systems stabilized under controlled conditions.

To achieve this goal, several requirements are necessary:

- The duration of the phenomenon to be observed must be suitable with the image recording time. If the system evolution is too fast, it will be impossible to record several images and observe this evolution. At the contrary, if the reaction kinetic is low, the time necessary for image recording will be too long and incompatible with experimentation. The high and low limits can be estimated ranging between 2 minutes and 48 hours.
- The system must remain stable under the environmental conditions and/or irradiation by the electron beam during the time necessary for image recording. In the case of easily degradable samples, it is necessary to adjust the imaging conditions (high voltage, beam current, aperture, working distance, detector bias…) constantly, as the sample environmental conditions are modified during the experiment. Thus, the effect of the electron beam on the sample morphology modifications must be verified. Some authors report that it can act as an accelerator (Popma, 2002) or inhibitor (Courtois et al., 2011) of the observed reactions.
- The image resolution must fit well with the size of details to be observed. Improvements in the image resolution have been achieved in the last decade thanks to the field effect emission guns. However, the presence of gas in the VP-SEM/ESEM chamber contributes to the incident electron beam scattering and subsequent degradation of the image resolution. Thus, the acquisition conditions must be adapted to the sample to be studied depending on the higher magnification to be reached.
- The gaseous environmental conditions in which the studied system evolutes (or can be stabilized) must be reproduced in the SEM/LV-SEM/ESEM chamber. The development of the ESEM offers real new opportunities in term of composition of the atmosphere surrounding the sample. The large field detector and the gaseous secondary electron detector (Stokes, 2008) have been developed specifically for imaging under "high pressure" conditions (up to 300Pa and 3000Pa respectively) whatever the gas composition (air, water, He, He+H_2 mixtures, O_2). Other detectors have been developed for very specific applications (high temperature under vacuum (Nakamura et al., 2002), EBSD at high temperature (Fielden, 2005)).

- The constraint in which the studied system evolutes (or can be stabilized) must also be reproduced in the microscope chamber. Some devices are commercialized by official sellers. Among them, we must report the Peltier stage for temperature control in the -10 to 60°C range, hot stages for temperature control up to 1500°C, stages for mechanical tests (Figure 1). Some authors have developed their own specific stages adapted to the problem to be treated (Fielden, 2005; Bogner et al., 2007). However, the development of miniaturized stages that can be positioned in the SEM chamber without creating perturbations on the incident electron beam can be really challenging. This will probably be a key in the development of *in situ* experimentation in the next years (Torres & Ramirez, 2011).

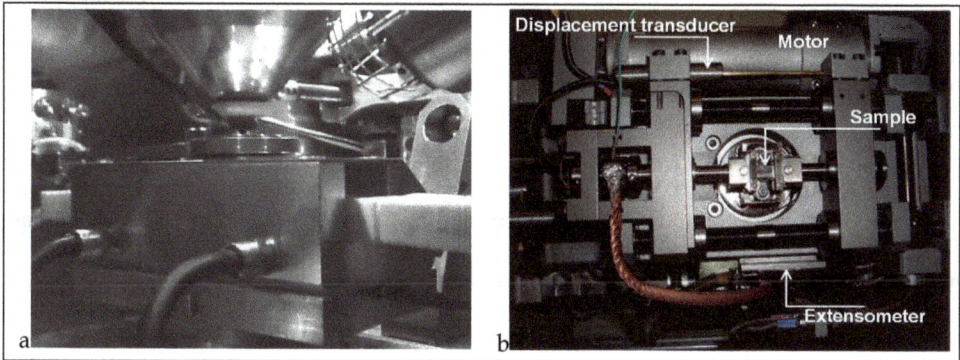

Fig. 1. a) hot stage (FEI) b) Hot tension/compression stage integrated into an SEM (Kammrath & Weiss Co.) (After Biallas & Maier, 2007 ; Gorkaya et al., 2010).

The basis of *in situ* experimentation in the SEM is the study of the morphological modifications of the sample under constraint. Thus, this requires recording of numerous high quality images for image post treatment and data extraction in order to characterize the reaction or matter properties. The sample size can vary from 1μm to 50mm, and the image resolution is in the 1-10nm range, depending on recording conditions. The images are SEM images, i.e. with a large depth of field and with grey level contrasts. In-SEM experimentation can be extended to a wide range of applications, corresponding to very different materials (plants (Stabentheiner et al., 2010), food (Thiel et al., 2002 ; James, 2009), paper (Manero et al., 1998), soft matter, polymers, metals, ceramics, solids, liquids...) or problems (plant behaviour, chemical reactivity, properties characterization, sintering, grain growth, corrosion...). In the literature, the main part of the data reported has been acquired using an environmental scanning electron microscope.

3. *In situ* mechanical tests

Boehlert (2011) have recently underlined the interest of performing *in situ* mechanical tests in the SEM and summarized it as follows. "*In situ* scanning electron microscopy is now being routinely performed around the world to characterize the surface deformation behavior of a wide variety of materials. The types of loading conditions include simple tension, compression, bending, and creep as well as dynamic conditions including cyclic fatigue with dwell times. These experiments can be performed at ambient and elevated temperatures and in different environments and pressures. Most modern SEMs allow for the adaptation of heating and mechanical testing assemblies to the SEM stage, which allows for tilting and rotation to optimal imaging conditions as well as energy dispersive spectroscopy X-ray capture. Perhaps some of the most useful techniques involve acquisition of electron backscatter diffraction (EBSD) Kikuchi patterns for the identification of crystallographic orientations. Such information allows for the identification of phase transformations and plastic deformation as they relate to the local and global textures and other microstructural features. Understanding the microscale deformation mechanisms is useful for modeling and simulations used to link the microscale to the mesoscale behavior. In turn, simulations require verification through *in situ* microscale observations. Together simulations and *in situ* experimental verification studies are setting the stage for the future of material science, which undoubtedly involves accurate prediction of local and global mechanical properties and deformation behavior given only the processed microstructural condition".

As a direct consequence of the great interest of the collected information, many different works from several scientific domains have been published for long. Thiel & Donald (1998) and Stabentheiner et al. (2010) describe the deformation of plants (carrots and leaves respectively) during room temperature tensile tests performed in the ESEM chamber. Similar tests are also reported with food (Stokes & Donald, 2000) and they are regularly performed on polymers (Poelt et al., 2010; Lin et al., 2011), composites (Schoßig et al., 2011) and metals (Boehlert et al., 2006; Gorkaya et al., 2007). Mechanical tests on metals, alloys and ceramics can also be performed at high temperature (Biallas & Maier, 2007; Chen & Boehlert, 2010). High temperature EDSB, developed by Seward et al. (2002), offers the possibility to observe phase transformations in materials as a function of temperature, as well as the direct visualization of the associated microstructural modifications (Seward et al., 2004).

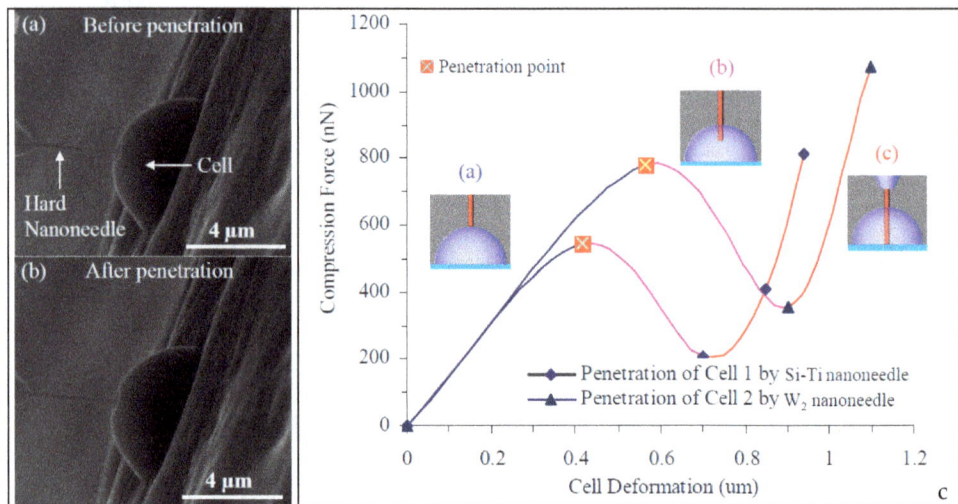

Fig. 2. (a) & (b) Single cell surgery without cell bursting using Si-Ti nanoneedle , (c) Force-cell deformation curve using Ti-Si and W2 nanoneedles at three different stages, i.e. (a) before penetration, (b) after penetration and (c) touching the substrate. (Ahmad et al., 2010).

Several recently developed techniques allow characterizing materials at the nanometer scale through both technological miniaturization and advancements in imaging and small-scale mechanical testing. Ahmad et al. (2010) have developed a coupled ESEM-atomic force microscope to characterize single cells mechanical properties (Figure 2). This ESEM-nanomanipulation system allowed determining effects of internal influences (cell size and growth phases) and external influence (environmental conditions) on the cell strength. Gianola et al. (2011) reports the development of a quantitative *in situ* nanomechanical testing approach adapted to a dualbeam focused ion beam and scanning electron microscope. *In situ* tensile tests on 75 nm diameter Cu nanowhiskers as well as compression tests on nanoporous Au micropillars fabricated using FIB annular milling are reported, the scientific question being the mechanical behaviour of nanosize materials. Both examples probably represent what will be the future of *in situ* mechanical tests using scanning electron microscopes.

4. *In situ* experimentation under wet conditions

4.1 Conditions for experimentation

Combination of the use of the ESEM and a Peltier stage with the development of specific detectors allows the possibility to control both specimen temperature and water pressure around the sample (Leary & Brydson, 2010). Water can be condensed or evaporated on the demand from the sample (Figure 3). This allows performing *in situ* experiments in a temperature-pressure domain that is reported on Figure 3a (dot zone). An easy to perform experiment, illustrated by a 6 images series, corresponding to the NaCl dissolution (during the increasing of the water pressure in the ESEM chamber and consecutive water condensation, at constant temperature) in water followed by the crystallization of NaCl (decrease of the water pressure) is reported on Figure 3b. This example corresponds to an "isothermal experiment". Another ways to work are to perform isobar experiments or to heat or cool a sample using a constant relative humidity (iso-RH experiments). These techniques allow the characterization of structural transitions of hydrated samples as a function of temperature (Bonnefond, 2011).

4.2 Biology and soft matter applications

This technique is particularly well adapted for the observation or experimentation on biological samples (Muscariello et al., 2005). Images of small and highly hydrated samples such as liposomes have been obtained by several authors (Perrie et al., 2007 ; Ruozi et al;, 2011) without any particular sample preparation. Perrie et al. (2007) have also been able to dynamically follow the hydration of lipid films and changes in liposome suspensions as water condenses onto, or evaporates from, the sample in real-time. The data obtained provides an insight into the resistance of liposomes to coalescence during dehydration, thereby providing an alternative assay for liposome formulation and stability (Perrie et al., 2010). However, Kirk et al. (2009) report that ESEM imaging of biological samples must remain combined with the classical techniques for sample preparation. Several works are specifically dedicated to *in situ* experimentation. Stabentheiner et al. (2010) state that "one unrivaled possibility of ESEM is the *in situ* investigation of dynamic processes that are impossible to access with CSEM where samples have to be fixed and processed". These authors have studied the anther opening that is a highly dynamic process involving several tissue layers and controlled tissue desiccation. This phenomenon can be observed because the sample is very stable under the ESEM conditions (Figure 4). Another recent study is relative to the closure of stomatal pores by Mc Gregor & Donald (2010). Even if the possibility for experimentation on biological samples is clearly demonstrated, the authors outline the fact that the electron beam damages are important even at low accelerating voltage (Zheng et al., 2009). Another surprising example that can be reported is the direct observation of living acarids available online: in the movie, colonies of acarids are directly observed in the ESEM chamber under several conditions (FEI movie).

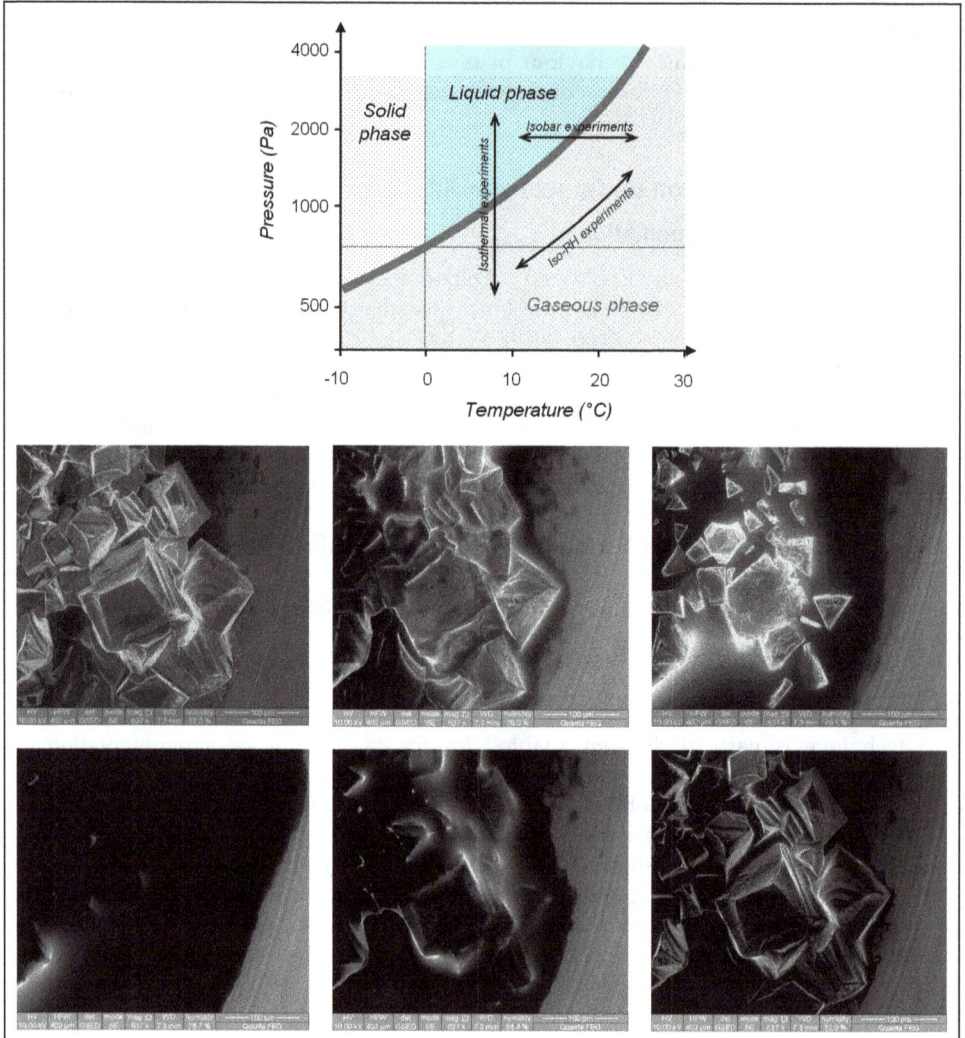

Fig. 3. (a) Simplified phase diagram for water indicating the ESEM domain (dot zone) and schemes to understand how isothermal or isobar experiments are performed.
(b) Solubilisation and crystallization of NaCl directly observed in the ESEM chamber.

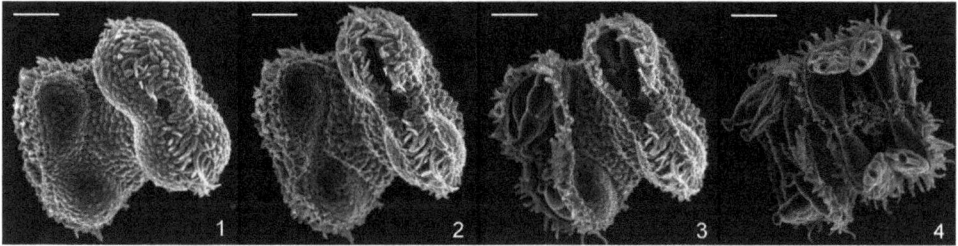

Fig. 4. *In situ* anther opening of C. angustifolia observed in LV-ESEM. 1) At the beginning, the valves of the anther are closed; 2) opening starts at the end of the stomium; 3) polyads are already seen; 4) opening proceeds till the valves are completely bent back and all eight polyads are presented (scale bar = 100μm). Time span from 1) to 3) was 25 min; 4) imaged 1 h after the start of the opening process (after Stabentheiner et al., 2010)

4.3 Applications on cements

Several works have been performed in order to study the reactivity of cement materials versus humidity. Hydration or dehydration (Sorgi & De Gennaro, 2007; Fonseca & Jennings, 2010; Camacho-Bragado et al., 2011) of phases have been followed and used to extract kinetic parameters (Montes-Hernandez, 2002 ; Montes & Swelling, 2005 ; Maison et al., 2009), as reported on Figure 5. In this work, the author uses ESEM image series to determine a three-step mechanism for bentonite aggregates evolution with relative humidity corresponding to an arrangement of particles followed by a particle swelling and a full destructuration. In SEM experiments are also used to characterize chemical reactivity (Camacho-Bragado et al., 2011). It has been recently used to characterize reaction of fly ash activated by sodium silicate by Duchene et al. (2010). These authors have determined very accurately the different steps of the reaction determining that the sodium silicate activator dissolves rapidly and begins to bond fly ash particles. Open porosity was observed and it was rapidly filled with gel as soon as the liquid phase is able to reach the ash particle. The importance of the liquid phase is underlined as a fluid transport medium permitting the activator to reach and react with the fly ash particles. The reaction products had a gel like morphology and no crystallized phase was observed.

4.4 Hydration and dehydration experiments

As previously reported for liposomes, new opportunities for the study of polyelectrolyte microcapsules versus their resistance to relative humidity and temperature modifications are opened and under consideration. The image series reported on Figure 6 clearly illustrate the possibility to image the native soft capsule at high relative humidity without any deformation. When decreasing the water pressure near the capsule, the object is deformed and do not shrink as observed when it is heated in water at temperature higher than 25°C (Basset et al., 2010). Thus, the walls of the object do not rearrange but collapse when submitted to a relative humidity decrease.

Similar tests have been performed on self-organized metal-organic framework compounds (Bonnefond, 2011). According to the image series reported on Figure 7, when the water pressure decreases, the size of sample remains constant up to a given water pressure (i.e. relative humidity) and for a transition pressure, the sample size decreases regularly. This can be associated to a local reorganisation in the sample that corresponds to a water loss associated to the sample collapsing The enthalpy of water ordering in the sample can be derived from the recorded image series as reported by Sievers et al.

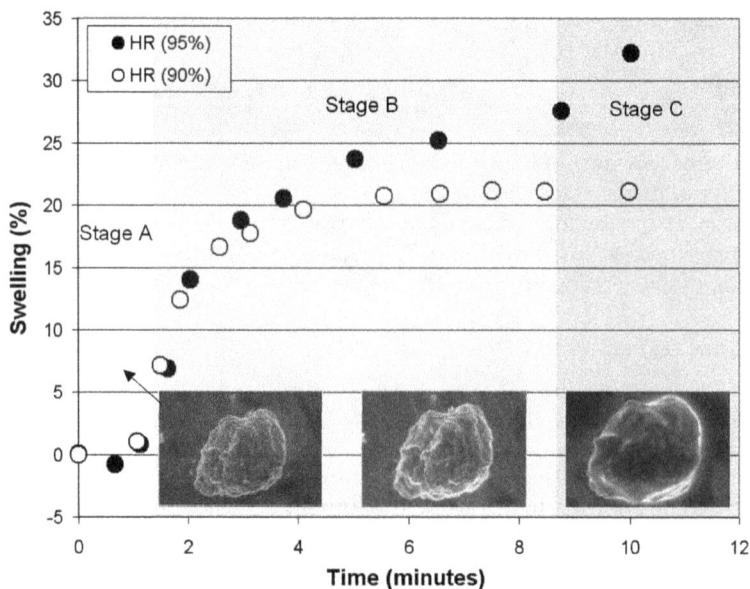

Fig. 5. Swelling kinetics of raw bentonite aggregates scale using ESEM-digital image analyses coupling (after Montes & Swelling, 2005).

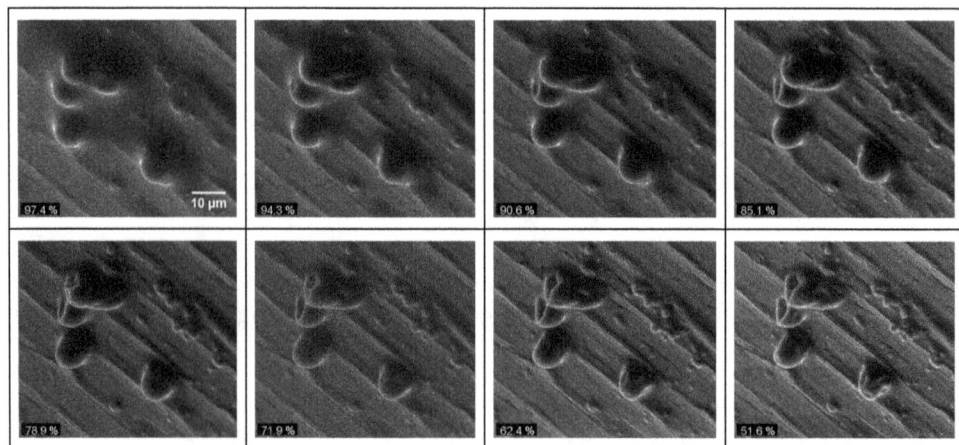

Fig. 6. ESEM micrographs of polyelectrolyte microcapsules suspended in double distilled water. Microcapsules were subjected to controlled dehydration in the ESEM sample chamber at T=5°C. At an operating pressure of 800Pa, vesicles appeared as spherical structures. (a) Gradual decrease of the operating pressure to 350 Pa showed regular deformation of the microcaspsules (b to h)

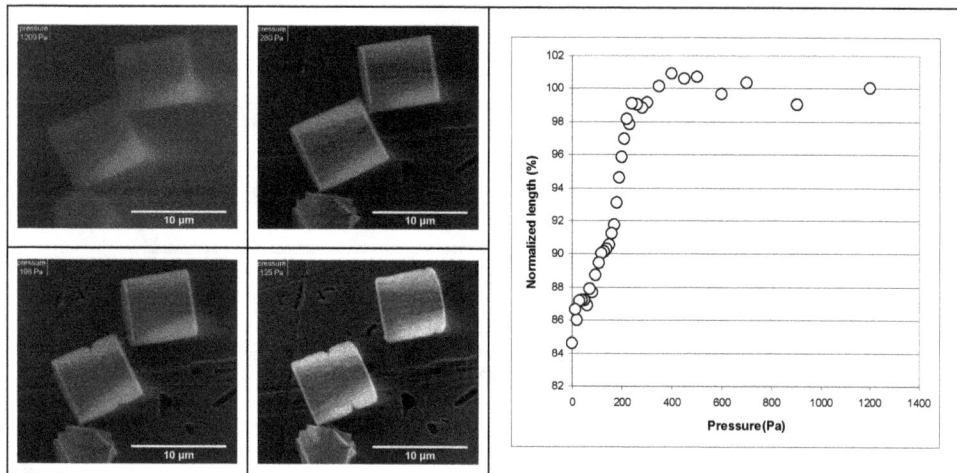

Fig. 7. Dehydration experiments performed on self-assembled organo-metallic compounds at T=22°C and corresponding size modification *versus* water vapour pressure (Bonnefond, 2011).

The effect of dehydration on lamellar bones was also studied by *in situ* ESEM experiments (Utku et al., 2008). The obtained results indicate that dehydration affects the dimensions of lamellar bone in an anisotropic manner in longitudinal sections, whereas in transverse sections the extent of contraction is almost the same in both the radial and tangential directions.

An original work on the heterogeneous ice nucleation on synthetic silver iodide, natural kaolinite and montmorillonite particles has been performed using the "increasing water pressure at constant temperature" (Zimmermann et al., 2007) in the temperature range of 250–270 K. Ice formation was related to the chemical composition of the particles. The obtained data are in very good agreement with previous ones obtained by diffusion chamber measurements (Figure 8).

4.5 Characterization of surface wetting properties

Characterization of the wetting properties of surfaces through the formation of microdroplets or nanodroplets is another important investigation field that can be explored using the ESEM. A recent review by Mendez-Vilas et al. (2009) has highlighted the main fundamental and applied results. Several strategies for the contact angle between water and the surface determination are reported (Stelmashenko et al., 2001; Stokes, 2001; Lau et al., 2003; Wei, 2004; Yu et al., 2006; Jung & Bhushan, 2008; Rykaczewski & Scott, 2011). The investigation of the hydrophobicity and/or hydrophilicity of a catalyst layer have been performed using ESEM for the first time by Yu et al. (2006). These authors have determined the micro-contact angle distribution as a function of the catalyst microstructure. Microdroplets growing and merging process was observed directly in the ESEM chamber by Lau et al. (2003).

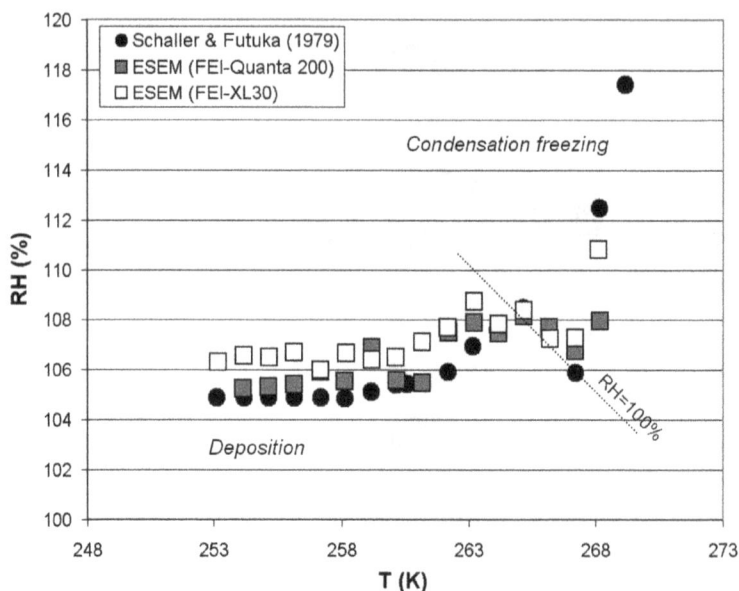

Fig. 8. Supersaturation *versus* temperature diagram for silver iodide (After Zimmermann et al., 2007).

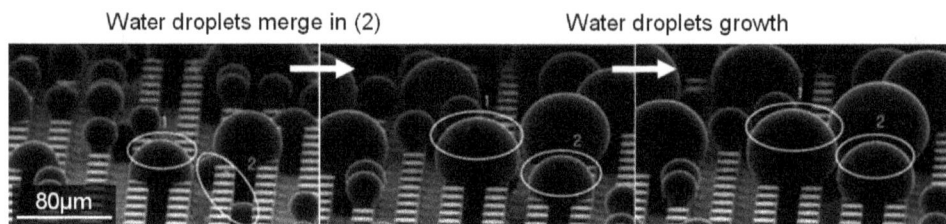

Fig. 9. Microdroplets growing and merging process under ESEM during increasing condensation by decreasing temperature. (After Jung & Bhushan, 2008)

4.6 Using the Wet-STEM mode

The development of the Wet-STEM by Bogner et al. (2005, 2007) allows observing samples in the transmission mode in the ESEM chamber, and more particularly, it offers the possibility to image directly nanoparticles dispersed in a few micrometer thin water film (Bogner et al., 2008), emulsions or vesicles (Maraloiu et al., 2010), without removing the liquid surrounding the objects of interest. One must keep in mind that images with soft matter, and more generally sample sensitive to the electron beam are very hard to obtain. Nevertheless, this technique also opens new research fields using *in situ* experimentation that only begin to be explored for wettability or deliquescence studies. By combining Wet-STEM imaging with Monte-Carlo simulation (Figure 10), Barkay (2010) have studied the initial stages of water nanodroplet condensation over a nonhomogeneous holey thin film. This study has shown a preferred water droplet condensation over the residual water film areas in the holes and has provided corresponding droplet shape and contact angle. On a similar way, Wise et al. (2008) have studied water uptake by NaCl particles prior to deliquescence by varying the relative humidity in the Wet-STEM environment (Figure 11).

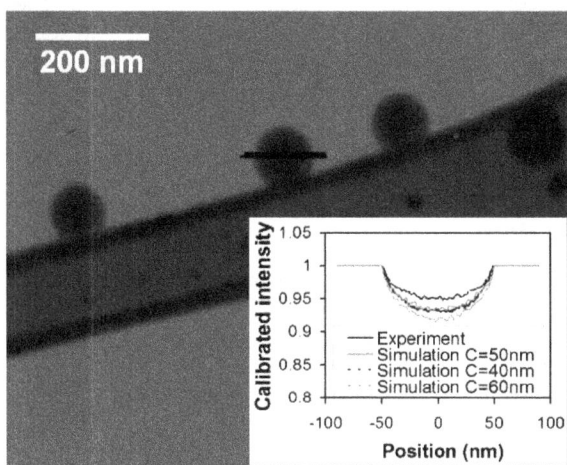

Fig. 10. Bright field image of 100 nm polystyrene latex spheres. Insert is the calibrated intensity corresponding to the dark line in the image (After Barkay (2010))

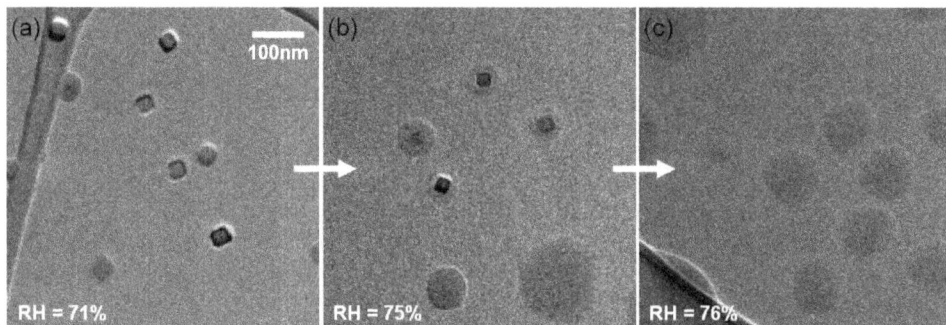

Fig. 11. ~40 nm NaCl particles as the RH was increased past the deliquescence point. Water uptake [(a) → (b)] prior to full deliquescence (c) is clearly observed. (After Wise et al., 2008)

4.7 Development of specific materials for experimentation

Several specific devices have been developed to characterize specific properties or reactions. Two of them will be shortly described below.

Chen et al. (2011) have developed an experimental platform that can be used to investigate chemical reaction pathways, to monitor phase changes in electrodes or to investigate degradation effects in batteries. They have performed *in situ* experiment runs inside a scanning electron microscope (SEM) and tracked the morphology of an electrode including active and passive materials in real time. This work has been used to observe SnO_2 during lithium uptake and release inside a working battery electrode.

Direct imaging of micro ink jets inside the ESEM chamber has been achieved using a specific device developed by Deponte et al. (2009), using a two-fluid stream consisting of a water inner core and a co-flowing outer gas sheath. ESEM images of water jets down to 700 nm diameter have been recorded. Details of the jet structure (the point of jet breakup, size and shape of the jet cone) can be measured. The authors conclude that ESEM imaging of liquid jets offers a valuable research tool for the study of aerosol production, combustion processes, ink-jet generation, and many other attributes of micro- and nanojet systems.

5. High temperature in the SEM

5.1 Application domains of HT-(E)SEM

Specific stages (and associated detectors) have been developed to heat samples up to 1500°C directly in the microscope chamber (Knowles & Evans, 1997; Gregori et al., 2001). The environmental scanning electron microscope (ESEM) equipped with this heating stage is an excellent tool for the *in situ* and continuous observation of system modifications involved by temperature. It allows recording image series of the morphological changes of a sample during a heat treatment with both high magnification and high depth of focus. The experiments can be carried out to observe the influence of all these parameters on the studied phenomenon under various conditions (heating rates, atmosphere compositions, variable pressure, final temperature and heating time). Images have been recorded up to 1400°C, with a decrease of the image resolution when the sample temperature increases (Podor et al., 2012). It is possible to work under vacuum (classical SEM) or under controlled atmosphere (H_2O, O_2, $He+H_2$, N_2, air...). Different types of studies have been reported, relative to corrosion of metals (Jonsson et al., 2011), oxidation of metals (Schmid et al., 2001a, 2001b ; Oquab & Monceau, 2001 ; Schmid et al., 2002 ; Abolhassani et al., 2003 ; Reichmann et al., 2008 ; Jonsson et al., 2009 ; Mège-Revil et al., 2009 ; Quémarda et al., 2009 ; Delehouzé et al., 2011), reactivity at high temperature (Maroni et al., 1999 ; Boucetta et al., 2010), phase changes (Fischer et al., 2004 ; Hung et al., 2007 ; Beattie & McGrady, 2009), hydrogen desorption (Beattie et al., 2009, 2011), redox reactions (Klemensø et al., 2006), microstructural modifications (Bestmann et al., 2005 ; Fielden, 2005 ; Yang, 2010), magnetic properties (Reichmann et al., 2011), sintering (Sample et al., 1996 ; Srinivasan, 2002 ; Marzagui & Cutard, 2004 ; Smith et al., 2006 ; Subramaniam, 2006 ; Courtois et al., 2011 ; Joly-Pottuz et al., 2011 ; Podor et al., 2012), thermal decomposition (Gualtieri et al., 2008 ; Claparède et al., 2011 ; Goodrich & Lattimer, 2011 ; Hingant et al., 2011), crystallisation (Gomez et al., 2009) in melts (Imaizumi et al., 2003 ; Hillers et al., 2007) and study of self-repairing – self-healing – properties of materials (Wilson & Case, 1997 ; Coillot et al., 2010a, 2010b, 2011) …

Even if numerous researchers are invested in HT-ESEM, only few of them have been successful in pursuing dynamic experiments at temperatures higher than 1100°C. Two recent studies report experiments performed at T=1350°C (Subramaniam, 2005) and 1450°C (Gregori et al., 2002). However, the resolution of the images remains poor (more than 1μm) mainly due to water cooling induced vibrations. Furthermore, the precision on the measure of the sample temperature remains poor (temperature differences up to 150°C with the expected temperature are sometimes measured). A recent device has been proposed by Podor et al. (2011) to overcome this difficulty.

A complete review specifically dedicated to *in situ* high temperature experimentation in the ESEM will be available soon. Several examples of *in situ* studies performed at high temperature in the ESEM chamber will be reported below, on the basis of original data acquired in our laboratory.

5.2 Investigation of the crystallization behaviour in silicate melts

The crystal growth and morphology during isothermal heating of glass melts can be directly observed using the hot stage associated with the ESEM. The image series reported on Figure 12 have been recorded during 10 minutes while heating the borosilicate melt sample isothermally at T=740°C. The development of large crystals in the melt rapidly yields to the complete crystallization of the melt. The crystal morphology presents cells filled with a second phase and the crystal formation yields to the deformation of the sample surface. Hillers et al. (2007) have used such data to quantify the variation of crystal length with time. They have established that the growth is only linear during the first minutes; afterward the growth rate decreases progressively with time.

This technique can also be used to determine the temperature of formation of the first crystals at the melt surface and to observe their formation. In the case of glass-ceramics, the density of nuclei as well as their size and shape development can be directly observed and used for crystallization kinetic determination (Vigouroux et al., 2011, in prep).

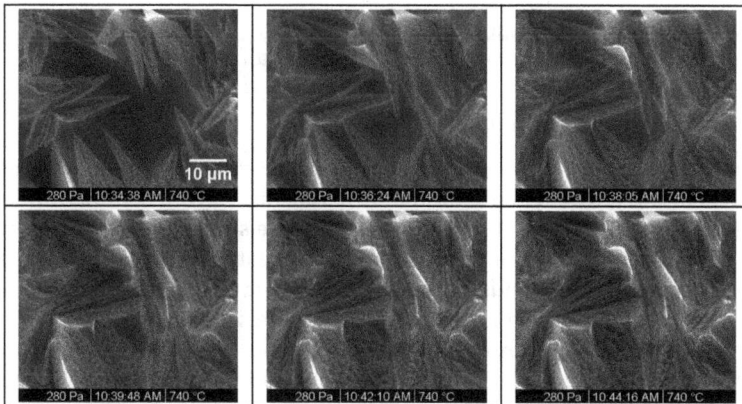

Fig. 12. Growth of crystals in a borosilicate melt during 10 minutes isothermal heat treatment at 740°C observed using the hot stage associated with the ESEM.

5.3 Decomposition of compounds

In situ thermal decomposition of composites, oxalates, oxides have been reported by several authors. Images of the heat treatment of a mixed uranium-cerium oxalate grain from 25°C to 1235°C are gathered on Figure 13. Morphological changes with temperature are directly linked with the oxalate decomposition as stated by Hingant et al. (2011) in the temperature range 25-500°C. The sample shrinkage observed when T>500°C is probably related with the first stage of the sintering process – i.e. beginning of bond formation between the nanograins and with the oxide grain growth (that can not be directly observed at this stage by HT-ESEM, but that is confirmed by X-Ray diffraction). Such a process has also been recently reported by Claparede et al. (2011) and Joly-Pottuz et al. (2011).

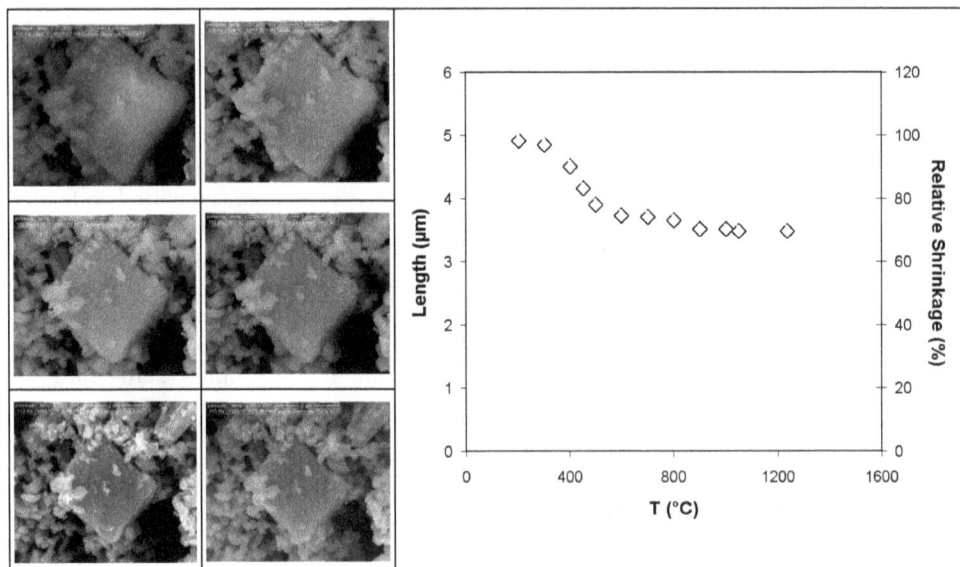

Fig. 13. Decomposition of a uranium-cerium mixed oxalate observed during *in situ* heating in the ESEM chamber and relative size and shrinkage modifications.

5.4 Study of sintering and grain growth

Several studies are relative to the sintering and grain growth processes in metals and ceramics. Depending on the system, the experiments have been performed in the temperature range 300-1450°C. The main interest of these studies is the possibility of direct observation of the individual grain behaviour during heat treatment. The example that is reported on Figure 14a corresponds to the heat treatment of the grain decomposed *in situ* (Figure 13). The image resolution is high enough to observe the nanograins growth inside the square plate agglomerate. Consequently, relative shrinkage and average grain diameter are extracted by image processing (Figure 14b). Assuming that the final density of the agglomerate is 99%, the sintering map is directly derived from these experimental data (Figure 14c). Thus, *in situ* sintering experiments can allow the establishment of the trajectories of theoretical sintering. Such data have never been already reported in previous studies, mainly due to the poor resolution of the recorded images.

The effect of the electron beam on sintering is controversy. Indeed, Popma (2002) noted that a local sintering stop was achieved by focusing the electron beam at a certain position during the *in situ* sintering experiments in the ESEM (performed on ZrO_2 nanolayers). On the contrary, Courtois et al (2011) performed experiments on the sintering of a lead phosphovanadate and concluded that the electric current induced by the electron beam was found to reduce the effective temperature of sintering by 50 to 150°C as well as to accelerate the kinetics of shrinkage of a cluster composed of sub-micrometric grains of material. Such effects were not evidenced in our study: the local sintering on sample surface zones that were not observed (i.e. exposed to the electron beam) was identical to the local sintering determined on the observed zone.

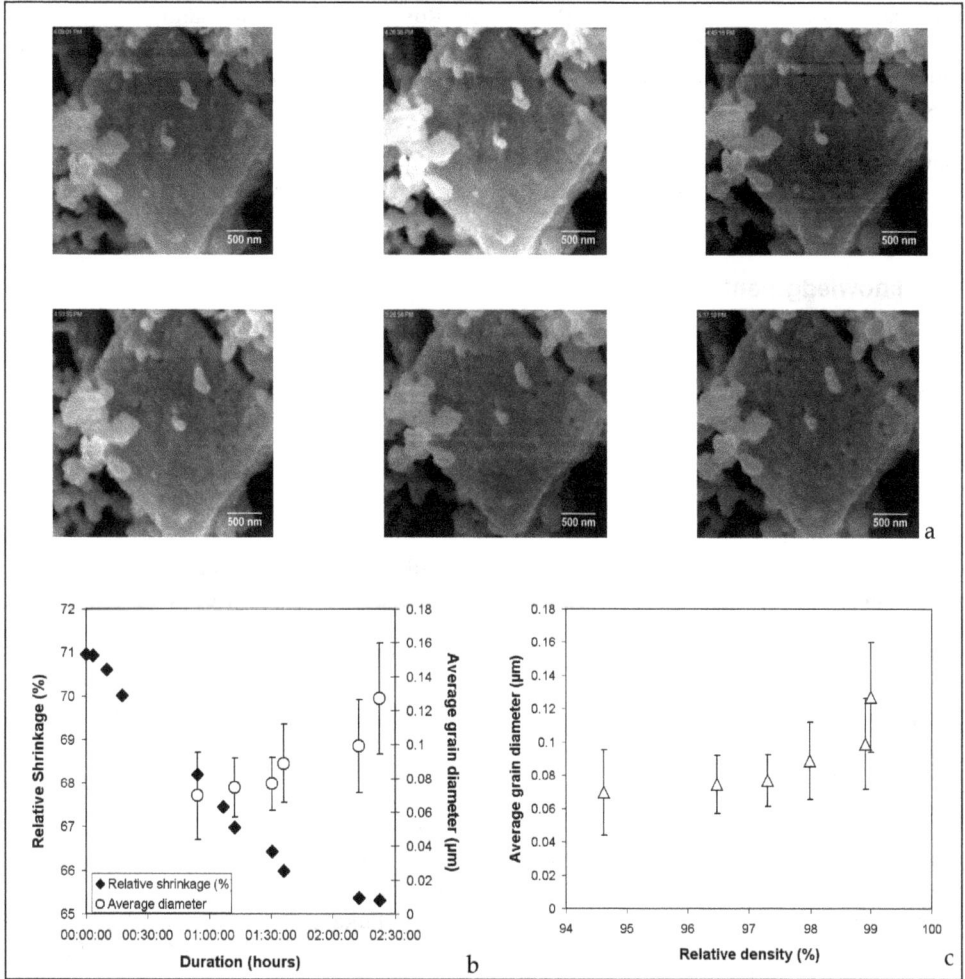

Fig. 14. (a) Sintering and grain growth of a uranium-cerium mixed oxide observed *in situ* in the ESEM chamber at T=1235°C, after 55', 70', 90', 95', 130', 140' (a). Corresponding Relative (b) Shrinkage and Average grain diameter versus duration and (c) derived sintering map - Grain growth versus densification rate –

6. Conclusions and perspectives

In situ scanning electron microscopy experimentation, that is generally associated with the use of the ESEM, allows the study of very different problems, the main limit being the availability of specific devices. Torres & Ramirez (2011) have written the best conclusion indicating that "the new generation of SEMs shows innovative hardware and software solutions that result in improved performance. This progress has turned the SEM into an extraordinary tool to develop more complex and realistic *in situ* experiments, achieving even at the subnanometer scale". In the near future, new SEM imaging modes, nanomanipulation and nanofabrication technologies (Miller & Russell, 2007 ; Romano-Rodriguez & Hernandez-Ramirez, 2007 ; Wich et al., 2011) will make possible to replicate more closely the conditions as the ones associated to the problems to be treated. *In situ* ESEM will probably be used to overcome technical and fundamental challenges in many scientific domains. The recent developments of a high temperature stage in the FIB (Fielden, 2008), a new tomography mode in the ESEM (Jornsanoh et al., 2011) and of the atmospheric scanning electron microscope (Nishiyama et al, 2010 ; Suga et al, 2011) can be cited as examples for this future.

7. Acknowledgment

The authors want to thank all the co-workers of the studies cited in this chapter, and more particularly F. Bonnefond, H. Boucetta, C. Dejugnat, T. Demars, A. Monteiro and L. Claparède for providing the samples and challenging projects.

8. References for videos

Reactivity of a salt with silicate melt at high temperature	http://www.dailymotion.com/icsmweb#videoId=xjknrt
Sintering of CeO$_2$ at T=1200°C	http://www.youtube.com/watch?v=4ijIUdQe3M4
Self-healing of a metal-glass composite at high temperature	http://www.dailymotion.com/icsmweb#videoId=xjknpp
Deformation of vesicles during dehydration	http://www.dailymotion.com/icsmweb#videoId=xjk75u
NaCl solubility and precipitation in water	http://www.dailymotion.com/icsmweb#videoId=xk22i9

9. References

Abolhassani, S., Dadras, M., Leboeuf, M. & Gavillet, D. (2003). In situ study of the oxidation of Zircaloy-4 by ESEM. *Journal of Nuclear Materials* 321, 70-77.

Ahmad, M.R., Nakajima, M., Kojima, S, Homma, M. & Fukada, T. (2010). *Single cell analysis inside ESEM – (ESEM)-nanomanipulator system* , InTech, ISBN 978-953-7619-93-0, "Cutting Edge Nanotechnology" 413-438.

Barkay, Z. (2010). Wettability study using transmitted electrons in environmental scanning electron microscope. *Applied Physic Letters* 96, 183109.

Basset, C., Harder, C., Vidaud, C. & Déjugnat, C. (2010). Design of Double Stimuli-Responsive Polyelectrolyte Microcontainers for Protein Soft Encapsulation. *Biomacromolecules* 11, 806–814.

Beattie, S.D. & McGrady, G.S. (2009). Hydrogen desorption studies of NaAlH$_4$ and LiAlH$_4$ by in situ heating in an ESEM. *International Journal of Hydrogen Energy* 34, 9151-9156.

Beattie, S.D., Langmi, H.W. & McGrady, G.S. (2009). In situ thermal desorption of H$_2$ from LiNH2-2LiH monitored by environmental SEM. *International Journal of Hydrogen Energy* 34, 376-379.

Beattie, S.D., Setthanan, U. & McGrady, G.S. (2011). Thermal desorption of hydrogen from magnesium hydride (MgH$_2$): An in situ microscopy study by environmental SEM and TEM. *International Journal of Hydrogen Energy* in press.

Bestmann, M., Piazolo, S., Spiers, C.J. & Prior, D.J. (2005). Microstructural evolution during initial stages of static recovery and recrystallization: new insights from in-situ heating experiments combined with electron backscatter diffraction analysis. *Journal of Structural Geology* 27, 447–457.

Biallas, G. & Maier, H.J. (2007). In-situ fatigue in an environmental scanning electron microscope – Potential and current limitations. *International Journal of Fatigue* 29, 1413–1425.

Boehlert, C.J., Cowen, C.J., Tamirisakandala, S., McEldowney, D.J. & Miracle, D.B. (2006).In situ scanning electron microscopy observations of tensile deformation in a boron-modified Ti–6Al–4V alloy. *Scripta Materialia* 55, 465–468.

Boehlert, C. J. (2011). In situ scanning electron microscopy for understanding the deformation behaviour of structural materials. Seminarios Internacionales de Fronteras de la Ciencia de Materiales. April 11th, 2011 (http://www.youtube.com/watch?v=wH3EYxT_ysM)

Bogner, A., Guimarães, A., Guimarães, R.C.O., Santos, A.M., Thollet, G., Jouneau, P.H. & Gauthier, C. (2008). Grafting characterization of natural rubber latex particles: wet-STEM imaging contributions. *Colloid and Polymer Science* 286, 1049–1059.

Bogner, A., Jouneau, P.H., Thollet, G., Basset, D. & Gauthier C. (2007). A history of scanning electron microscopy developments: Towards "wet-STEM" imaging. *Micron* 38, 390–401.

Bogner, A., Thollet, G., Basset, D., Jouneau, P.H. & Gauthier, C., (2005). Wet STEM: A new development in environmental SEM for imaging nano-objects included in a liquid phase *Ultramicroscopy* 104, 290-301.

Bonnefond, F. (2011). *Etude in situ de la déshydratation de composés organométalliques*. Master 1 thesis (30p.)

Boucetta, H., Schuller, S., Ravaux, J & Podor, R. (2010). *Etude des mécanismes de formation des phases cristallines RuO₂ dans les verres borosilicate de sodium*. Proceeding of Matériaux 2010 (18-22 oct Nantes, France)

Camacho-Bragado, G.A., Dixon, F. & Colonna, A. (2011). Characterization of the response to moisture of talc and perlite in the environmental scanning electron microscope. *Micron* 42, 257-262.

Chen, D., Indris, S., Schulz, M., Gamer, B. & Mönig, R. (2011). In situ scanning electron microscopy on lithium-ion battery electrodes using an ionic liquid. *Journal of Power Sources* 196, 6382–6387.

Chen, W., Boehlert, C.J. (2010). The 455°C tensile and fatigue behavior of boron-modified Ti–6Al–2Sn–4Zr–2Mo–0.1Si(wt.%). *International Journal of Fatigue* 32, 799-807.

Claparède, L., Clavier, N., Dacheux, N., Moisy, P., Podor, R. & Ravaux, J. (2011). Influence of crystallization state and microstructure on the chemical durability of cerium-neodynium mixed dioxides. *Inorganic Chemistry*, 50[18], 9059–9072.

Coillot, D., Podor, R., Méar, F.O. & Montagne, L. (2010a). Characterisation of self-healing glassy composites by high-temperature environmental scanning electron microscopy (HT-ESEM). *Journal of Electron Microscopy* 59, 359-366.

Coillot, D., Méar, F.O., Podor, R. & Montagne, L. (2010b). Autonomic Self-Repairing Glassy Materials. *Advanced Functional Materials* 20(24), 4371-4374.

Coillot, D., Méar, F.O., Podor, R. & Montagne, L. (2011). Influence of the Active Particles on the Self-Healing Efficiency in Glassy Matrix. *Advanced Engineering Materials* 13, 426-435.

Courtois, E., Thollet, G., Campayo, L., Le Gallet, S., Bidault, O. & Bernard, F. (2011). In situ study of the sintering of a lead phosphovanadate in an Environmental Scanning Electron Microscope. *Solid State Ionics* 186, 53–58.

Delehouzé, A., Rebillat, F., Weisbecker, P., Leyssale, J.M., Epherre, J.F., Labrugère C. & Vignoles G.L. (2011). Temperature induced transition from hexagonal to circular pits in graphite oxidation by O_2. *Applied Physics Letters* 99, 044102.

DePonte, D.P., Doak, R.B., Hunter, M., Liu, Z., Weierstall, U. & Spence, J.C.H. (2009). SEM imaging of liquid jets. *Micron* 40, 507–509.

Donald, A.M. (2003). The use of environmental scanning electron microscopy for imaging wet and insulating materials. *Nature Materials* 2, 511-516.

Duchene, J., Duong, L., Bostrom, T. & Frost, R. (2010). Microstructure study of early in situ reaction of fly ash geopolymer observed by ESEM. *Waste Biomass Valorisation* 1, 367–377.

FEI movie (1998) http://www.dailymotion.com/video/xirinx_acariens-cannibales-les-envahisseurs-invisibles_animals

Fielden, I.M. (2005). *Investigation of microstructural evolution by real time SEM of high temperature specimens*. PhD thesis Sheffield Hallam University (170p).

Fielden, I.M. (2008). In-Situ Focused Ion Beam (FIB) microscopy at high temperature. *Electron Microscopy and Analysis Group*.

Fischer, S., Lemster, K., Kaegi, R., Kuebler, J. & Grobety, B. (2004). In situ ESEM observation of melting silver and inconel on an Al_2O_3 powder bed. *Journal of Electron Microscopy* 53, 393-396.

Fonseca, P.C. & Jennings, H.M. (2010). The effect of drying on early-age morphology of C–S–H as observed in environmental SEM. *Cement and Concrete Research* 40, 1673–1680.

Gianola, D.S., Sedlmayr, A., Mönig, R., Volkert, C.A., Major, R.C., Cyrankowski, E., Asif, S.A.S., Warren, O.L., & Kraft, O. (2011). In situ nanomechanical testing in focused ion beam and scanning electron microscopes. *Review of Scientific Instruments* 82, 063901.

Gómez, L.S., López-Arce, P., Álvarez de Buergo, M. & Fort, R. (2009). Calcium hydroxide nanoparticles crystallization on carbonates stone: dynamic experiments with heating/cooling and Peltier stage ESEM. *Acta Microscopica* 18, 105-106.

Goodrich, T.W. & Lattimer, B.Y. (2011). Fire Decomposition Effects on Sandwich Composite Materials. *Composites A*, doi:10.1016/j.compositesa.2011.03.007

Gorkaya, T., Burlet, T., Molodov, D.A. & Gottstein, G. (2010). Experimental method for true in situ measurements of shear-coupled grain boundary migration. *Scripta Materialia* 63, 633–636.

Gregori, G., Kleebe, H.J., Siegelin, F. & Ziegle, G.(2002). In situ SEM imaging at temperatures as high as 1450°C. *Journal of Electron Microscopy* 51, 347-52.

Gualtieri, A.F., Lassinantti Gualtieri, M. & Tonelli, M. (2008). In situ ESEM study of the thermal decomposition of chrysotile asbestos in view of safe recycling of the transformation product. *Journal of Hazardous Materials* 156, 260-266.

Hillers, M., Matzen, G., Veron, E., Dutreilh-Colas, M. & Douy, A. (2007). Application of In Situ High-Temperature Techniques to Investigate the Effect of B_2O_3 on the

Crystallization Behavior of Aluminosilicate E-Glass. *Journal of the American Ceramic Society* 90, 720-726.

Hingant, N., Clavier, N., Dacheux, N., Hubert, S., Barré, N., Podor, R. & Aranda, L. (2011). Preparation of morphology controlled $Th_{1-x}U_xO_2$ sintered pellets from low-temperature precursors. *Powder Technology* 208, 454–460.

Hung, J.H.H., Chiu, Y.L., Zhu, T. & Gao, W. (2007). In situ ESEM study of partial melting and precipitation process of AZ91D. *Asia-Pacific Journal of Chemical Engineering*, Special Issue: Special issue for the Chemeca 2006 John A Brodie Medal Nominated Papers. Volume 2, Issue 5, pages 493–498, September/October 2007.

Imaizumi, K., Matsuda, N. & Otsuka, M. (2003). Coagulation/phase separation process in the silica/inorganic salt systems (1) – observation of state transformation – *Journal of Materials Science* 38, 2979 – 2986.

James, B. (2009). Advances in "wet" electron microscopy techniques and their application to the study of food structure. *Trends in Food Science & Technology* 20, 114-124.

Joly-Pottuz, L., Bogner, A., Lasalle, A., Malchere, A., Thollet, G. & Deville, S. (2011). Improvements for imaging ceramics sintering in situ in ESEM. *Journal of Microscopy*, 244, 93-100.

Jonsson, T., Folkeson, N., Svensson, J.E., Johansson, L.G., & Halvarsson M. (2011). An ESEM in situ investigation of initial stages of the KCl induced high temperature corrosion of a Fe–2.25Cr–1Mo steel at 400 °C. *Corrosion Science* 53, 2233-2246.

Jonsson, T., Pujilaksono, B., Hallström, S., Ågren, J., Svensson, J.E., Johansson, L.G. & Halvarsson, M. (2009). An ESEM in situ investigation of the influence of H_2O on iron oxidation at 500°C. *Corrosion Science* 51, 1914-1924.

Jornsanoh, P., Thollet, G., Ferreira, J., Masenelli-Varlot, K., Gauthier, C. & Bogner, A. (2011). Electron tomography combining ESEM and STEM: A new 3D imaging technique *Ultramicroscopy*, doi:10.1016/j.ultramic.2011.01.041

Jung, Y.C. & Bhushan, B. (2008). Wetting behaviour during evaporation and condensation of water microdroplets on superhydrophobic patterned surfaces *Journal of Microscopy* 229, 127-140.

Kirk, S.E., Skepper, J.N. & Donald, A.M. (2009). Application of environmental scanning electron microscopy to determine biological surface structure. *Journal of Microscopy* 233, 205–224.

Klemensø, T., Appel, C.C. & Mogensen, M. (2006). In Situ Observations of Microstructural Changes in SOFC Anodes during Redox Cycling. *Electrochemical and Solid-State Letters* 9, A403-A407

Knoll, M. (1935). Aufladepototentiel une Sekündaremission elektronbestrahlter Körper. Zeitschrift fur technische Physik 16, 467-475.

Knowles, R. & Evans, B. (1997). *High temperature specimen stage and detector for an ESEM.* Patent WO 97/07526

Lau, K.K.S., Bico, J., Teo, K.B.K., Chhowalla, M., Amaratunga, G.A.J., Milne, W.I., McKinley, G.H. & Gleason, K.K. (2003). Superhydrophobic Carbon Nanotube Forests. *Nano Letters* 3, 1701-1705.

Leary, R. & Brydson, R. (2010). Characterisation of ESEM conditions for specimen hydration control. *Journal of Physics: Conference Series* 241, 012024.

Lin, T., Jia, D. & Wang, M. (2010). In situ crack growth observation and fracture behavior of short carbon fiber reinforced geopolymer matrix composites. *Materials Science and Engineering* A 527, 2404–2407.

Maison, T., Laouafa, F., Fleureau, J.M. & Delalain, P. (2009). *Analyse aux échelles micro et macroscopique des mécanismes de dessiccation et de gonflement des sols argileux.* Proceeding of the 19ème Congrès Français de Mécanique Marseille, 24-28 août 2009.

Manero, J.M., Masson, D.V., Marsal, M. & Planell, J.A. (1998). Application of the Technique of Environmental Scanning Electron Microscopy to the Paper Industry. *Scanning* 21, 36–39.

Maraloiu, V.A., Hamoudeh, M., Fessi, H. & Blanchin, M.G. (2010). Study of magnetic nanovectors by Wet-STEM, a new ESEM mode in transmission. *Journal of Colloid and Interface Science* 352, 386–392.

Maroni, V.A., Teplitsky, M. & Rupich M.W. (1999). An environmental scanning electron microscope study of the AgrBi-2223 composite conductor from 25 to 840°C. *Physica* C 313, 169–174.

Marzagui, H. & Cutard, T. (2004). Characterisation of microstructural evolutions in refractory castables by in situ high temperature ESEM. *Journal of Materials Processing Technology* 155-156, 1474-1481.

McGregor, J.E. & Donald, A.M. (2010). ESEM imaging of dynamic biological processes: the closure of stomatal pores. *Journal of Microscopy* 239, 135–141.

Mège-Revil, A., Steyer, P., Thollet, G., Chiriac, R., Sigala, C., Sanchéz-Lopéz, J.C. & Esnouf, C. (2009). Thermogravimetric and in situ SEM characterisation of the oxidation phenomena of protective nanocomposite nitride films deposited on steel. *Surface & Coatings Technology 204*, 893–901

Mendez-Vilas, A., Belen Jodar-Reyes, A., & Gonzalez-Martin, M.L. (2009). Ultrasmall Liquid Droplets on Solid Surfaces: Production, Imaging, and Relevance for Current Wetting Research. *Small* 5(12), 1366–1390.

Miller, M.K. & Russell, K.F. (2007). Atom probe specimen preparation with a dual beam SEM/FIB miller. *Ultramicroscopy* 107, 761-766.

Montes-H., G. (2005). Shrinkage measurements of bentonite using coupled environmental scanning electron microscopy and digital image analysis. *Journal of Colloid and Interface Science* 284, 271–277.

Montes-Hernandez, G. (2002). *Etude expérimentale de la sorption d'eau et du gonflement des argiles par microscopie électronique à balayage environnementale (ESEM) et l'analyse digitale d'images.* Thèse de 3éme cycle 162pp.

Muscariello, L., Rosso, F., Marino, G., Giordano, A., Barbarisi, M., Cafiero, G. & Barbarisi, A. (2005). A Critical Overview of ESEM Applications in the Biological Field. *Journal of Cellular Physiology* 205, 328–334.

Nakamura, M., Isshiki, T., Tamai, M. & Nishio, K. (2002). *Development of a new heating stage equipped thermal electron filter for scanning electron microscopy.* Proceeding of the 15th International Congress on Electron Microscopy Durban, South Africa.

Nishiyama, H., Suga, M., Ogura, T., Maruyama, Y., Koizumi, M., Mio, K., Kitamura, S. & Sato, C. (2010). Atmospheric scanning electron microscope observes cells and tissues in open medium through silicon nitride film. *Journal of Structural Biology* 169, 438-449.

Oquab, D. & Monceau, D. (2001). In-situ SEM study of cavity growth during high temperature oxidation of β-(Ni, Pd)Al. *Scripta Materialia* 44, 2741-2746.

Perrie, Y., Ali, H., Kirby, D.J., Mohammed, A.U.R., McNeil, S.E. & Vangala A. (2010). *Environmental Scanning Electron Microscope Imaging of Vesicle Systems in "Liposomes: Methods and Protocols, Volume 2: Biological Membrane Models"*, Methods in Molecular Biology 606, 319-331.

Perrie, Y., Mohammed, A.U.R., Vangala, A. & McNeil, S.E. (2007). Environmental Scanning Electron Microscopy Offers Real-Time Morphological Analysis of Liposomes and Niosomes. *Journal of Liposome Research* 17, 27-37.

Podor, R., Clavier, N., Ravaux, J., Claparéde, L., Dacheux, N. & Bernache-Assollant, D. (2012). Dynamic aspects of cerium dioxide sintering: HT-ESEM study of grain growth and pore elimination. *Journal of the European Ceramic Society*, 32, 353-362.

Podor, R., Pailhon, D., Ravaux, J. & Brau, H.P. (2011). *Porte-échantillon à thermocouple intégré*. Demande de brevet français déposée le 21 juillet 2011 sous le n° 11 56612.

Poelt, P., Zankel, A., Gahleitner, M., Ingolic, E. & Grein, C. (2010). Tensile tests in the environmental scanning electron microscope (ESEM) - Part I: Polypropylene homopolymers. *Polymer* 51, 3203-3212.

Popma, R.L.W. (2002). *Sintering characteristics of nano-ceramics*, PhD thesis, University of Groningen.

Proff, C., Abolhassani, S., Dadras, M.M. & Lemaignan, C. (2010). In situ oxidation of zirconium binary alloys by environmental SEM and analysis by AFM, FIB, and TEM. *Journal of Nuclear Materials* 404, 97–108.

Quémarda, L., Desgranges, L., Bouineau, V., Pijolat, M., Baldinozzi, G., Millot, N., Nièpce, J.C. & Poulesquen, A. (2009). On the origin of the sigmoid shape in the UO_2 oxidation weight gain curves. *Journal of the European Ceramic Society* 29, 2791–2798.

Reichmann, A., Poelt, P., Brandl, C., Chernev, B. & Wilhelm, P. (2008). High-Temperature Corrosion of Steel in an ESEM With Subsequent Scale Characterisation by Raman Microscopy. *Oxidation of Metals* 78, 257-266.

Reichmann, A., Zankel, A., Reingruber, H., Pölt, P. & Reichmann, K. (2011). Direct observation of ferroelectric domain formation by environmental scanning electron microscopy. *Journal of the European Ceramic Society* , 31[15], 2939-2942.

Romano-Rodriguez, A. & Hernandez-Ramirez, F. (2007). Dual-beam focused ion beam (FIB): A prototyping tool for micro and nanofabrication. *Microelectronic Engineering* 84,789–792.

Ruozi, B., Belletti, D., Tombesi, A., Tosi, G., Bondioli, L., Forni, F. & Vandelli, M.A. (2011). AFM, ESEM, TEM, and CLSM in liposomal characterization: a comparative study. *International Journal of Nanomedicine* 6, 557–563.

Rykaczewski, K. & Scott, J.H.J. (2011). Methodology for Imaging Nano-to-Microscale Water Condensation Dynamics on Complex Nanostructures. *ACSNano* 5[7], 5962-5968.

Sample, D.R., Brown, P.W. & Dougherty, J.P. (1996). Microstructural evolution of copper thick films observed by environmental scanning electron microscopy. *Journal of the American Ceramic Society* 79, 1303-1306.

Schaller, R.C., Fukuta, N. (1979). Ice nucleation by aerosol particles: Experimental studies using a wedge-shaped ice thermal diffusion chamber. *Journal of Atmospheric Sciences* 36, 1788–1802.

Schmid, B., Aas, N., Grong, Ø. & ØDegard, R. (2001). High-temperature oxidation of nickel and chromium studied with an in-situ environmental scanning electron microscope. *Scanning* 23, 255-266.

Schmid, B., Aas, N., Grong, Ø. & ØDegard, R. (2001). In situ environmental scanning electron microscope observations of catalytic processes encountered in metal dusting corrosion on iron and nickel. *Applied Catalysis* A 215, 257-270.

Schmid, B., Aas, N., Grong, Ø. & ØDegard, R. (2002). High-Temperature Oxidation of Iron and the Decay of Wüstite Studied with in situ ESEM. *Oxidation of Metals* 57, 115-130.

Schoßig, M., Zankel, A., Bieröge, C., Pölt, P. & Grellmann, W. (2011). ESEM investigations for assessment of damage kinetics of short glass fibre reinforced thermoplastics – Results of in situ tensile tests coupled with acoustic emission analysis. *Composites Science and Technology* 71, 257-265

Seward, G.G.E., Prior, D.J., Wheeler, J., Celotto, S., Halliday, D.J.M., Paden, R.S. & Tye, M.R. (2002). High-Temperature Electron Backscatter Diffraction and Scanning Electron Microscopy Imaging Techniques: In-situ Investigations of Dynamic Processes. *Scanning* 24, 232–240.

Seward, G.G.E., Celotto, S., Prior, D.J., Wheeler, J. & Pond R.C. (2004). In situ SEM-EBSD observations of the hcp to bcc phase transformation in commercially pure titanium. *Acta Materialia* 52, 821-832.

Sievers, T.K., Bonnefond, F., Demars, T., Genre, C., Meyer, D., Podor, R. Vapour pressure dependent size of coordination polymer network meso-particles. *Advanced Materials* (submitted)

Smith, A.J., Atkinson, H.V., Hainsworth, S.V. & Cocks, A.C.F. (2006). Use of a micromanipulator at high temperature in an environmental scanning electron microscope to apply force during the sintering of copper particles. *Scripta Materialia* 55, 707-710.

Sorgi, C. & De Gennaro, V. (2007). Analyse microstructurale au MEB environnemental d'une craie soumise à chargement hydrique et mécanique. *Comptes-rendus Geosciences* 339, 468–481.

Srinivasan, N.S. (2002). Dynamic study of changes in structure and morphology during the heating and sintering of iron powder. *Powder Technology* 124, 40-44.

Stabentheiner, E., Zankel, A. & Pölt, P. (2010). Environmental scanning electron microscopy (ESEM) — a versatile tool in studying plants. *Protoplasma* 246:89–99.

Stelmashenko, N.A., Craven, J.P., Donald, A.M., Terentjev, E.M. & Thiel, B.L. (2001). Topographic contrast of partially wetting water droplets in environmental scanning electron microscopy. *Journal of Microscopy* 204, 172-183.

Stokes, D.J. & Donald A.M. (2000). In situ mechanical testing of dry and hydrated breadcrumb in the ESEM. *Journal of Materials Science* 35, 599-607.

Stokes, D.J. (2001). Characterisation of Soft Condensed Matter and Delicate Materials Using Environmental Scanning Electron Microscopy (ESEM). *Advanced Engineering Materials* 3, 126-130.

Stokes, D.J. (2008). *Principles and practice of variable pressure/environmental scanning electron microscopy (VP/ESEM)*, John Wiley & Sons Ltd, The Atrium, Southern Gate, Chichester, West Sussex, UK.

Subramaniam S. (2006). *In Situ High Temperature Environmental Scanning Electron Microscopic Investigations of Sintering Behavior in Barium Titanate.* PhD thesis, University of Cincinnati, Cincinnati USA.

Suga, M., Nishiyama, H., Konyuba, Y., Iwamatsu, S., Watanabe, Y., Yoshiura, C., Ueda, T. & Sato, C. (2011). The Atmospheric Scanning Electron Microscope with open sample space observes dynamic phenomena in liquid or gas. *Ultramicroscopy,* doi:10.1016/j.ultramic.2011.08.001

Thiel, B.L. & Donald, A.M. (1998). In situ Mechanical Testing of Fully Hydrated Carrots (Daucus carota) in the Environmental SEM. *Annals of Botany* 82: 727-733.

Thiel, B.L., Stokes, D.J. & Donald, A.M. (2002). Application of Environmental Scanning Electron Microscopy to the Study of Food Systems. *Microscopy and Microanalysis* 8, 960-961.

Torres, E.A. & Ramirez, A.J. (2011). In situ scanning electron microscopy. *Science and technology of welding and Joining* 16(1), 68-78.

Utku, F.S., Klein, E., Saybasili, H., Yucesoy, C.A. & Weiner, S. (2008). Probing the role of water in lamellar bone by dehydration in the environmental scanning electron microscope. *Journal of Structural Biology* 162, 361-367.

Vigouroux, H., Fargin, E., Le Garrec, B., Dussauze, M., Rodriguez, V., Adamietz, F., Ravaux, J., Podor, R., Vouagner, D., De Ligny, D. & Champagnon, B. (2011). *Phase Separation and Crystallization Mechanism in $LiNbO_3$-SiO_2 Glasses.* International conference on the chemistry of glasses, 4-8 sept 2011, Oxford (UK).

Vigouroux, H., Fargin, E., Le Garrec, B., Dussauze, M., Rodriguez, V., Adamietz, F., Ravaux, J., Podor, R., Vouagner, D., De Ligny, D. & Champagnon, B. In-Situ Study of $LiNbO_3$ crystallization in lithium niobium Silicate glass ceramic. (in prep).

von Ardenne, M. (1938a). Das Elektronen-Rastermikroskop. Praktische Ausführung. *Zeitschrift fur technische Physik* 19, 407-416.

von Ardenne, M. (1938b). Das Elektronen-Rastermikroskop. Theoretische Grundlagen. *Zeitschrift fur Physik* 109, 553-572.

Wei, Q.F. (2004). Surface characterization of plasma-treated polypropylene fibers. *Materials Characterization* 52, 231-235.

Wich, T., Stolle, C., Luttermann, T. & Fatikow, S. (2011). Assembly automation on the nanoscale. *CIRP Journal of Manufacturing Science and Technology,* doi:10.1016/j.cirpj.2011.03.003

Wilson, B.A. & Case, D.E. (1997). In situ microscopy of crack healing in borosilicate glass. *Journal of Materials Science* 32, 3163-3175.

Wise, M.E., Martin, S.T., Russell, L.M. & Buseck, P.R. (2008). Water uptake by NaCl particles prior to deliquescence and the phase rule. *Aerosol Science and Technology* 42(4), 281-294.

Yang, J. (2010). In-situ High Resolution SEM Imaging with Heating Stage. Scanning Electron Microscopes (SEM) from Carl Zeiss.

Yu, H.M., Ziegler, C., Oszcipok, M., Zobel, M. & Hebling, C. (2006). Hydrophilicity and hydrophobicity study of catalyst layers in proton exchange membrane fuel cells. *Electrochimica Acta* 51, 1199-1207.

Zheng, T, Waldron, K.W. & Donald, A.M. (2009). Investigation of viability of plant tissue in the environmental scanning electron microscopy. *Planta* 230, 1105–1113.

Zimmermann, F., Ebert, M., Worringen, A., Schutz, L. & Weinbruch, S. (2007). Environmental scanning electron microscopy (ESEM) as a new technique to determine the ice nucleation capability of individual atmospheric aerosol particles. *Atmospheric Environment* 41, 8219–8227.

3

Gaseous Scanning Electron Microscope (GSEM): Applications and Improvement

Lahcen Khouchaf

Université Lille - Nord de France, Ecole des Mines de Douai, Douai, France

1. Introduction

The imaging and the microanalysis of hydrated and insulating materials using electron beam probe methods (Conventional SEM, Auger Electron Microscopy,...) are very limited by the necessity to keep the sample under high vacuum and the presence of the charge effect. In this case the sample must be coated except when the experiment is performed at very low energy in order to avoid the charge phenomenon. Studies of vegetable and biological samples are almost impossible without degradation. In Conventional SEM (Pressure = 10^{-5} mbar in the specimen chamber) image quality and microanalysis results are strongly related to the size of the electron beam, the accelerating voltage and the nature of the sample. A large description of different aspects on SEM/EDS exists in the literature (Newbury et al, 1986; Goldstein et al, 1992).

In order to overcome the high vacuum in the specimen chamber different types of microscopes with the possibility to introduce different gases inside the sample chamber are now available (Danilatos, 1980, 2009, Carlton, 1997, Wight, 2001). Depending on the pressure value in the specimen chamber different names are given in the literature such as ESEM: Environmental Scanning Electron Microscope, LVSEM: Low Vacuum Scanning Electron Microscope, HPSEM: High Pressure Scanning Electron Microscope, VPSEM: Variable Pressure Scanning Electron Microscope, CPSEM: Controlled pressure Scanning Electron Microscope and depending on the maximum pressure attainable in the specimen chamber (Danilatos 1988, Khouchaf & Vertraete, 2002, 2004; Khouchaf et al., 2006, 2007, 2010, 2011; Kadoun et al, 2003; Gilpin, 1994; Carlton, 1997; Doehne, 1997; Newbury 2002; Gauvin, 1999; Bolon, 1991; Wight, 2001). But all these microscopes differ from CSEM by the capability to introduce the gas as an environment unlike High vacuum in CSEM and the use of gaseous detection system such as Gaseous Secondary Electron Detector (GSED). Indeed, all these microscopes may be called Gaseous Scanning Electron Microscope (GSEM).

Unlike CSEM, with GSEM image quality and microanalysis results are strongly related to the size of the electron beam, the accelerating voltage, the nature of the sample and depend on the pressure in the chamber and the kind of gas used. GSEM showed the enormous use in several fields using materials. The electron beam scattering tends to decrease the resolution. Different correction methods were developed (Bilde-Sorensen et al, 1996; Doehne, 1996-1997; Le Berre et al, 1997; Gauvin et al, 1999) but weren't satisfactory for many reasons such as the time and the difficulty of implementation.

In this chapter different applications using GSEM will be given in the first part. In the second part an introduction of some physical phenomena related to the scattering of the primary electron beam with the gas and their consequence on the image quality and on the microanalysis results is given. The focus here is to present the potential, the limitation and some way to optimize the use of GSEM.

2. Applications of ESEM

GSEM allows imaging and analysis of many types of materials without any preparation with the presence of a gaseous environment. Gas atoms or molecules interact with the primary electron beam and produce positive ions. The presence of the positive ions allows neutralization the negative charge on the surface of the insulating sample. Gas serves also for detection (Danilatos, 1988).

Different types of gases may be introduced such as: N_2, O_2, Ar, He, H_2O...... As an illustration, we give two examples. By introduction of gases, vacuum incompatible materials, and dynamic surface modification may be studied. At pressures in excess of 500 Pa, water can be condensed in situ enabling characterization of hydrated materials.

In this study, the experiments were performed in environmental 'wet' mode using an Environmental Scanning Electron Microscope (ESEM) « ElectroScan 2020 » equipped with EDS Microanalysis system « Oxford Linkisis ». The electron source is a tungsten filament. The energy of the electron beam used was 20 kV with an emission current of 49 µA. The condenser is also fixed at 43 % value and the diameter of the projection aperture is 50 µm. Secondary electrons were detected using a long gaseous secondary electron detector (GESD)at a working distance of 19 mm is used in order to reduce the skirt beam phenomena. The chamber pressure is varied by introducing gas.

The sample is embedded in epoxy resin and polished.

For the estimation of the unscattered fraction and the skirt radius an ESEM electron flight simulator software was used (Electron Flight Simulation Software, version 3.1-E).

2.1 Observation of vegetable

Plant material is insulating and has a fragile structure. Its observation using an electron beam is a delicate operation. Its structure is degraded under the electron beam. Some plants contain Stomata (or epidermis) the structures deposited on the outer leaf skin layer. They consist of two cells, called guard cells that surround a tiny pore called a stoma. Stomata allow communication between the internal and external environments of the plant.

Figure 1a shows an image of a plant after the coating operation in order to neutralize the negative charges. The image is obtained at 20 kV and high vacuum using the Gaseous Secondary Electron Detector (GSED) detector. This specimen is formed by stomata which are tiny openings or pores, found mostly on the underside of a plant leaf and used for gas exchange. The image below shows an elongated and irregular structure.

The same observation is made without coating under a gaseous environment of water vapor (Fig.2b). In this case, the morphological aspect and the structure of the plant are very different compared to the image in Fig. 1a. The sample kept its structure. We can observe the stomata in their natural state. In order to study mechanism of exchange in the plant, it is necessary to keep the sample in its natural state. This is possible by using GSEM. It is interesting to underline that despite the scattering phenomena, the image kept its quality. That will be explained below.

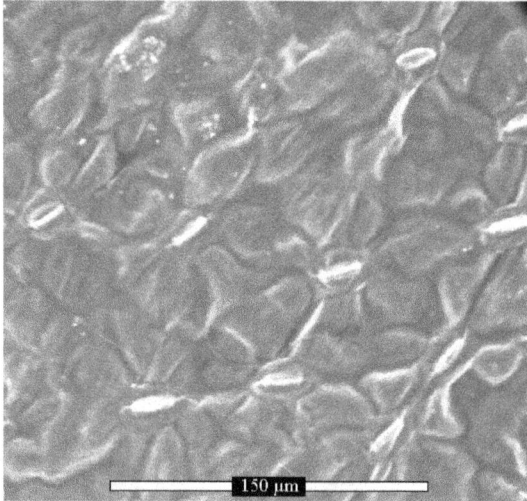

Fig. 1a. ESEM micrographs of a plant before the coating process.

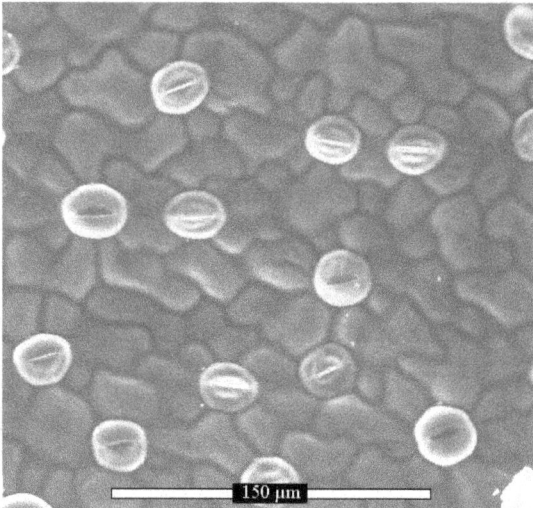

Fig. 1b. ESEM micrographs of a plant before the coating process.

2.2 Detection of calcium potassium inside SiO₂ Framework

Figure 2 shows an example of a flint aggregate subjected to attack by Alkali-Silica Reaction (ASR). ASR is a physicochemical process which takes place during the degradation of concrete. Observation of the ASR effects using Scanning Electron Microscope (SEM) is important because the attack of the aggregate is heterogeneous on the microscopic scale. When using CSEM, sample preparation prior to imaging is required, which may lead to an alteration of the true surface morphology or even the creation of artifacts.

The use of GSEM overcomes these problems and gives direct imaging of the samples in their natural state. The image in Fig. 2 shows the presence of different degraded zones (1 to 8) affected by chemical reaction. These zones have different micronic sizes and under the given conditions, the effect of the beam skirt is not the same. If the volume of the generation of X-rays is lower than the size of the zone then the effect of the skirt may be neglected. If the volume of the generation of X-rays is higher than the size of the zone then the effect of the skirt may be taken into account.

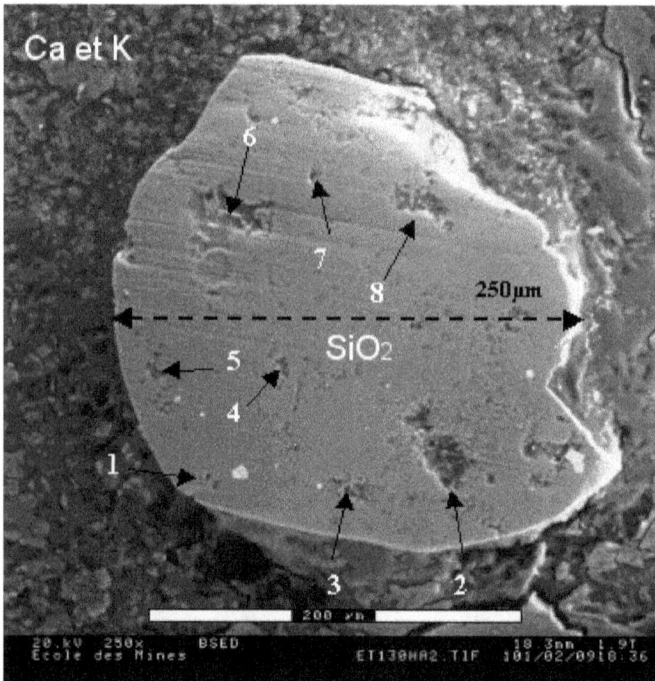

Fig. 2. ESEM micrographs of flint aggregate after reaction at (GSED, P=532 Pa).

The images below (Figures 3) show the effect of calcium cations during the degradation of concrete. Figure 3a presents flint aggregate after 30 hours of reaction with the presence of calcium and potassium and figure 3b without calcium. With the presence of calcium, it is interesting to note that the reaction has not finished particularly in the center of the aggregate (Fig. 3a).

However, when the calcium is removed, the grain is fragmented to small grain clearly separated by showing that the mechanism is different when calcium is present (Fig. 3b). Despite a high pressure (532 Pa) and high accelerating voltage 20 kV, again the resolution of the image is not bad.

Fig. 3. ESEM micrographs of (a) flint aggregate after reaction at (GSED, P=532 Pa) with the presence of calcium, (b) flint aggregate after reaction (GSED, P=545.3 Pa) without calcium.

3. Description of the beam skirt

Experiments above were performed with insulating materials without any preparation and without the coating procedure by introducing a gaseous environment into the specimen chamber. A part of primary electron beam interacts with the atoms or molecules of the gas.

The average collision number with particle gas per electron is given by the equation below:

$$m = \sigma_t \times n \times L \qquad (1)$$

Where

σ_t: scattering cross section is specific to each gas molecule
n: gas particle number/volume
L: Working distance (distance between the final aperture PLA1 and the surface of the sample see Figure 4). The Gas Path Length GPL is introduced and corresponds to the distance that electrons have to travel through the gas to reach the sample. m may be expressed as below (Danilatos, 1988):

$$m = \frac{\sigma_T . L . P}{k . T} \qquad (2)$$

where σ_T is the total cross section of the gas, L the gas path length, P the gas pressure, k the Boltzmann constant and T the temperature.

Using the number m it is possible to define three different scattering regimes corresponding to:

Minimal scattering: $m < 0.05$
Partial scattering: $0.05 < m < 3$
Plural scattering: $m > 3$

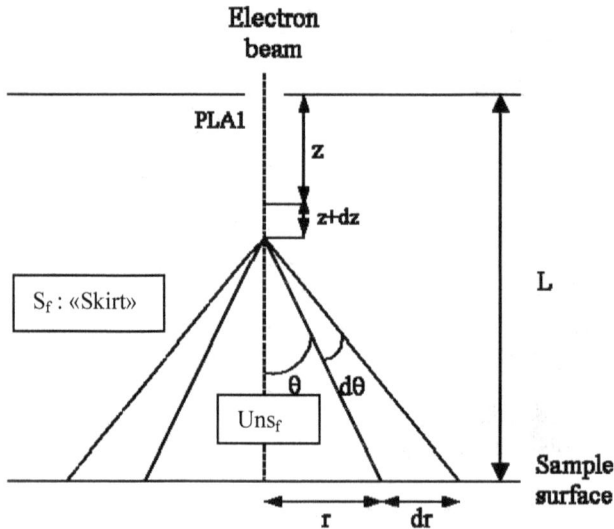

Fig. 4. An electron after PLA1 aperture of the ESEM, moves along the axis of PLA1, undergoes a collision at a distance between z and z + dz in an angle of $\theta + d\theta$; it is then scattered and arrives at the surface of the sample in an annulus between r and r + dr.

After interaction between electron and gas, the primary electron beam is divided into two parts (Fig. 4) called "scattered fraction: S_f" which corresponds to the elastic scattering by the gas atoms or molecules and a second part called "unscattered fraction: Uns_f". The "unscattered fraction" of the electron beam can be written by the equation below when a simple mode of scattering is considered (Danilatos, 1988):

$$Uns_f = \exp(-\frac{P \times L \times \sigma_t}{k \times T})$$
(3)

P: Pressure in the specimen chamber
L: Working distance (WD, distance between the final aperture PLA1 (Fig. 1) and the surface of the sample). In this study the Gas Path Length GPL is introduced and corresponds to the distance that electrons have to travel through the gas to reach the sample.
σ_t: total scattering cross section is specific to each gas molecule
k: Boltzman constant
T: Temperature in Kelvin.

Elastic scattering leads to the enlargement of the primary electron beam to form a skirt producing the generation of X-rays which are not representative of the zone of interest for X-ray microanalysis. Different correction methods have been developed in order to take into account the contribution of the skirt (Bilde-Sorensen et al, 1996; Doehne, 1996-1997; Le Berre et al, 1997; Gauvin et al, 1999). Up to now these methods have not been successful. Danilatos, 1988) introduced the radius rs which represents the radius containing 90% of the incident beam) as below:

$$r_s = \frac{364.Z}{E} \cdot \left(\frac{P}{T}\right)^{\frac{1}{2}} \cdot GPL^{\frac{3}{2}}$$
(4)

where r_s is the skirt radius, Z the gas atomic number, E the incident beam energy, P the pressure, T the temperature and GPL the gas path length.

Considering the condition used in $ 2.2, the simulations of the electron beam scattering were performed using the Electron Flight Simulator software (Figure 5a). In this case the unscattered fraction is about 85.7% with a rs of 26 μm. Worst case X-ray Gen radius means the approximate region where X-ray signals will be generated and given just as indication. Based on our previous sudy using helium gas (Khouchaf et al, 2004, 2007, 2011), the same simulation under helium gas (Fig1.b) leads to Uns$_f$ of 97.6%, rs < 1 μm showing a good improvement of the conditions with a gas having a low average atomic number.

Some authors (John F. Mansfield) have suggested that quantitative analysis is possible with ESEM (with water vapour as the standard gas) only under very restrictive conditions such as : short working distances between 6 mm and 7.2mm, gas path length between 1.2mm and 2.2mm in the 70 to 350Pa range at high accelerating voltage of 30 kV. From these conditions it is easy to notice that the values of working distance, gas path length and pressure must be very low when the accelerating voltage decreases. Using conditions close to that given by Mansfield we perform a simulation by means of Electron Flight Simulator (Fig. 6a and 6b).

At P= 70 Pa and GPL=1mm, Uns$_f$ and r_s are close to 95.5% and <1 μm respectively. When the pressure increases to 350 Pa and GPL to 2mm (Fig. 6b), Uns$_f$ decreases to 93.5% and r_s increases to 2 μm.

Parameters suggested by Mansfield consider a high accelerating voltage of 30 kV which is conform with excitation of heavy or metallic elements and which we can study by using

CSEM. In addition, high voltage leads to large volume of interaction and a degradation of the resolution by generation of unwanted x-rays. More applications in GSEM require a low accelerating voltage which is possible by recent microscope with a FEG gun.

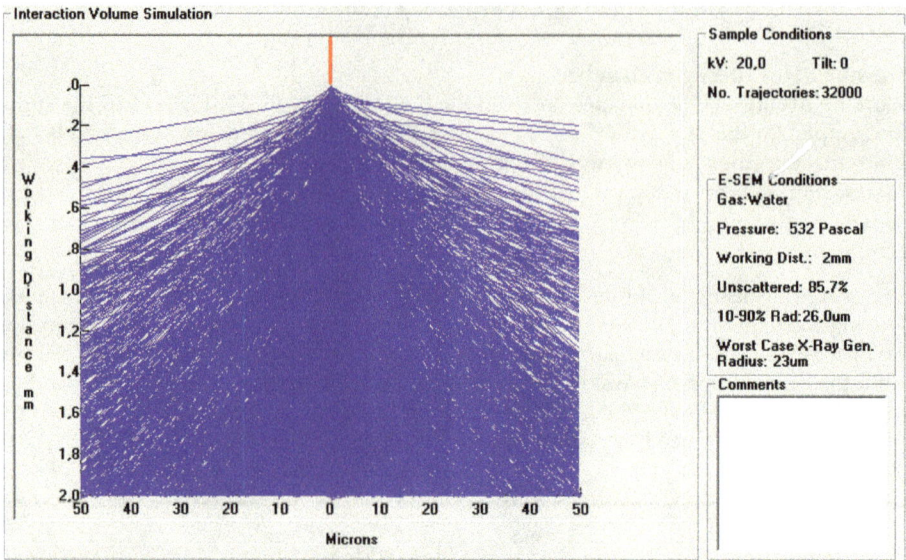

Fig. 5a. Monte Carlo simulation using Electron Flight Simulator of the electron beam scattering under water vapor, V=20kV, P= 532 Pa, GPL = 2mm.

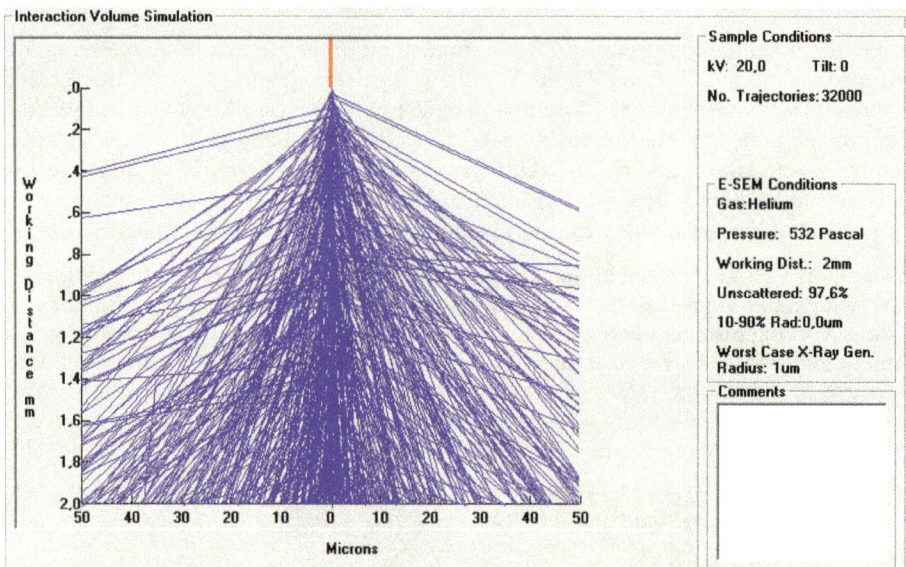

Fig. 5b. Monte Carlo simulation using Electron Flight Simulator of the electron beam scattering under helium, V=20kV, P= 532 Pa, GPL = 2mm.

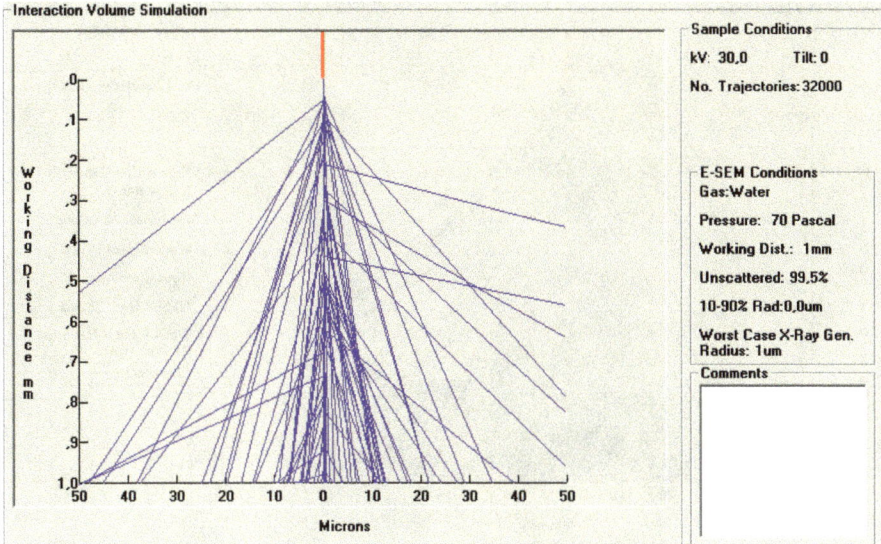

Fig. 6a. Monte Carlo simulation using Electron Flight Simulator of the electron beam scattering under water vapor, V=30kV, P= 70 Pa, GPL = 1mm.

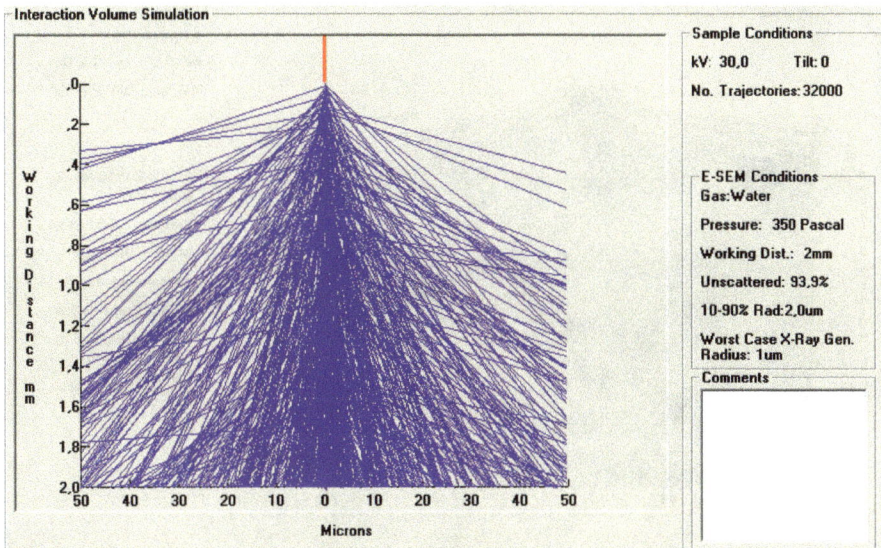

Fig. 6b. Monte Carlo simulation using Electron Flight Simulator of the electron beam scattering under water vapor, V=30kV, P= 350 Pa, GPL = 2mm.

Let us consider the conditions above at low accelerating voltage of 5 kV under water vapor (Fig7a and 7b). At P= 70 Pa and GPL=1mm, Uns_f and r_s are close to 93.2% and 12 μm respectively, but at P= 350 Pa and GPL=2mm, Uns_f and r_s are close to 47% and 109.3 μm respectively . It is easy to conclude that the quality of the results will be affected.

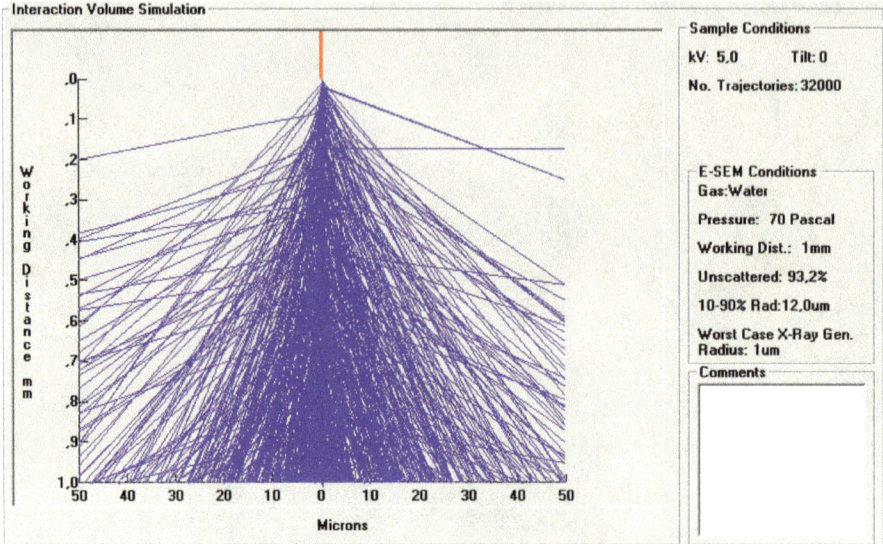

Fig. 7a. Monte Carlo simulation using Electron Flight Simulator of the electron beam scattering under water vapor, V=5kV, P= 70 Pa, GPL = 1mm.

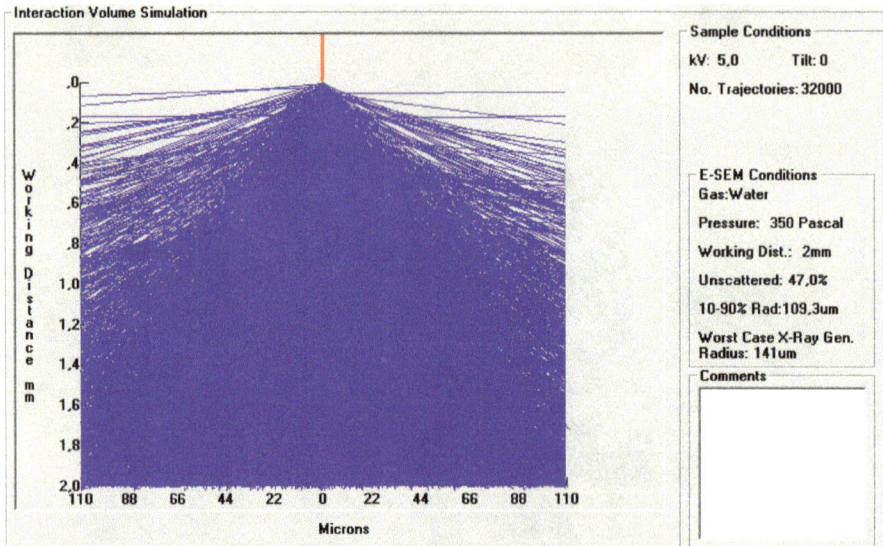

Fig. 7b. Monte Carlo simulation using Electron Flight Simulator of the electron beam scattering under water vapor, V=5kV, P= 350 Pa, GPL = 2mm.

As given by equation 4, the value of rs depends on the gas introduced, the incident energy, the pressure, the temperature and the working distance. Indeed, the good way to minimize the beam skirt phenomena is to optimize different parameters used during X-ray microanalysis. With equation 3 this leads to choose a gas with a low average atomic number, to increase the incident beam energy, to reduce the pressure and the gas path length, to increase the temperature. Unfortunately, these conditions lead to a significant limitation to use GSEM. For example for ESEM, the standard gas is water vapor and the best results are obtained by helium (ref Khouchaf). Increasing the incident beam leads to a minimization of the beam skirt by decreasing the resolution and degrading the fragile materials.

Based on the conditions and parameters such as the bem energy, the pressure, the gas, the GPL it's not sufficient to define the limit of the use of GSEM. In order to obtain the best results it is also necessary to take into account the value oft he average number of scattering events per electron m. The best results will be obtained with a minimal scattering regime corresponding to m<0.05. This suggests the use a gas with a low average atomic number (Khouchaf et al, 2011).

 Unfortunately most new microscopes use gases with a high average atomic number such as (N_2, air, H_2O vapor). The improvement of the results can also be reached by increasing the temperature. One way is to use a gas with a low average atomic number such as helium. Figures (8a and 8b) below show the electron Flight Simulator spectra obtained under helium environment. The results may be compared to those in figures 6a and 6b.

At P= 70 Pa and GPL=1mm, Uns_f and r_s are close to 99.8% and <1 µm respectively and at P= 350 Pa and GPL=2mm, Uns_f and r_s are close to 99.1% and <1 µm respectively. It is easy to conclude that the quality of the results will be improved.

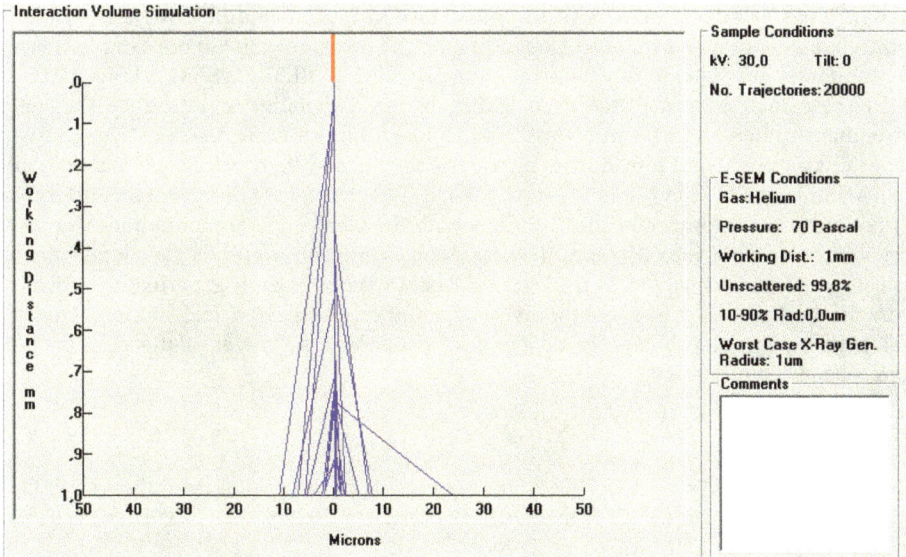

Fig. 8a. Monte Carlo simulation using Electron Flight Simulator of the electron beam scattering under helium, V=5kV, P= 70 Pa, GPL = 1mm.

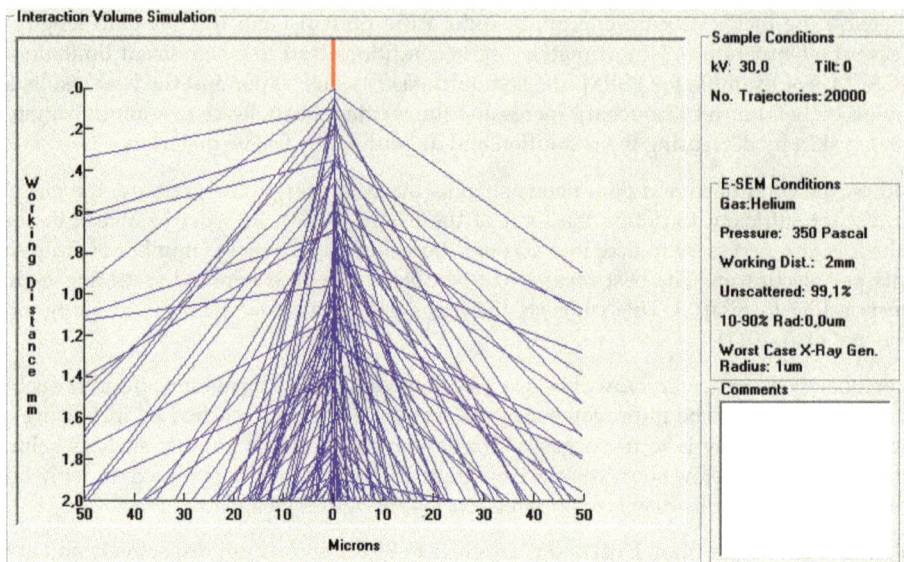

Fig. 8b. Monte Carlo simulation using Electron Flight Simulator of the electron beam scattering under helium, V=5kV, P= 350 Pa, GPL = 2mm.

4. Conclusion

Different types of microscopes with the possibility to introduce different gases inside the sample chamber are now available. Depending on the pressure value in the sample chamber different names are given in the literature such as ESEM, LVSEM, HPSEM, VPSEM, CPSEM. But all these microscopes differ from CSEM by the capability to introduce the gas as environment unlike High vacuum in CSEM. Indeed, all these microscopes work under a gaseous environment and introducing a gaseous detection system in this way may be called gaseous Scanning Electron Microscope (GSEM). In this chapter we demonstrate and confirm the possibility to perform interesting studies with the GSEM if some limitations due to the beam skirt are taken into account. The different correction methods developed are not satisfactory. The good way is to find the best parameters for each experiment in order to obtain the best results based on the average number of collision m. Another way is to develop new microscopes capable of avoiding (isolating) the travel of the electron beam across gaseous environment.

5. References

Bilde-Sorensen, JB & Appel, CC. (1996). Improvements of the spatial resolution of XEDS in the environmental SEM. In *EUREM, 11th Euro Congress on EM*, Dublin, Ireland, Published by EUREM 96 on CD-ROM

Carlton, PA. (1997). The effect of some instrument operating conditions on the x-ray microanalysis of particles in the environmental scanning electron microscope, *Scanning*, 19, pp.(85-91)

Danilatos, G.D. (1980). An atmospheric scanning electron microscope (ASEM). *Scanning*, 3, pp.(215-217)

Danilatos, G.D. (1988). *Foundations of Environmental Scanning Electron Microscopy*, Advances in Electronics and Electron Physics. 71. Academic Press, ISBN 0120146711pp. 109–250

Danilatos, G.D. (2009). Optimum beam transfer in the environmental scanning electron microscope. Journal of Microscopy, 234, 1, pp.(26–37)

Doehne, E. (1996). A new correction method for energy-dispersive spectroscopy analysis under humid conditions. *Scanning*, 18, pp.(164–165)

Doehne, E. (1997). A new correction method for high resolution energy dispersive X-ray analyses in the environmental scanning electron microscope. *Scanning*, 19, pp.(75–78)

Fletcher, A.L.; Thiel, B.L.; & Donald, A.M. (1997). Amplification measurements of potential imaging gases in environmental SEM. *J. Phys*, D 30, pp.(2249–2257)

Gauvin, R. (1999). Some theoretical considerations on X-ray microanalysis in the environmental or variable pressure scanning electron microscope. *Scanning*, 21, pp(388–393)

Goldstein, J.; Newbury, D.; Echlin, P.; Joy, C.; Roming, A.D.; Lyman, E.; & Lifshin, E. (1992). *Scanning Electron Microscopy and X-ray Microanalysis*, 3rd ed. Plenum Press, ISBN 0-306-44175-6, New York , USA

Kadoun, A. ; Belkorissat, R. ; Khelifa, B. ; & Mathieu, C. (2003). Comparative study of electron beam–gas interaction in an SEM operating at pressures up to 300 Pa.*Vacuum*, 69, pp.(537- 543)

Khouchaf, L.; Blondiaux, J.; Hedouin, V.; Gosset, D. ; Dürr, J. & Flipo, R.-M. (2000). La Microscopie Electronique à Balayage Environnementale équipée en Microanalyse X : son utilisation en Pathologie Osseuse Humaine. Perspectives et limites. *Journal. Phys. IV*, 10, pp.(551-559)

Khouchaf, L. & Verstraete J. (2002). X-ray microanalysis in the environmental scanning electron microscope (ESEM): Small size particles analysis limits. *J. Phys. IV*, 12, 6, pp.(341-346)

Khouchaf, L. & Verstraete, J. (2004). J. Electron scattering by gas in the Environmental Scanning Electron Microscope (ESEM): Effects on the image quality and on the X-ray microanalysis. *J. Phys IV*, 118, pp.(237-243)

Khouchaf, L.; Boinski, F. (2007). Environmental Scanning Electron Microscope study of SiO_2 heterogeneous material with helium and water vapor, *Vacuum*, 81, pp.(599-603)

Khouchaf, L.; Mathieu C. & Kadoun, A. (2010). Electron microbeam changes under gaseous environment: CP-SEM case and microanalysis limits, *Microscopy: Science, Technology, Applications and Education*, A. Méndez-Vilas and J. Díaz (Eds.), FORMATEX, ISBN-13: 978-84-614-6190-5, Badajoz, Spain

Khouchaf, L.; Mathieu C. & Kadoun, A. (2011). Microanalysis results with low Z gas inside Environmental SEM. *Vacuum*, 86, pp.(62-65)

Le Berre, J.F; Demopoulos, G.P; Gauvin, G. (2007). Skirting: A Limitation for the Performance of X-ray Microanalysis in the Variable Pressure or Environmental Scanning Electron Microscope. *Scanning*, 29, pp.(114-122)

Mansfield, J.F. (2000). X-ray microanalysis in the environmental SEM. a challenge or a contradiction, *Mikrochim Acta*,132, pp(137–143)

Mansour, O.; Aidaoui, K.; Kadoun, A.; Khouchaf, L. & Mathieu, C. (2010). Monte Carlo simulation of the electron beam scattering under gas mixtures in an HPSEM at low energy. *Vacuum*, 84, pp.(458463)

Newbury, D. (2002). X-Ray Microanalysis in the Variable Pressure (Environmental) Scanning Electron Microscope. *J. Res. Natl. Inst. Stand. Technol.*,107, pp.(567–603)

Newbury, D.; Joy, C.; Echlin, P.; Fiori, C. & Goldstein, J. (1986). *Advanced Scanning Electron Microscopy and X-ray Microanalysis*, Plenum Press, ISBN 0-306-42140-2, New York , USA

Stokes, D. J.; (2008). *Principles and practice of Variable Pressure/Environmental scanning electron microscopy (VP-ESEM)*, John Wiley & Sons, Ltd. ISBN: 978-0-470-06540-2, Cornwall, UK

Wight, A. (2001). Experimental data and model simulations of beam spread in the environmental scanning electron microscope. *Scanning*, 23, pp.(320–327)

4

Some Applications of Electron Back Scattering Diffraction (EBSD) in Materials Research

Zhongwei Chen[1], Yanqing Yang[1,*] and Huisheng Jiao[2]
[1]State Key Laboratory of Solidification Processing,
Shaanxi Materials Analysis & Research Center,
Northwestern Polytechnical University, Xi'an,
[2]Oxford Instruments Shanghai Office, Shanghai,
P.R. China

1. Introduction

Electron Back Scattering Diffraction (EBSD) is a technique based on the analysis of the Kikuchi pattern by the excitation of the electron beam on the surface of the sample in a scanning electron microscope (SEM). The crystal structure, orientation and correlative information can be acquired by the technique. EBSD has a unique advantage in the determination of the crystal orientation and microstructure compared with the traditional analysis methods. It can observe the grain boundary types, misorientations, and the distribution of them, and the statistical measurement and quantitative analysis also can be carried out. Therefore, the quantitative relationship between grain boundary structure, orientation, texture and material properties can be established. Consequently, it has been a very important experimental technique in materials science and engineering.

This chapter presents a few examples of applying EBSD to characterize the microstructure of different materials including steels and molybdenum sheets after rolling and heat treatment, and casting aluminum alloys in order to reveal the formation mechanism of microstructure during solidification.

2. Applications of EBSD in steel and molybdenum

In this section, we summarize the applications of EBSD in microstructure characterization, including second phase identification, texture analysis of steel and molybdenum after different heat treatments and rolling, in understanding the microstructure change during the cold and hot work processing, and the properties of the materials. EBSD also has been used to measure the Kurdjumov-Sachs orientation relationships between austenite and ferrite in stainless steel. The orientation relationships between the particles and the matrix show a spread around the Kurdjumov-Sachs relationship; the close packed planes in the FCC and BCC phases are usually parallel to or nearly parallel to each other. A model has

---

* Corresponding Author

been proposed for the interfacial structure in a duplex stainless steel based on the topological theory to explain the deviation angles. In order to verify the prediction of the distribution of the orientation relationship between BCC/FCC structures, EBSD provides a convenient approach for orientation relationship determination and fast data collection.

2.1 Microstructure characterization of steel after different heat treatments and rolling

Steel's mechanical properties and corrosion resistance are seriously affected by inclusions and precipitates at grain boundaries and inside grains. For example, in ferritic stainless steel, the precipitation of chromium carbides and nitrides at grain boundaries causes local depletion of chromium in the surrounding material, resulting in a much-reduced corrosion resistance [Kim *et al.*, 2010; Park *et al.*, 2006]. This will lead to a dramatic drop in the strength and stiffness of the materials. In order to prevent the formation of chromium carbide and nitride, stronger nitride and carbide formation elements are added to form more stable precipitates.

It is shown that, in this research, a number of precipitates have been formed in a ferritic stainless steel intended for a high temperature application during the steel making process. The precipitates vary in size from 50µm down to a few 10s of nanometers. The larger precipitates can be easily identified using a combination of EBSD and EDS chemical analyses, but the smaller ones make any chemical analysis problematical. However the superior resolution of the EBSD technique, coupled with a knowledge of the probable composition of the precipitates, makes their identification using EBSD alone relatively simple.

The sample was mechanical polished and the final polish with colloidal silica. Zeiss Supra 55 VP SEM was used for EBSD and EDS analysis with beam current around 1-2nA and 20kV. A combination of imaging and "point and click" phase identification (with an integrated EDS-EBSD system) has been used to identify and then map the larger particles; however, these are commonly intragranular particles and, as such, are less damaging to the properties of the steel. The finest particles lie on the grain boundaries – these have the potential to cause intergranular corrosion, and EBSD analysis alone was necessary for their identification. Different mapping step size was used for mapping coarse inclusions in grains and fine precipitates at grain boundaries, 0.5µm and 20nm respectively.

2.1.1 Phase identification of coarse precipitates

Imaging using the Nordlys forescatter system allows both orientation contrast images and atomic number contrast images to be collected, shown in Fig.1, simply by switching from the bottom to the top forescatter diodes. These clearly show 2 types of precipitates within the grains – a large precipitate (type 1) and a number of smaller precipitates (type 2).

The integrated EBSD+EDS system allows 1-click phase identification. From a single point, an EDS spectrum is collected along with the diffraction pattern (EBSP); the spectrum peaks are identified and this chemistry is used to search a phase database (or several databases) to find all matching phases. The EBSP is then indexed and the best matching phase is determined, shown in Fig.2. Here was used to identify the 2 types of precipitate: type 1 precipitates are aluminum nitrides (AlN - hexagonal), and the type 2 precipitates are chromium carbides ($Cr_{23}C_6$ - cubic).

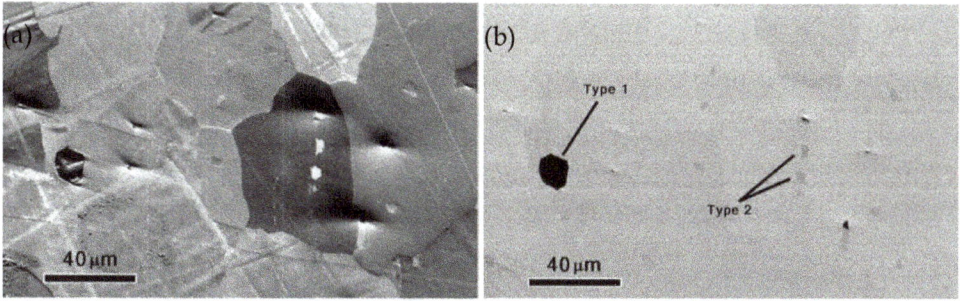

Fig. 1. Forescatter images. (a) orientation contrast image collected using the lower forescatter diodes, showing grains and surface topography; (b) atomic number contrast image collected using the top diodes, showing 2 types of precipitate in the ferrite matrix – types 1 and 2.

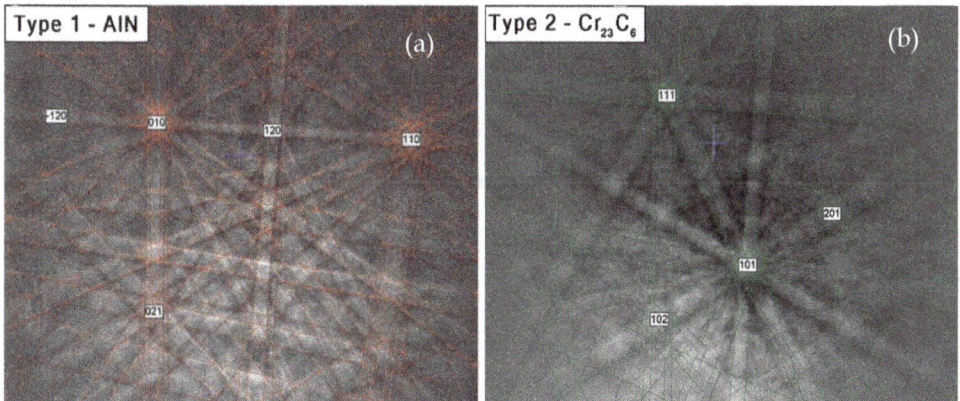

Fig. 2. Indexed EBSPs from the 2 types of precipitate. (a) type 1, indexed as hexagonal AlN; (b) type 2, indexed as cubic $Cr_{23}C_6$.

With the identity of the precipitates now known, it is possible to map the area and to show the distribution of the phases. The CHANNEL5 EBSD system can discriminate between the 2 precipitates and the ferrite matrix, on the basis of crystallography alone. The resulting phase map is shown in Fig.3.

2.1.2 Identification of small grain boundary precipitates

A closer look at the microstructure shows that many of the grain boundaries have small, elongated precipitates, less than 200 nm across. Unlike the coarse AlN and $Cr_{23}C_6$ precipitates, their location at the boundaries could cause intergranular corrosion, and as such it is important to identify them (see Fig. 4).

The size of these particles makes chemical analysis by EDS problematic, as the signal will predominantly originate from the steel matrix. Therefore EBSD is the ideal technique to identify such precipitates. It is expected that these precipitates are either carbides or nitrides, and so matching phases that fit the chemistry (Fe, Cr)(N, C) were used to index the EBSPs. In all cases the precipitates were identified as having a hexagonal M_7C_3 structure –

$(Fe,Cr)_7C_3$. An automated EBSP map along one of these boundaries was collected, and this showed that some of the precipitates have a distinct crystallographic relationship with one of the neighbouring grains: $(0001)_{carbide} | | (111)_{ferrite}$ – a basal orientation relationship.

Fig. 3. Phase map showing the same area imaged in Fig. 1. Blue = ferrite, Red = AlN and Green = $Cr_{23}C_6$. Black lines represent grain boundaries.

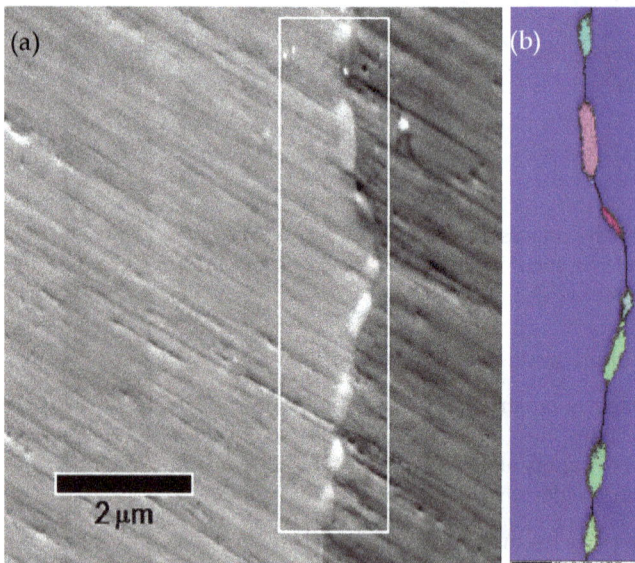

Fig. 4. (a) Forescatter orientation contrast map of grain boundary precipitates: white box marks EBSD analysis area; (b) Phase and orientation map of the boundary zone. Ferrite is shown in blue, with the $(Fe,Cr)_7C_3$ precipitates colored according to their orientation.

2.2 Microtexture analysis in molybdenum sheets

EBSD is an ideal technique for microtexture analysis. With the development of the speed of detectors, macro texture analysis is also possible. Comparison with XRD, in characterizing texture EBSD provides not only the types and of percentages of textures, but also the microstructure information. In this section, microtextures of a cross-rolled molybdenum were analyzed using EBSD technique. Molybdenum and its alloys are used in a variety of markets, including the electronics, materials processing and aerospace industries. There are a number of different properties that make molybdenum so attractive, notably its strength at high temperature, high stiffness, excellent thermal conductivity and low coefficient of thermal expansion [Cockeram *et al.*, 2005].

As with any metal, the physical characteristics of molybdenum can be tailored to suit particular applications. This is done using specific machining or metalworking procedures. One example of this is found in the aerospace industry, where molybdenum's strength and stiffness at high temperatures make it the ideal material for space satellite components. Sheets of molybdenum are cross-rolled in order to further enhance its properties [Oertel *et al.*, 2008]. Cross-rolling involves rolling the original sheet both parallel and perpendicular to its length, producing a specific texture (defined as the {001}<110> texture).

This research looks at a specific case in which, during the production of a molybdenum dish, undesirable surface ripples were observed. Obviously these ripples would have a damaging effect on the dish's performance, and therefore it was decided that the microstructural and crystallographic textural characteristics of the molybdenum sheet should be investigated in order to deduce the cause of the ripple formation. Samples were mechanically polished; final electro-polishing with 10% sulphuric acid/methanol electrolyte at -25°C, 55V applied voltage.

The microstructure of the sample is shown in Fig.5, a low magnification backscattered electron micrograph. It is clear that the sample has partially recrystallized, producing large, strain free grains. The recrystallized and unrecrystallized (i.e. deformed) fractions are arranged in alternate bands parallel to the final rolling direction (RD). The scale bar represents 50 μm.

Fig. 5. Forescatter orientation contrast image

The results of the EBSD analysis are shown in this orientation map, Fig.6. The color scheme reflects the orientation (see the inverse pole figure, inset), with green color showing points with the crystallographic <110> direction aligned with the rolling direction. Grain boundaries (>10°) are marked in black, with subgrain boundaries (2-10°) in grey. Note that in this and all other maps, the final rolling direction is horizontal and the scale bar represents 300 μm.

There are many ways that the CHANNEL 5 data processing software can be used to characterize the extent of deformation and recrystallization in a sample such as this.

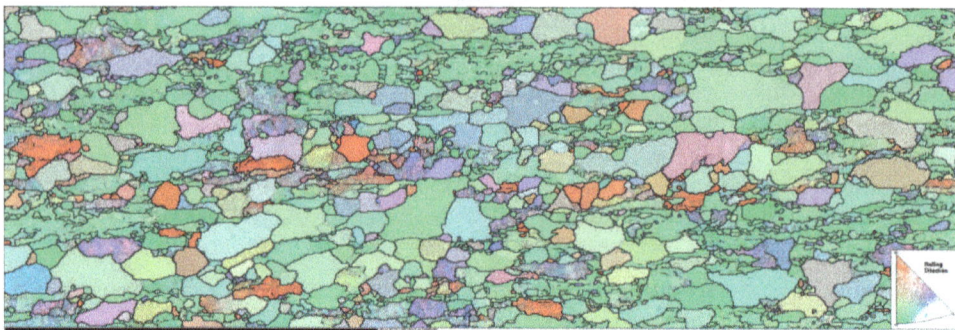

Fig. 6. Orientation map showing the orientation of each grain

In average misorientation map of Fig. 7(a) the misorientation between all the points in each grain has been calculated, and assigned a color (as defined in the legend, inset). The blue color indicates where grains have very little internal misorientation - these typically have been recrystallized. The green, yellow and red colors represent progressively increasing levels of internal misorientation, indicating no recrystallization. The software can also determine the recrystallized fraction automatically, as shown in Fig.7(b). Here the recrystallized grains are marked in blue and make up 54% of the total area. Unrecrystallized grains are colored red. It is clear from the orientation map (Fig.6) that this sample has a strong texture. This can be represented in pole figures or in orientation distribution functions (ODFs) in Fig.8.

In Fig. 8(a) the contoured pole figures show the complex nature of this texture. As observed in the orientation map, there is a strong alignment of the <110> axes with the rolling direction. However, there are also significant {100} and {111} textures. In Fig. 8(b) plotting the texture in an ODF clearly illustrates the texture characteristics. There is a strong fibre texture that splits into 2 branches, as well as other less important texture components. The colors are the same as in Fig. 7(b), showing that there is little difference between the orientation of recrystallized (blue) and non-recrystallized grains (red).

With a complex texture such as this, there are many individual "texture components" that can be used to describe parts of the overall texture. There are 2 main texture components that account for most of the texture in this sample, but if this sample has been fully cross-rolled, then a strong {001}<110> texture would be expected. In Fig.9, the inset shows the color scheme (up to a maximum 20° deviation) and the background grains are colored according to the pattern quality.

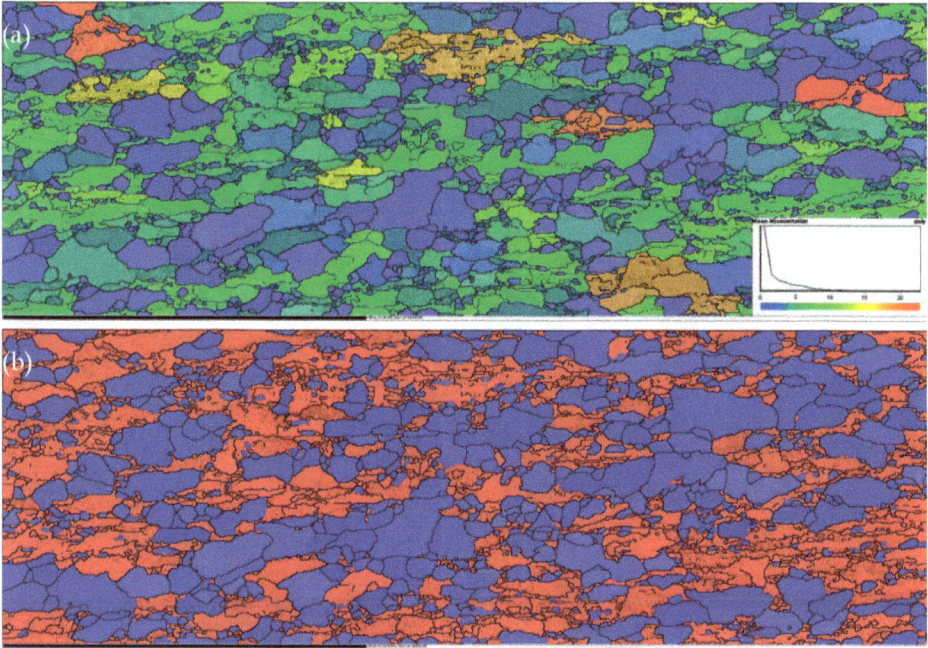

Fig. 7. (a) The average misorientation map and (b) recrystallised fraction map.

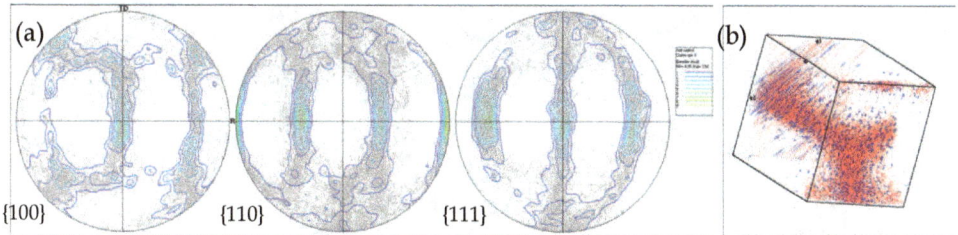

Fig. 8. (a) The contoured pole figures and (b) ODF showing texture characterisitics.

Cross-rolling should produce a dominant {001}<110> texture in BCC-metals (such as molybdenum). However this texture accounts for less than 25% of the area in this sample, with {110} and {111}-fibre textures more dominant. This indicates that the sheet has not been fully cross-rolled, and this would explain the formation of undesired ripples on the sheet surface.

2.3 Measuring the Kurdjumov-Sachs orientation relationships between austenite and ferrite in stainless steel

The interface between FCC and BCC crystals can be found in many important metallic alloys. The orientation relationships between these two phases show a spread around the Kurdjumov-Sachs relationship (K-S OR), Nishiyama-Wassermann relationship or other relationship; the close packed planes in the FCC and BCC phases are usually parallel to or nearly parallel to each other.

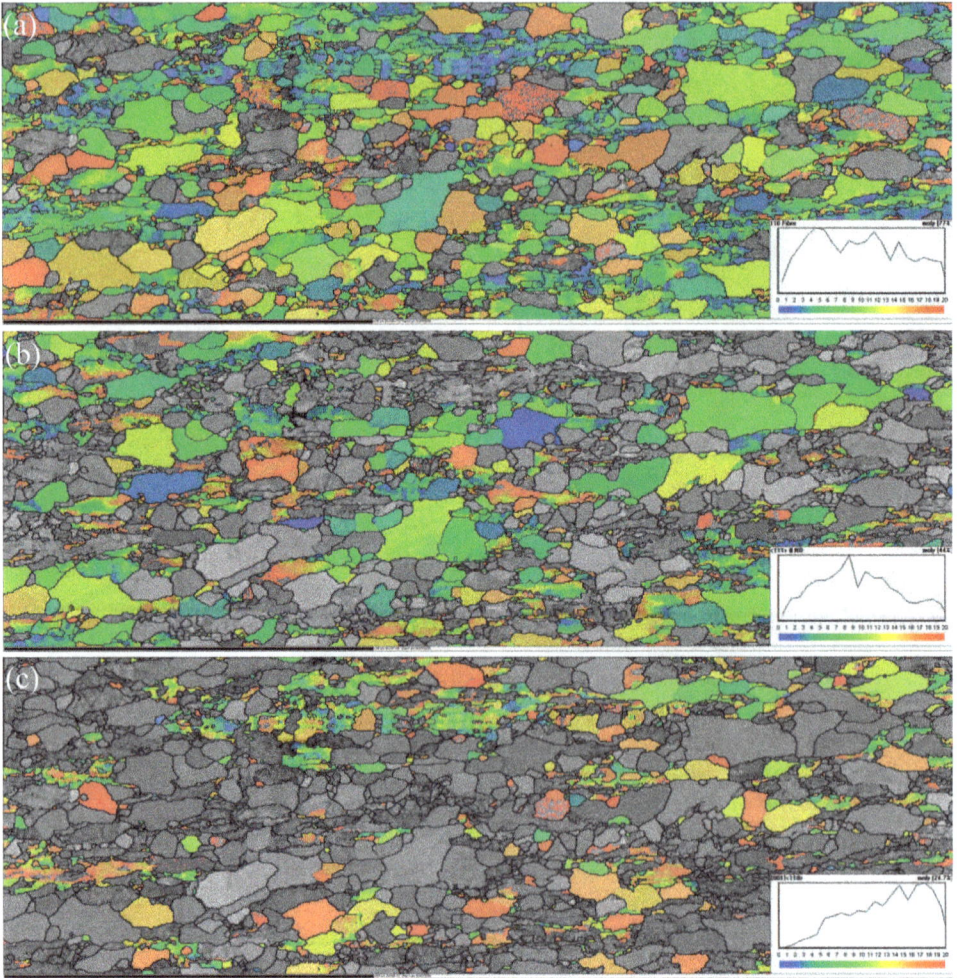

Fig. 9. (a) {110}-fibre texture (<110> parallel to RD) - 77% of the area; (b) {111}-fibre texture (<111> parallel to normal direction) - 44% of the area; (c) {001}<110> texture (a subset of the {110}-fibre texture) - 24.7% of the area.

In order to understand the morphology and structure of such interphase boundaries, different theoretical models have been proposed, for example, the O-lattice theory [Bollmann,1970], Invariant line model [Dahmen, 1982], CSL/DSC model [Balluffi, 1982], and the structural ledge model [Hall *et al.*, 1972]. However, these models can only explain or predict interfacial structure in part; in particular, interfacial defects have not always been accurately characterized in previous studies. Recently, by considering the symmetry of the bicrystals, the topological theory was developed for characterizing the parameters of interfacial defects [Pond, 1989]. A model has been proposed for the interfacial structure in a duplex stainless steel based on the topological theory [Jiao *et al.*, 2003]. The FCC/BCC interfaces have been characterized as arrays of interfacial defects superimposed on reference bicrystal structures. This model predicted deviation angle ranged from 0.3° to 4.97° from the ideal K-S OR. In order to verify the prediction of the distribution of the orientation relationship between BCC/FCC structures, EBSD has been used for orientation relationship determination and fast data collection.

In this research, a sample of Zeron-100 duplex stainless steel was heat treated at 1400°C for 30 minutes and followed by water quenching, and 10 seconds at 1000°C for precipitating. Sample was electropolished with a solution of 10 wt% oxalic acid in H_2O with a voltage of 10~15V at room temperature for 1~3 minutes.

Fig. 10 shows the forescatter orientation contrast image of the duplex stainless steel. The EBSD data was collected from the center area of the image. In Fig.11, it can be seen there are two phases, FCC γ phase is in red and BCC α matrix is in blue. For the γ phase we can see that there are three types: small particles, large particles in α grain and at grain boundaries. From the pole figures in Fig.12 the orientation relationship (OR) between FCC and BCC phases was determined as K-S OR. In Fig.11(b) the K-S OR interface boundaries are plotted in white. If the deviation from K-S OR over 7° the interface boundaries are in black. From the map we can see that the most of particles shows a K-S OR to matrix. However, the particles at the grain boundaries only show K-S OR with one side, which is because of the particle is nucleated from one grain. The large particles inside the α grain show non K-S OR with the matrix; that because these particles are retained during homogenization.

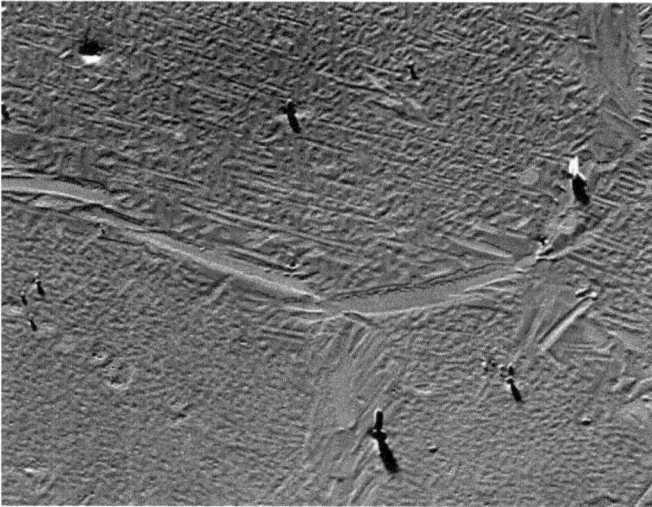

Fig. 10. Forescatter orientation contrast image of the duplex stainless steel. The EBSD data was collected from the center area of the image.

Although these two phases show a good consistent with K-S OR, there is always a small deviation from the ideal OR; that means the {111} plane in γ phase is not exactly parallel to {110} plane in BCC matrix. From the above distribution, it is found that most the particles show a 1.5° away the K-S OR and with a range of 0.2° to 5.4°, as shown in Fig. 13, which is in a good agreement with the prediction from the topological interfacial model.

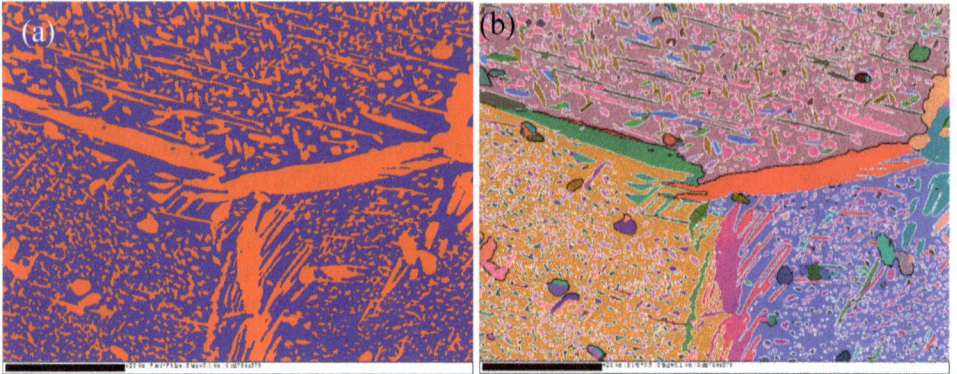

Fig. 11. (a) Phase map and (b) orientation map of the sample.

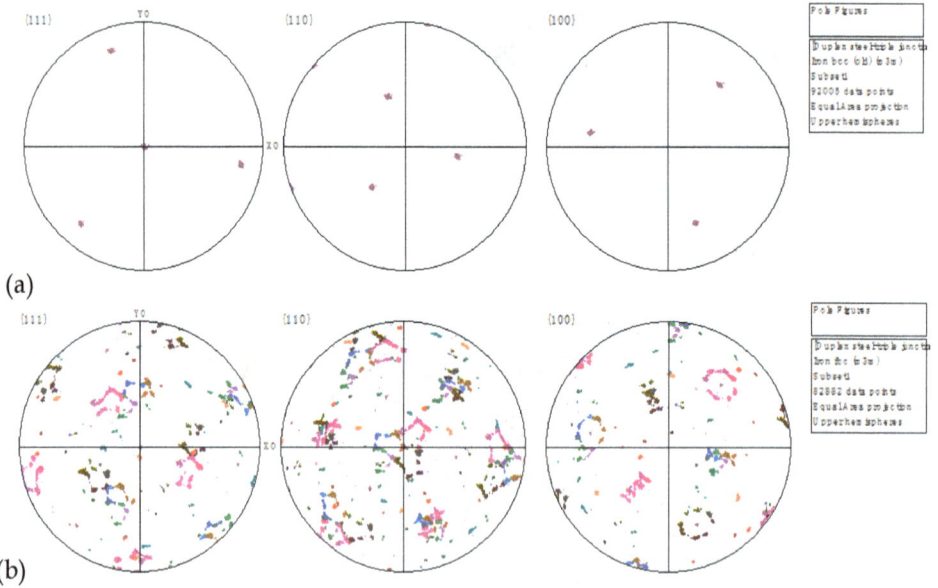

Fig. 12. Pole figures from one α grain (a) and γ particles in this grain (b). From these pole figures it is found that these two phases fall into K-S OR.

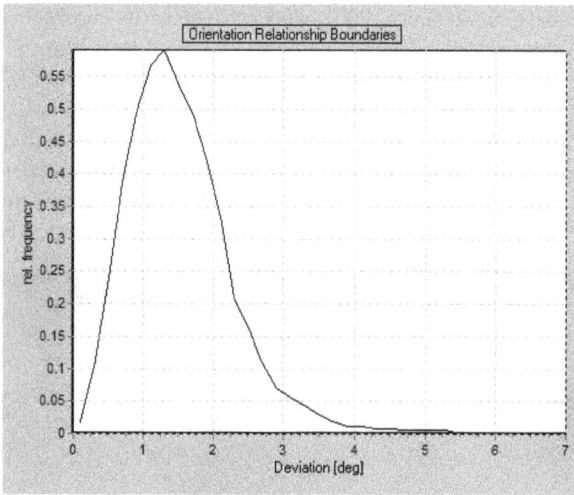

Fig. 13. The distribution of the deviation angle from K-S OR.

3. EBSD analysis in the field of solidification

3.1 Preparation of EBSD samples by ion etching

Both electrolytic polishing and ion etching can be used to prepare EBSD samples, however, the electrolyte for Al-Si alloy is relatively difficult to be prepared. On the other hand, for A357 alloy, electrolytic polishing of α-Al dendrites is faster than that of the eutectic Si due to the difference in electrochemical property, and then the prominency of eutectic Si phase is visible, which influences the surface roughness awfully. As a result, the reflection of the backscatter electron can not be received by the screen, inducing a low calibration rate, consequently the sample could not be analyzed [Nogita & Dahle, 2001]. Ion etching, as a new technique for EBSD sample preparation, is suitable for eliminating the surface stress layer of most materials, and the etching speed could be selected according to the etching voltage, ion beam current, geometrical shape and materials of the samples. Therefore, ion etching is chosen for EBSD sample preparation of particular materials which contain some hard brittle phases in microstructure.

Fig.14 shows the Kikuchi pattern of A357 alloy sample prepared by mechanical buffing and ion etching respectively. The difference between these two pictures is obvious. Rheology on the surface of the sample took place due to mechanical buffing, therefore no Kikuchi pattern could be observed, as shown in Fig. 14(a). Since the electron beam effects only 1-2μm deep on the surface of the sample, so ion etching must be performed to remove the surface stress layer in order that Kikuchi pattern could be observed. Fig.14(b) shows clear Kikuchi pattern.

3.2 Misorientation of secondary eutectic phase with primary phase in modification alloys

A357 aluminum alloy is casting alloy with the coarse α-Al dendrites and plate-like eutectic silicon in cast microstructure, of which the volume fraction of the eutectic silicon phase is more than 50% [Heiberg & Arnberg, 2001] and influences its mechanical properties. The coarse primary α-Al dendrite could be refined and equiaxed by adding proper quantities of Ti and B into the alloy [Easton & St John, 2001; Shabestari & Malekan, 2010], while Sr is a good modificator, which improves the mechanical properties of Al-Si alloy by changing the morphology of the eutectic Si [Chen & Zhang, 2010; Martínez et al., 2005]. In the section, EBSD investigation of A357 alloy has been used to analyze the misorientation of eutectic phase with primary phase in A357 aluminum alloy with and without Sr modification and to study the nucleation and growth mechanism of the eutectic solidification by Sr modification.

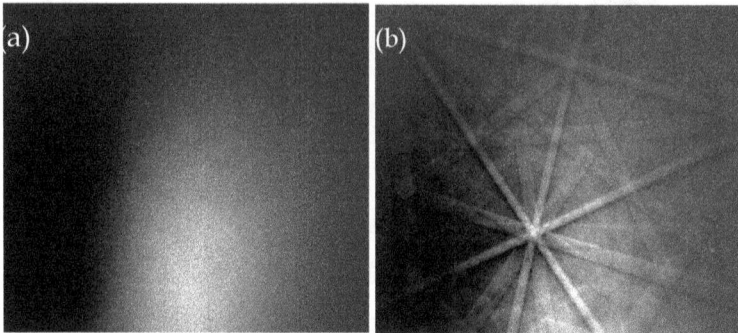

Fig. 14. Kikuchi pattern of unmodified A357 alloy samples prepared by (a) mechanical polishing and (b) ion etching.

For EBSD samples of the cast A357 alloys, the ion etching was carried out for the EBSD sample preparation by Gatan 682 ion etching equipment. EBSD tests were performed on ZEISS SUPRA55 field emission gun scanning electron microscope (FEG-SEM) with HKL channel 5 backscattered electron diffraction camera. The parameter of FEG-SEM is set as following: acceleration voltage of 20kV, working distance of 21.0 mm.

Fig.15 shows the microstructure of the unmodified and modified A357 alloy. The coarse α-Al dendrites and the plate-like Al-Si eutectic both can be observed clearly, as shown in Fig.15 (a). The coarse plate-like eutectic Si phases were transformed to fine fibrous eutectic by Sr modification, as shown in Fig.15 (b). It is concluded that there are approximately 480 grains in a visual field of the unmodified A357 sample, of which the average grain size is 21.0μm, while about 1362 grains can be found in the same scanning region of the sample which was modified and the average grain size is 7.4μm, according to the EBSD statistical results. The average grain size of the sample decreases by nearly two times, indicating that the microstructure can be refined by Sr modification.

Fig.16 shows the crystal misorientation and its distribution with and without Sr modification. Low-angle grain boundaries (LAGBs) are dominate in the unmodified sample and few high-angle grain boundaries (HAGBs) can be observed, while both of them in the modified sample can be found that they distribute evenly and occupy certain proportion respectively, as shown in Fig.16(d). Fig.17 shows the polar figures of the samples with and without Sr modification. Crystal orientation of the sample without Sr modification is relatively concentrated, which basically tends to be two kinds of crystal orientation, represented as red and purple respectively, as shown in Fig.17 (a), while crystal orientation in the modified sample is changed and tends to be scattered, as shown in Fig.17 (b).

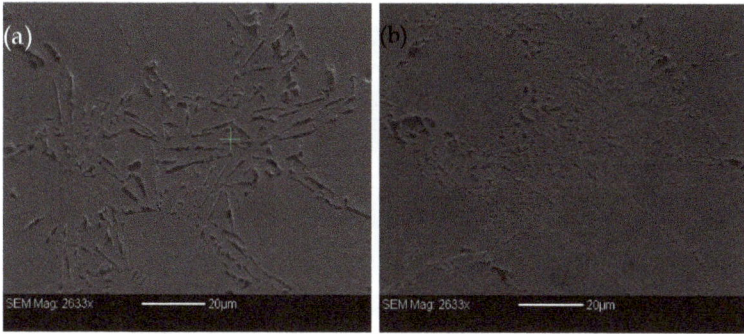

Fig. 15. Microstructure of samples (a) unmodified and (b) modified.

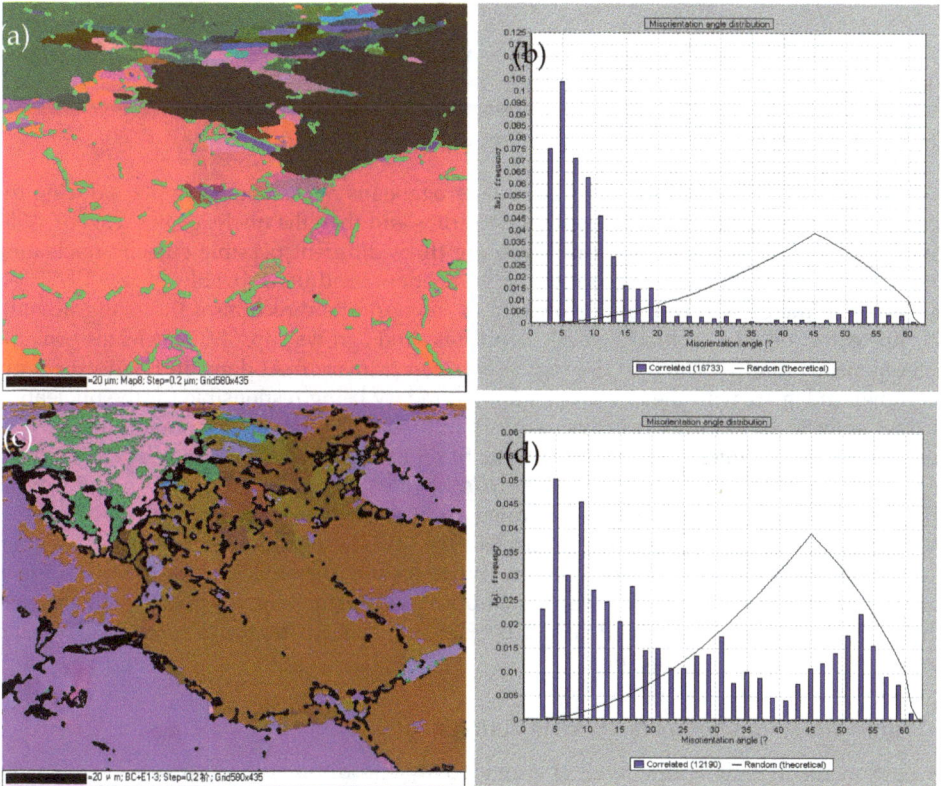

Fig. 16. Misorientation angle distribution for the Al grains. (a) and (b) unmodified; (c) and (d) modified.

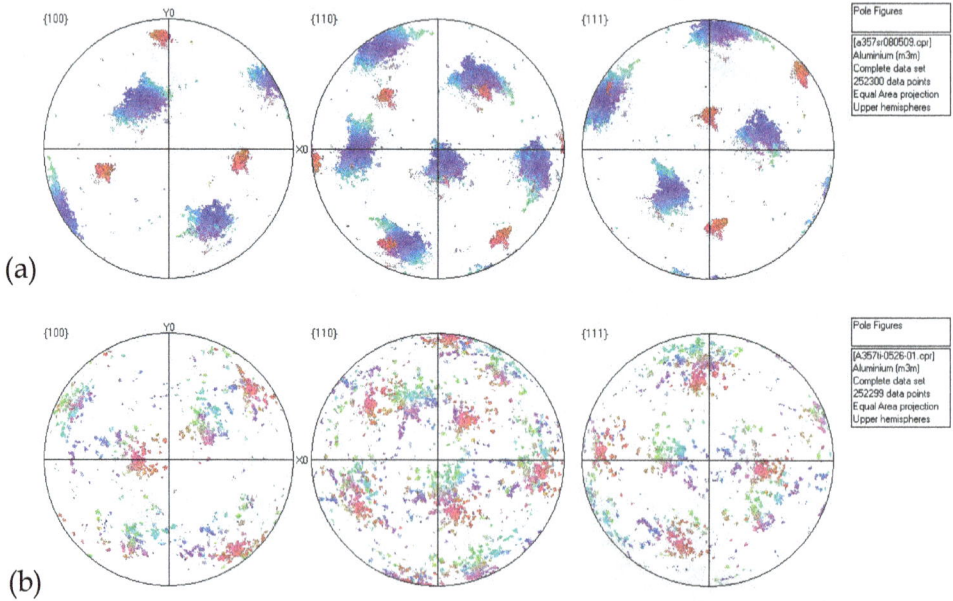

Fig. 17. Pole figure maps of samples. (a) unmodified and (b) modified.

A357 alloy is hypoeutectic Al-Si alloy, and heterogeneous nucleation takes place during the solidification process of the primary α-Al dendrites and then the nuclei grows up. For A357 alloy, recent work has indicated that there are three different possible eutectic nucleating and growth modes depending on the solidification conditions [Dahle et al., 1997]: (1) nucleates on the casting wall and grows up in the opposite direction of thermal gradients, (2) nucleates on the primary α-Al dendrites and grows up, (3) nucleates on the heterogeneous nuclei located in the region between primary α-Al dendrites and grows up. As shown in Fig.16, with the low-angle grain boundary being responsible for the nucleation of the eutectic Si on the primary α-Al dendrites, the high-angle grain boundary is responsible for the nucleation of the eutectic Si on the heterogeneous nuclei located in the region between primary α-Al dendrites. It is obvious that the eutectic Si in the sample without Sr modification nucleates and grows up in mechanism (2) and those in the sample modified nucleates and grows up in mechanism (3). The results indicate that nucleation mechanism of the eutectic Si phase changes due to the modification, leading to the change of the growth patterns. Therefore, misorientation of secondary eutectic phase with primary phase in casting alloys can be detected by EBSD and used to study the nucleation and growth mechanism of modification.

3.3 Agglomeration of primary crystals during solidification

In suction casting, the extra force is provided by the differential pressure between the melting chamber and the suction chamber. Thus, suction casting is successful in preventing casting defects by means of suction force, and a high cooling rate in suction casting is generated due to the use of the Cu-mold. Although the primary solidification under the intensive force has been comprehensively understood [Chen et al., 2009], not much attention has been paid to the solidification of the remaining liquid in the die-casting mould. However, the secondary solidification of the remaining liquid plays an important role in determining the final microstructure and corresponding mechanical properties [Stangleland et al., 2004; Vernède et al., 2006]. The secondary solidification was also found in rheo-diecasting (RDC) process of semi-solid metal (SSM) technology [Fan et al., 2005; Hitchcock et al., 2007]. Therefore, it is necessary to understand the secondary solidification behavior of the remaining liquid in relation to the final microstructure.

Agglomeration of primary crystals in the mushy zone during solidification is an important phenomenon influencing many aspects of casting processes and often affects the rheology of partially solid alloys and the microstructure of the as-cast component. Microstructural characterization of AlFeSi specimens produced by suction casting and die casting has been performed using EBSD. EBSD is used to study the agglomeration of equiaxed crystals in suction casting by analyzing the grain misorientation.

Typical as-cast microstructures produced by the two casting techniques are shown in Figs. 18(a) and 18(b). The primary grains have quasi-equiaxed morphology in two cases. From Fig. 18(b), it can be seen that in located region of the relatively large primary α-Al dendrites, there are fine α-Al particles, contributed by the solidification of the remaining liquid in the located region of α1 dendrites, which is referred to as "secondary solidification".

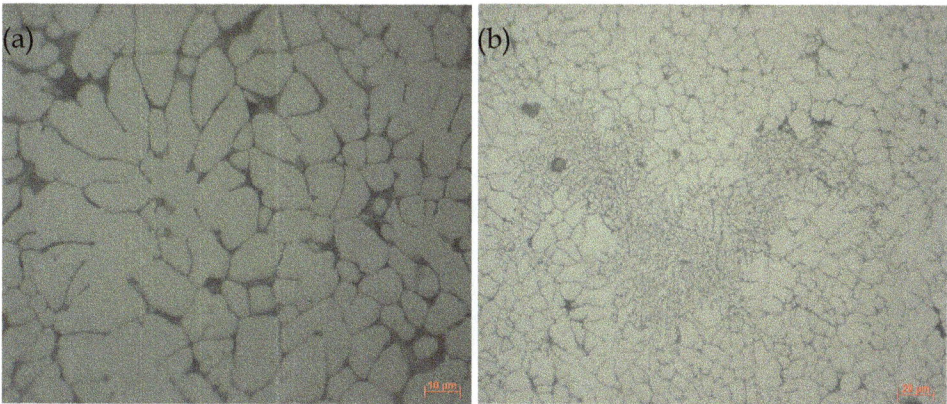

Fig. 18. Optical metallographs of AA8011 alloy. (a) die casting sample and (b) central region in 2 mm diameter sample in suction casting.

Similar to previous research [Otarawanna et al., 2010], low-angle grain boundaries (LAGBs) and high-angle GBs (HAGBs) are defined here as boundaries with a misorientation between 5° and 15°, and more than 15°, respectively. Three types of GBs − LAGBs, HAGBs and coincidence-site-lattice GBs (CSL-GBs) − were determined after result extrapolation in each EBSD map. Fig. 19 shows extrapolated EBSD maps by LAGBs and HAGBs in each of the samples investigated. It shows that the fraction of low-energy GBs (LAGBs) is higher in the suction casting sample than in the die casting sample.

By producing samples with different casting methods where the solidifying alloy experiences different levels of external mechanical stresses, the effects of mechanical stresses applied during solidification can be assessed. The length percent of low-energy GBs shown in Fig. 19 is associated with crystal agglomeration during solidification. The high pressure applied on the solidifying alloy during suction casting is likely to promote crystal collisions and result in a strong degree of crystal agglomeration.

When two growing primary crystals impinge with one another, GB formation depends on the interfacial energy of the potential new GB. Coalescence or bridging is the transformation of two impinging solidification fronts into a solid bridge [Rappaz et al., 2003] and in this case a new agglomerate can form. Therefore, coalescence can be considered as the disappearance of two solid–liquid interfaces, each with interfacial energy $\gamma_{s/l}$, and the formation of a GB with interfacial energy γ_{gb} [Mathier et al., 2004]. When two solidifying crystals impinge on one another, bridging occurs readily if $\gamma_{gb} < 2\gamma_{s/l}$ [Rappaz et al., 2003]. On the other hand, some energy is required to form a new boundary if $\gamma_{gb} > 2\gamma_{s/l}$. γ_{gb} is a function of the misorientation between the two impinging crystals. $\gamma_{gb} < 2\gamma_{s/l}$ occurs if the misorientation is less than 15° (LAGBs) [Mathier et al., 2004]. In this case, there is an attractive force to bring the two crystals together and coalescence occurs as soon as the two interfaces are close enough. Only GBs with $\gamma_{gb} < 2\gamma_{s/l}$ are thought to form after a collision of two crystals [Sannes et al., 1996]. If not, a liquid film is stable and the colliding crystals bounce back. In suction casting samples, the fraction of low-energy GBs among in-cavity solidified grains is significantly higher than in diecasting specimens. This is attributed to the increased number of crystal collisions during suction casting, which promotes agglomeration of favorably oriented crystals.

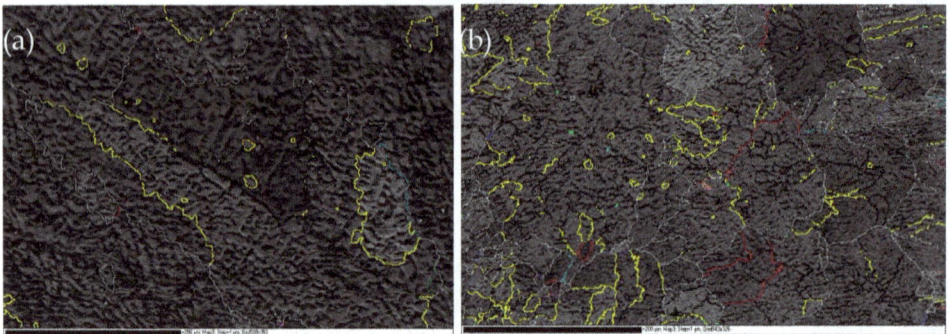

Fig. 19. EBSD micrographs by GB lines in color. White—HAGBs, Yellow—LAGBs, other color-- CSL-GBs. (a) die casting sample, (b) central region in 2 mm diameter sample in suction casting

4. Conclusions

Some examples of applications of EBSD to materials processing are presented in the chapter. It is shown that combined with SEM imaging and EDS composition analyzing, EBSD is very powerful for materials research. Phase identification, microtexture characterization and crystallographic orientation relationship determination are carried out in different steels and molybdenum sheet after heat treating and rolling in order to understand the microstructure evolution during the hot processing. It is also shown that EBSD analysis on casting aluminum alloys provides invaluable insight into mechanisms of nucleation and growth in modification during solidification, and agglomeration of equiaxed crystals in secondary solidification.

5. Acknowledgement

The authors gratefully acknowledge the financial support from the Research Fund of the State Key Laboratory of Solidification Processing (No. 42-QP-009) and the Fundamental Research Fund of Northwestern Polytechnical University (No. JC200929).

6. References

Balluffi, R. W.; Brokman, A. & King, A. H. (1982). CSL/DSC Lattice model for general crystal boundaries and their line defects. *Acta Metallurgica*, Vol.30, No.8, (August 1982), pp. 1453-1470, ISSN 0001-6160

Bollmann, W. (1970). *Crystal Defects and Crystalline Interfaces*, Springer, ISBN 978-0387050577, New York, UAS

Chen, Z. W. & Zhang, R. J. (2010). Effect of strontium on primary dendrite and eutectic temperature of A357 aluminum alloy. *China Foundry*, Vol.7, No.2, (May 2010), pp. 149-152, ISSN 1672-6421

Chen, Z. W.; He, Z. & Jie, W. Q. (2009). Growth restriction effects during solidification of aluminium alloys. *Transactions of Nonferrous Metals Society of China*, Vol.19, No.2, (April. 2009), pp. 410-413, ISSN 1003-6326

Cockeram, B. V.; Smith, R. W. & Snead, L. L. (2005). Tensile properties and fracture mode of a wrought ODS molybdenum sheet following fast neutron irradiation at temperatures ranging from 300°C to 1000°C. *Journal of Nuclear Materials*, Vol.346, No.2-3, (November 2005), pp. 165-184, ISSN 0022-3115

Dahmen, U. (1982). Orientation relationships in precipitation systems. *Acta Metallurgica*, Vo.30, No.1, (January 1982), pp.63-73, ISSN 0001-6160

Dahle, A. K.; Hjelen, J. & Arnberg, L. (1997). Formation of eutectic in hypoeutectic Al-Si alloys, *Proceedings of 4th Decennial International Conference on Solidification Processing*, pp. 527-530, ISBN 0-9522507-2-1, Sheffield, UK, July 7-10, 1997

Easton, M. A. & St John, D. H. (2001). A model of grain refinement incorporating alloy constitution and potency of heterogeneous nucleant particles. *Acta Materialia*, Vol.49, No.10, (June 2001), pp. 1867-1878, ISSN 1359-6454

Fan, Z.; Liu, G. & Hitchcock, M. (2005). Solidification behaviour under intensive forced convection, *Materials Science and Engineering: A*, Vol.413-414, (December 2005), pp. 229-235, ISSN 0921-5093

Hall, M. G.; Aaronson, H. I. & Kinsma, K. R. (1972).The structure of nearly coherent fcc:bcc boundaries in a CuCr alloy, *Surface Science*, Vol.31, (June 1972), pp. 257-274, ISSN 0039-6028

Heiberg, G. & Arnberg L. (2001). Investigation of the microstructure of the Al-Si eutectic in binary aluminum-7 wt% silicon alloys by electron backscatter diffraction (EBSD). *Journal of Light Metals*, Vol.1, No.1, (February 2001), pp. 43-49, ISSN 1471-5317

Hitchcock, M.; Wang, Y. & Fan, Z. (2007). Secondary solidification behavior of the Al-Si-Mg alloy prepared by the rheo-diecasting process. *Acta Materialia*, Vol.55, No.5, (March 2007), pp. 1589-1598, ISSN 1359-6454

Jiao, H. S.; Aindow, M. & Pond, R. C. (2003). Precipitate orientation relationships and interfacial structures in duplex stainless steel Zeron-100. *Philosophical Magazine*, Vol.83, No.16, (June 2003), pp. 1867-1887, ISSN 1478-6435

Kim, J. K.; Kim, Y. H.; Lee, J. S. & Kim K. Y. (2010). Effect of Chromium Content on Intergranular Corrosion and Precipitation of Ti-stabilized Ferritic Stainless Steels. *Corrosion Science*, Vol.52, No.5, (May 2010), pp. 1847-1852, ISSN 0010-938X

Martínez, D. E. J.; Cisneros, G. M. A. & Valtierra, S. (2005). Effect of strontium and cooling rate upon eutectic temperatures of A319 aluminum alloy. *Scripta Materialia*, Vol.52, No.6, (March 2005), pp. 439-443, ISSN 1359-6462

Mathier, V.; Jacot, A. & Rappaz, M. (2004). Coalescence of equiaxed grains during solidification. *Modeling and Simulation in Materials Science and Engineering*, Vol.12, No.3, (May 2004), pp. 479-490, ISSN 0965-0393

Nogita, K. & Dahle, A. K. (2001). Eutectic solidification in hypoeutectic Al–Si alloys: electron backscatter diffraction analysis. *Materials Characterization*, Vol.46, No.4, (April 2001), pp. 305-310, ISSN 1044-5803

Oertel, C. G.; Huensche, I.; Skrotzki, W.; Knabl, W.; Lorich, A. & Resch, J. (2008). Plastic anisotropy of straight and cross rolled molybdenum sheets. *Materials Science and Engineering: A*, Vol.483-484, (June 2008), pp. 79-83, ISSN 0921-5093

Otarawanna, S.; Gourlay, C. M.; Laukli, H. I. & Dahle, A. K. (2010). Agglomeration and bending of equiaxed crystals during solidification of hypoeutectic Al and Mg alloys. *Acta Materialia*, Vol.58, No.1, (January 2010), pp. 261-271, ISSN 1359-6454

Park, C. J.; Ahn, M. K. & Kwon, H. S. (2006). Influences of Mo substitution by W on the precipitation kinetics of secondary phases and the associated localized corrosion and embrittlement in 29% Cr ferritic stainless steels. *Materials Science and Engineering: A*, Vol.418, No.1-2, (February 2006), pp. 211-217, ISSN 0921-5093

Pond, R. C. (1989). Line defects in interfaces, In: *Dislocations in Solids, Vol.8*, Nabarro, F.R.N. & Hirsh, J. P., (Eds.), pp. 1-62, Elsevier, ISBN 0-444-70515-5, Amsterdam, the Netherland

Rappaz, M.; Jacot, A. & Boettinger, W. J. (2003). Last-stage solidification of alloys: Theoretical model of dendrite-arm and grain coalescence. *Metallurgical and Materials Transactions A*, Vol.34, No.3, (March 2003), pp. 467-479, ISSN 11661-003-0083-3

Sannes, S.; Arnberg, L. & Flemings, M. C. (1996). Orientational relationships in semi-solid Al-6.5wt%Al, In: *Light Metals 1996*, Hale, W., (Ed.), pp. 795-798, TMS, ISSN 1096-9586, Warrendale, PA, USA

Shabestari, S. G. & Malekan, M. (2010). Assessment of the effect of grain refinement on the solidification characteristics of 319 aluminum alloy using thermal analysis. *Journal of Alloys and Compounds*, Vol.492, No.1-2, (March 2010), pp. 134-142, ISSN 0925-8388

Stangleland, A.; Mo, A.; Nielsen, Ø.; M'Hamdi, M. & Eskin, D. (2004). Development of thermal strain in the coherent mushy zone during solidification of aluminum alloys. *Metallurgical and Materials Transactions A*, Vol.35, No.9, (September 2004), pp. 2903-2915, ISSN 11661-004-0238-X

Vernède, S.; Jarry, P. & Rappaz, M. (2006). A granular model of equiaxed mushy zones: Formation of a coherent solid and localization of feeding. *Acta Materialia*, Vol.54, No.15, (September 2006), pp. 4023-4034, ISSN 1359-6454

5

Dopant Driven Electron Beam Lithography

Timothy E. Kidd

Physics Department, University of Northern Iowa, Cedar Falls, IA, USA

1. Introduction

The scanning electron microscope (SEM) can be used for far more than just obtaining images. It has a long tradition of being used to directly manipulate a sample to create various surface structures. The scanning coils within the microscope can be utilized for directing the electron beam in a controlled manner rather than simply raster across the surface as is used in imaging. By focusing the electron beam on a given area of the sample, it can be used to induce various localized changes to the surface of a material with a high degree of precision. There are several established techniques by which an electron beam can be used to create patterned structures upon a surface, the most common of which is electron beam lithography. Electron beam lithography is a multi-step process in which a sacrificial polymer layer is first deposited onto the sample that can achieve feature sizes down to ten nanometer length scales (Broers et al. 1996; Liu et al. 2002). The electron beam can also be used to locally induce or break bonds to pattern nanostructures (Mendes et al. 2004) or simply burn material away from selected areas of the sample (Egerton et al. 2004). In essentially every case, the electron beam interacts with the surface to locally alter or break chemical bonds to form patterned surface structures with very high precision.

In the present study, we have developed a new method by which the electron beam can be used to create patterned surfaces (Kidd et al. 2011). The discovery was quite by accident, occurring during some standard studies of layered dichalcogenide crystals. These highly two dimensional materials have intriguing electronic and chemical bonding characteristics which have made them of great interest for the study of novel electronic phase transitions (Wilson et al. 1975; Sipos et al. 2008) and potential use in a variety of alternative energy applications (Whittingham 1976; Kline et al. 1981; Chen et al. 2003). The systems we had chosen for study were doped crystals, as doping these materials can be used to induce dramatic changes in their electronic and chemical behaviors (Levy 1979; Friend and Yoffe 1987). The particular sample of interest at the time was Cu_xTiSe_2, owing to a recent discovery of a superconducting phase in the system in competition with the charge density wave ground state of pure $TiSe_2$ (Morosan et al. 2006).

The SEM was being used to examine the microstructure of single crystal samples while energy dispersive x-ray spectrometry (EDX), via a spectrometer attached to our microscope, was used for measuring the chemical composition of the sample. These measurements were meant to do nothing more than determine the homogeneity of our samples and how much copper was successfully incorporated as a dopant. Interesting anomalies were quickly discovered, however. The samples had an inhomogeneous appearance in the SEM even after surface layers were removed, although EDX measurements showed the chemical composition was essentially homogenous. The EDX measurements were then taken over longer time periods, to try to detect any subtle inhomogeneity in the sample stoichiometry. These longer measurements induced further changes in the appearance of the sample, and there were signs that the copper concentration might actually be varying with the duration of the EDX measurement itself. After some time, we hypothesized that perhaps the samples were contaminated in some way, so we attempted to drive off any contamination by heating the measured crystals in a vacuum oven. This resulted in the creation of a multitude of sub-micron and nanoscale copper iodide crystallites upon the sample surface, except for where the EDX measurements were performed. In areas of the sample that had been exposed for long durations to the electron beam, the surface was unchanged.

Through further experimentation, the process was refined so that the size and density of the crystallites could be controlled by varying exposure dosages and the temperature at which samples were heated after exposure. The iodine was found to come from within the sample itself. Iodine is used as a catalyst for growing large single crystals of $TiSe_2$ and other dichalcogenides, and some remains trapped in various nooks or cracks long after the samples are grown. This iodine is then released during heating and draws copper ions from within the bulk of the $TiSe_2$ to its surface to react and form the CuI crystallites (Jaegermann et al. 1996). The electron beam radiation serves to reduce the amount of copper arriving at the surface in the exposed portions of the surface, allowing one to devise a patterned array of sub-micron structures. Unlike traditional electron beam lithography, with its required sacrificial polymer coating, no intermediate processing is required. The sample is simply exposed to the electron beam in a controlled fashion and then reacted with some quantity of a halogen gas like iodine in a simple two-step process. This lithographic technique is essentially one derived solely from controlling the mobility of the dopant ions stored within the sample to reach the surface for reaction.

2. Experimental methods

The Cu_xTiSe_2 samples used in this study were synthesized using a technique proven for growing large single crystals of pure $TiSe_2$ (Balchin 1976; Kidd et al. 2002). The basic process is to heat the starting elemental powders (Alfa Aesar, >99.5% purity) with iodine as a catalyst in a sealed and evacuated silica ampoule. The original intent for these samples was to study the superconducting ground state which emerges upon doping with copper (Morosan, Zandbergen et al. 2006). In an attempt to dope the samples with a higher concentration of Cu than was found in the literature, the copper was included with the initial growth of Ti, Se, and I powders in a single step growth process rather than the standard multi-step method of first growing $TiSe_2$ powders and then later incorporating the copper dopants.

To put it mildly, high doping concentrations were not achieved using this modified single-step method. Instead, the copper reacted strongly with iodine in the growth ampoule to form a CuI film which coated the surfaces of very lightly doped $TiSe_2$ (Figure 1). The samples had a stoichiometry of $Cu_{0.04}TiSe_2$, with copper concentrations varying by ±0.01 as determined by EDX. This was much less than the goal of doping concentrations greater than 10%, and in fact our samples showed no signs of superconductivity down to 3K.

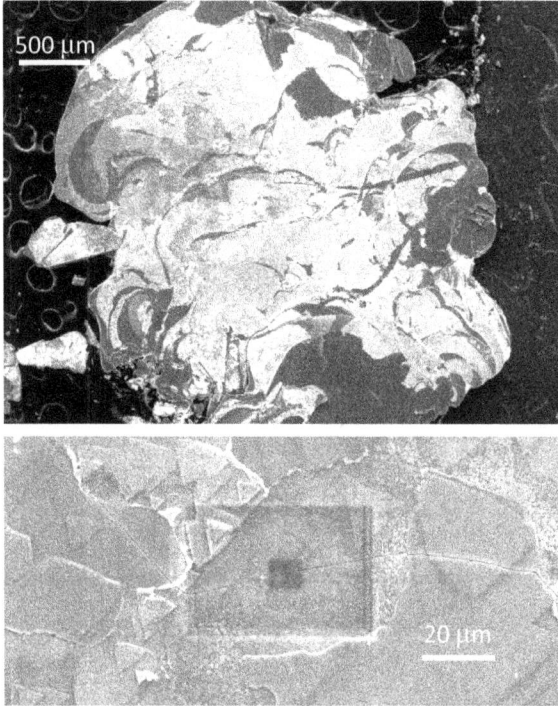

Fig. 1. SEM image of the CuI film coating as a grown Cu_xTiSe_2 single crystal. Most of the surface is covered with a CuI film, which appears much brighter in the SEM image than surrounding $TiSe_2$. The lower panel reveals triangular structures typical of the CuI coating. The darkened rectangular regions are areas in which higher magnification imaging was attempted. The focused beam burned away the CuI to reveal the underlying $TiSe_2$ surface.

The CuI surface layer was not a completely uniform coating, with portions seemingly chipped away during sample handling. It appeared much brighter than uncovered areas, and typically was composed of triangular features aligned over a local area of the sample. This CuI overlayer was very sensitive to electron beam exposure. High magnification imaging would often burn away a portion or all of the CuI layer within less than a minute of imaging. Rectangular areas in which the CuI layer was thus burned away can be seen in the lower panel of Figure 1.

The SEM measurements were performed using a TESCAN Vega II microscope with an attached Bruker Quantax EDX spectrometer. SEM images and EDX data shown here were taken using 20kV beam voltages. The samples were manipulated in air for short periods and stored long term in a dry box. The inert $TiSe_2$ crystals showed almost no signs of chemical decomposition months after they were grown. To expose clean portions of the samples and remove the CuI films, samples were exfoliated with Scotch tape to remove the uppermost surface layers. This process had to be repeated several times in some cases, as CuI inclusions could be found within some crystals. These inclusions, which were typically very thin CuI films, grew as the crystals were formed creating weak points within the crystals where exfoliation would occur but not provide a clean dichalcogenide surface.

3. Metastable surface inhomogeneity

SEM and EDX measurements were performed primarily on exfoliated Cu_xTiSe_2 single crystals. Interestingly, even in samples which showed no evidence of CuI via EDX after this process, the crystals appeared inhomogeneous in SEM images as in Figure 2. The apparent brightness of the surface varied in different locations. This was quite unexpected as pure $TiSe_2$ samples never showed any such inhomogeneity, and x-ray diffraction measurements indicated the samples were of a single phase. EDX measurements were then taken over different sections of the sample to try to correlate the apparent differences in brightness to differences in the local chemical composition. According to EDX measurements, the samples were essentially homogenous. Furthermore, high magnification SEM measurements did not reveal any substantial differences between the brighter and darker areas of the sample.

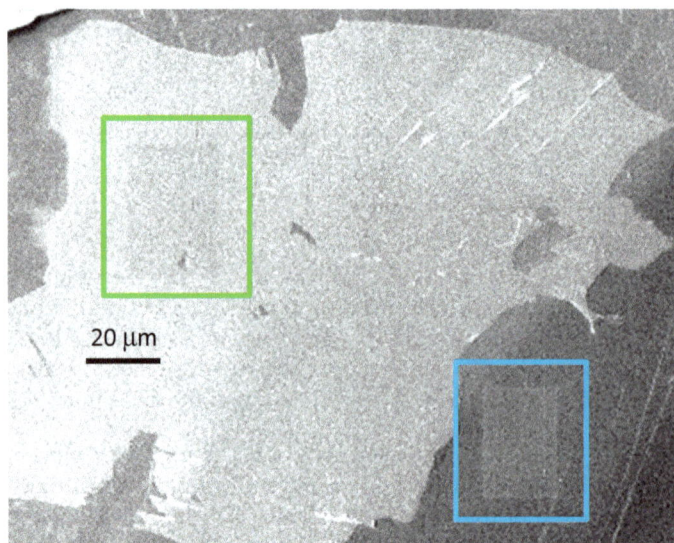

Fig. 2. SEM image taken from the inhomogeneous appearing surface of a freshly exfoliated single crystal of $Cu_{0.04}TiSe_2$. Various areas of the surface appeared more or less bright in the SEM. The colored boxes outline two areas exposed to relatively high electron beam radiation during long duration EDX measurements. The green square outlines an area from the brighter appearing portion of the surface that became darker after exposure. The blue square outlines a section from the darker area of the surface that became brighter after electron beam exposure.

To obtain more precise stoichiometry values, EDX measurements were made over longer exposure times than usual. Measurements were taken for half an hour or more, in which a relatively small portion of the surface was continuously exposed to electron beam radiation. While these measurements did not provide any conclusive results regarding the local stoichiometry, they did have a significant impact on the appearance of the sample surface (Figure 2). Essentially, long term electron beam exposure caused areas which initially appeared brighter to become darker and areas which were initially darker to become brighter.

The process could even be reversed to some extent in that an area could be switched back to its original appearance by a second exposure as seen in Figure 3. In this image, the sample areas appeared relatively dark until the portion in the upper half of the picture was exposed to long term electron beam radiation and became brighter. After this, a smaller portion of the sample was re-exposed to the beam. At this time, the brighter section reverted to its original appearance and the bottom half of the surface, which was not initially exposed to the beam for a long duration, was not changed at all. The edge of this second exposure is visible in the image, however. In general, it was often possible to effectively burn away a region made brighter by the beam, but we were not able to make a darkened area return to a more bright state by a second exposure.

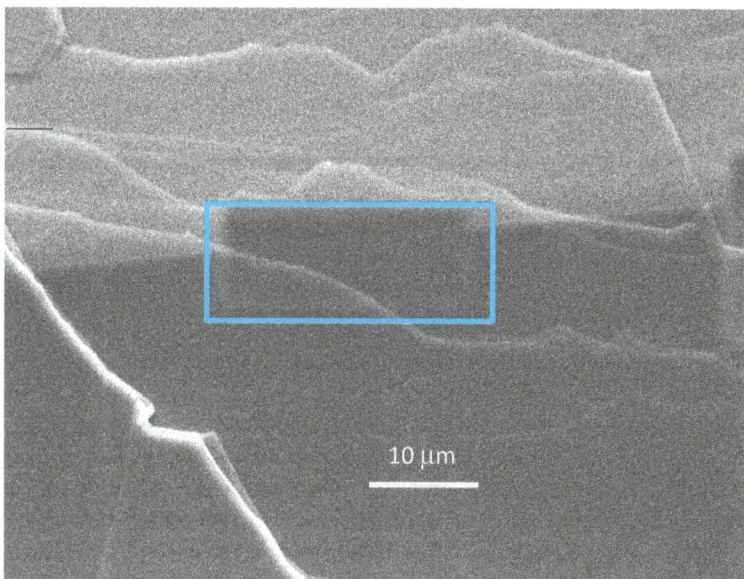

Fig. 3. SEM images of an area exposed to long term electron beam radiation multiple times. This area of the sample was originally one of the less bright regions of the surface. After long term EDX measurements, the top half of this image became brighter. A second long term EDX measurement was taken at the border between the original surface and the brightened area, as highlighted by the colored rectangle. Within this region, the area that became brighter in the initial exposure reverted to its original darker appearance after the second exposure.

These results indicate that the apparent brightness of the sample is most likely related to a phenomenon confined to the immediate surface of the sample. EDX measurements include x-rays generated by inelastically scattered electrons which can easily penetrate several hundred nanometers into the sample. Therefore, stoichiometry variations confined to a few surface layers are essentially undetectable by EDX. Furthermore, it is obvious that electron beam radiation can alter the apparent brightness of the sample as viewed in the SEM. When the beam is focused on a small area of the sample in high magnification measurements, this area would receive a relatively high dose of radiation even for short duration measurements. Therefore it is not surprising that significant differences could not be found when comparing different portions of the sample as the surface is likely changing during the measurement process itself.

These measurements in themselves show that the variation in brightness is a surface phenomenon sensitive to electron beam exposure, but do not reveal the source of the surface inhomogeneity. The apparent brightness in an SEM image can be influenced by many factors. The relative height, density, and conductivity of a given sample region can all play a role in determining how bright a particular area appears in an SEM image.

In this case, it appears that the brightness could be due to local changes in the sample density or surface conductivity. Copper ions have significant mobility within the $TiSe_2$ solid (Gunst et al. 2000), so that any local heating or other beam interactions could potentially cause the copper ions to migrate and/or aggregate to alter the local electron density in response to beam exposure. On the other hand, insulating copper iodide films appear not only on the surface of as-grown crystals, but could also be found after exfoliations interspersed within the crystal bulk. Even though EDX measurements did not detect significant iodine levels, the brighter areas could arise from an extremely thin CuI layer. This appears to be the more likely explanation. Copper ions assume a positive oxidation state as dopants within the layers of $TiSe_2$. If the beam caused them to migrate and cluster together in some manner, this would not be a stable configuration as the positive ions would repulse each other through Coulomb interactions to return to their equilibrium positions. The changes induced by beam exposure were stable for days if not weeks indicating equilibrium must be attained soon after the beam exposure is ended.

If the apparent changes in brightness arise to a thin CuI layer, there must be a source of iodine somewhere throughout the sample. In fact, a small concentration of iodine does remain interspersed throughout the sample from the growth process, where iodine is used to catalyze the growth of large single crystals. Also, iodine was found in significant amounts within cracks or at the edges of the crystals. These areas also tended to contain larger concentrations of copper, making them a mix of copper iodide and pure iodide in areas of the sample less affected by exfoliation. Due to charging effects in these insulating materials, the iodine rich particles appear much brighter than the surrounding dichalcogenide surface.

The presence of microscale iodide crystals was unexpected given the element's volatility. The samples used in this study were grown several weeks prior to the SEM measurements, making it likely that any surface iodine should have already sublimated. Therefore, the iodine seen here must have been stable while trapped inside various sample defects. This iodine would then be exposed after the exfoliation process to become more reactive.

Assuming the bright areas appear so due to a thin coating of CuI, they would become darker as this insulating material is burned away during the electron beam exposure. At the same time, the radiation from the electron beam would enhance the mobility of copper ions to flow towards the surface through local heating and/or the accumulation of negative charge. Another effect of the electron beam exposure would be to volatilize any exposed iodine. Any copper ions migrating to the surface would react with this iodine to become trapped as a very thin CuI layer, creating a region appearing more bright after exposure. This hypothesis is consistent with the fact that the surface was mostly stabilized within a day or so after exfoliation. While it was still possible to burn away brighter regions, it was no longer possible to reverse this process or induce darker regions to become bright on the aged samples. This is exactly what one would expect as the iodine available for reaction immediately after exfoliation would simply sublimate away so that there would be nothing left for copper ions to react with to form apparently bright regions as seen by SEM.

4. Formation of CuI crystals

The initial EDX measurements indicated that there could be variation in the copper concentration related to electron beam exposure, although these variations were not very significant given the precision of the measurement. If this were true, it could mean that the electron beam exposure could be used to create localized areas of superconductivity within the Cu_xTiSe_2 surface and induce interesting interfacial properties between charge density wave and superconducting regions of the sample. (Morosan, Zandbergen et al. 2006; Barath et al. 2008). To better attain information about the copper distribution, it was deemed important to first remove the iodine. Copper reacts strongly with any iodine present in the system, making iodine a contaminant that can strongly affect the distribution of the dopant copper ions within the TiSe$_2$ layers. Given the volatility of iodine, the decision was made to gently heat one of our exfoliated samples in a vacuum oven. For convenience and consistency, a sample in which EDX measurements had been performed was chosen. A low temperature, only 60°C, was used as this should be sufficient to vaporize any exposed iodine and it was not so high as to damage the carbon tape used to mount the sample.

The results were quite surprising, as seen in Figure 4. Not all of the iodine exposed by exfoliation sublimed from the sample during heating. In fact, it appeared as if the majority reacted with dopant copper ions to form a surface covered in CuI. That copper and iodine reacted was not completely unexpected. Halogen gas exposure has been shown to draw out high mobility dopants like copper or silver (Jaegermann, Pettenkofer et al. 1996). However, the relatively small amount of iodine seen in our measurements of the exfoliated surface did not lead us to expect the near totality with which the surface was covered after heating.

There are additional oddities concerning the copper iodine reaction of this surface. The region shown in Figure 4 is the same area as shown in Figure 3 after heat treatment. The portions of the sample that had turned relatively bright after e-beam exposure were almost completely free of CuI after heating. Other exposed areas can be seen as rectangular features in the image. The smaller region formed by the secondary exposure also had less CuI formation. However, some rectangular regions appear to have a more uniform CuI coverage than unexposed areas of the sample. These regions received a relatively small dose of electron beam radiation during the time in which images were taken, such as that shown in Figure 3.

The truly novel discovery was that rectangular portions of the surface remained CuI free. These rectangular areas could be identified as the regions on the sample which received very high electron beam exposure levels during the EDX studies. This discovery led to a more systematic series of measurements designed to illuminate whether controlled electron beam exposure could be used to create a patterned array of CuI features in a process similar to, but simpler than, standard electron beam lithography.

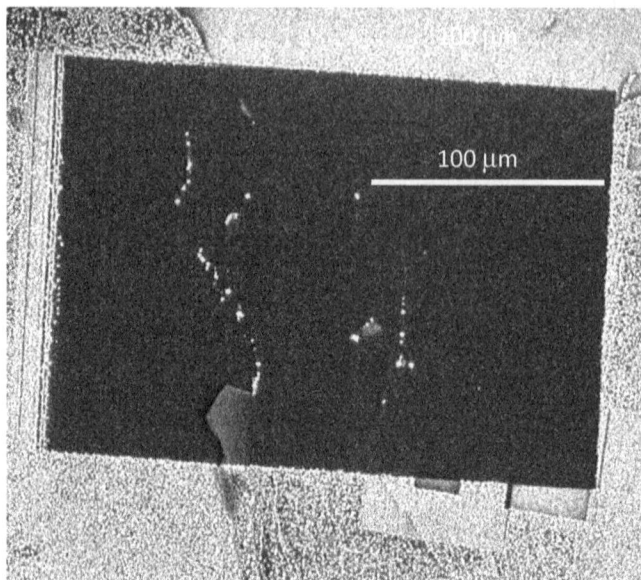

Fig. 4. SEM images taken from the area shown in Figure 3 after the sample was heated in a vacuum oven at 60°C for 24 hours. Iodine released from the sample during heating reacted with copper from within the sample to form CuI crystallites over the majority of the sample surface. The formation of CuI was reduced or eliminated in areas of the sample in which long term EDX measurements were taken.

5. Dopant driven lithography

To determine the effects of electron beam exposure, comparisons were made between a series of different exposure levels made upon a single Cu_xTiSe_2 crystal. After the exposures were performed, the samples were transferred to the vacuum oven and heated to temperatures between 60-100°C. The iodine exposure levels were varied by including various amounts of iodine in the vicinity of the sample when it was loaded into the vacuum oven.

Within a given sample, the results were very consistent. Figure 5 shows SEM images taken from a Cu_xTiSe_2 crystal that had been exposed to four different electron radiation doses before being heated in a vacuum oven. The surface was exposed to three ring patterns with dosages of 2.5, 5, and 10 mC/cm². Near the center of each ring a small circular area was also exposed to a dosage of roughly 50mC/cm². This surface was exfoliated multiple times to attain a mostly homogenous surface to minimize any effects from a potential pre-existing CuI film upon the surface. These efforts were successful in that no significant changes were induced by the electron beam exposure itself. However, the radiation exposure patterns were still easily identifiable after the heat treatment. Even at low magnification levels, it can be seen that the CuI coverage is significantly reduced by the electron beam exposure.

The influence of the electron beam exposure is more obvious at higher magnification as seen in the area exposed with a 5 mC/cm^2 dose in Figure 6. After heating, the surface is coated with small CuI crystallites, which are reduced in size and density in areas exposed to higher levels of electron radiation. Significant changes were difficult to detect in doses less than 1mC/cm^2.

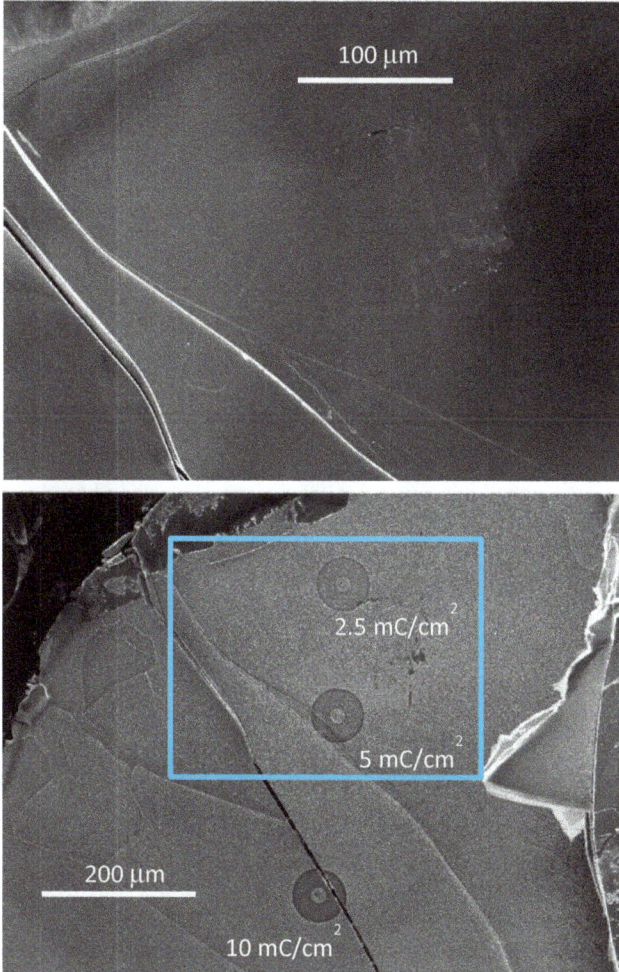

Fig. 5. SEM images before and after an irradiated sample was heated to 60°C in a vacuum oven for 24 hours. The top panel was taken immediately after selected areas of the surface were exposed to electron beam radiation. The surface showed no signs of damage from the electron beam exposure. The image in the bottom panel was taken after the heat treatment induced CuI formation. The blue rectangle indicates the area initially imaged in the upper panel, and the radiation exposure levels for each pattern are indicated on the image. Near the center of each ring, small circular areas were exposed to approximately 50 mC/cm^2 of electron beam radiation.

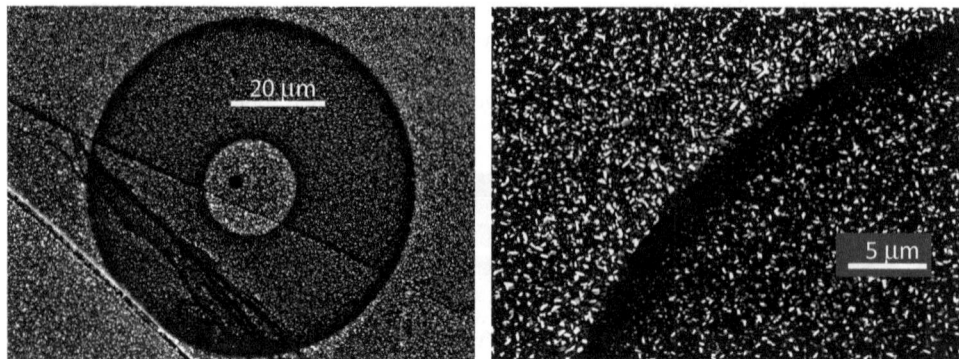

Fig. 6. SEM images taken after irradiation and vacuum heat treatment at 60°C. The left panel is an overview of an area exposed to 10mC/cm² of radiation in a ring pattern. The right panel is a magnified view of the exterior ring edge. CuI crystallites are smaller and fewer in number in the irradiated area. The boundary region of the exposed area received a larger dose of electron radiation, resulting in even less CuI crystallites at the edge of the pattern.

The edges of the exposed ring were relatively dark in the SEM images, with much less CuI crystallites. This is attributed to systematic errors in the coil control program which cause the beam to linger longer at edges in the prescribed pattern. The small circle offset from the center of the ring was exposed to a much higher dosage of electron beam radiation. Through additional measurements concerning multiple samples and treatment temperatures, it was found that a dosage of 50mC/cm² was consistently sufficient to completely suppress CuI formation.

It was difficult to determine how the chemical composition of the surface was altered by various exposure doses or iodine reaction temperatures. The signal for iodine was small and copper consistently measured between three and four percent of the overall composition. Iodine could not be detected at all in areas exposed to high enough doses to completely suppress CuI formation, but even in these areas the copper concentration was not significantly different from the rest of the sample. These results indicate the CuI formation was confined to the surface and that the overall copper doping concentration was not substantially altered within the exposed areas.

As shown in Figure 7, the CuI crystallites are typically sub-micron structures. The size and density of these crystallites are inversely proportional to the amount of electron radiation exposure. These results show that not only can the exposure process be used to direct where the CuI crystallites are formed, but also to control their size and average spacing.

Depending on the radiation dosage and iodine reaction temperature, crystallite sizes ranged from ten nanometers to a few microns. While not particularly uniform, the size of most crystallites within a given exposure region did not deviate by more than a factor of two. The crystallite size was dependent on both the exposure level and temperature at which the iodine exposure occurred within the vacuum oven. Higher processing temperatures led to the formation of larger crystallites (Figure 8). While the smaller crystallites appeared more rounded and randomly shaped, larger structures often showed clear symmetry respective of an underlying crystal structure. The variation in appearance is a size-effect related phenomenon. As the size of the crystallite decreases, atoms residing at the surface represent a larger and larger percentage of the material. Surface atoms have a relatively high energy state, and when the crystallite becomes small enough, minimizing the total number of surface atoms by attaining a more rounded structure becomes more energetically favorable than maintaining the proper bonding angles representing the crystal symmetry found in larger features.

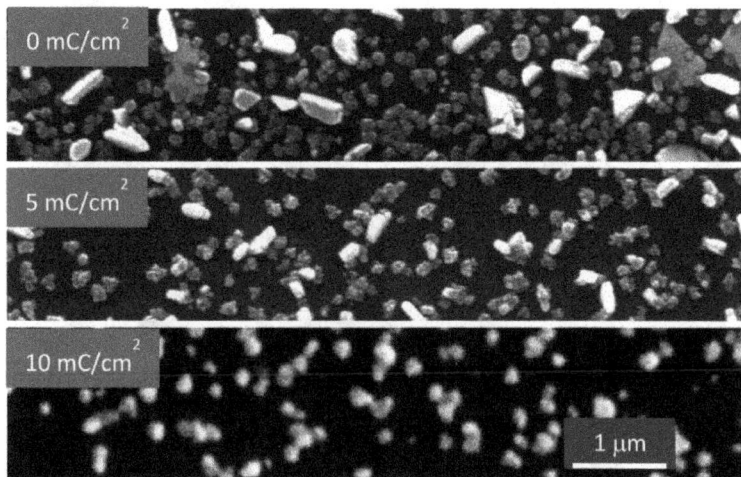

Fig. 7. SEM images comparing the effects of radiation exposure on the form of CuI crystallites formed after heating at 60°C. Each panel is shown at the same scale and was taken from different areas of the same sample with the indicated radiation exposure dose.

The lateral resolution of the process is theoretically expected to be on the order of the electron beam spot used during the exposure process. However, in practice the beam spot can be made to be much smaller than the average CuI crystallite size. Thus the lateral resolution of this technique is defined more by the size of the crystallites to be obtained rather than intrinsic exposure parameters.

There were significant variations found in the size and density of crystallites formed on different TiSe$_2$ crystals. This is attributed to variations in the amount of iodine available for reaction. It was found that including excess iodine seemed to have a much smaller effect than variations seen from sample to sample, indicating that the CuI reaction is dominated by iodine emerging from the TiSe$_2$ sample itself. While iodine is a volatile element in ambient conditions, it can remain present for years trapped between layers of inside defects of a TiSe$_2$ crystal. When the sample is exfoliated, trapped iodine becomes available for reaction, especially when the sample is heated, even at temperatures below 100°C. Iodine loaded in proximity to the crystal likely becomes volatile and spreads quickly throughout the vacuum oven, minimizing its effectiveness in forming CuI. Iodine trapped in the nooks and crannies of the sample will take more time to fully volatilize, and be in closer proximity to the surface, giving it more time to react with the copper dopant ions before it dissipates away from the sample. This hypothesis also supports the relationship between crystallite size and reaction temperature. At higher temperatures, the copper ions have a higher mobility which allows a larger number to reach the surface in a given time. As there is likely a limit to how long the chemical reaction takes place, heating to a higher temperature would allow more copper ions to take part in the reaction to form larger structures.

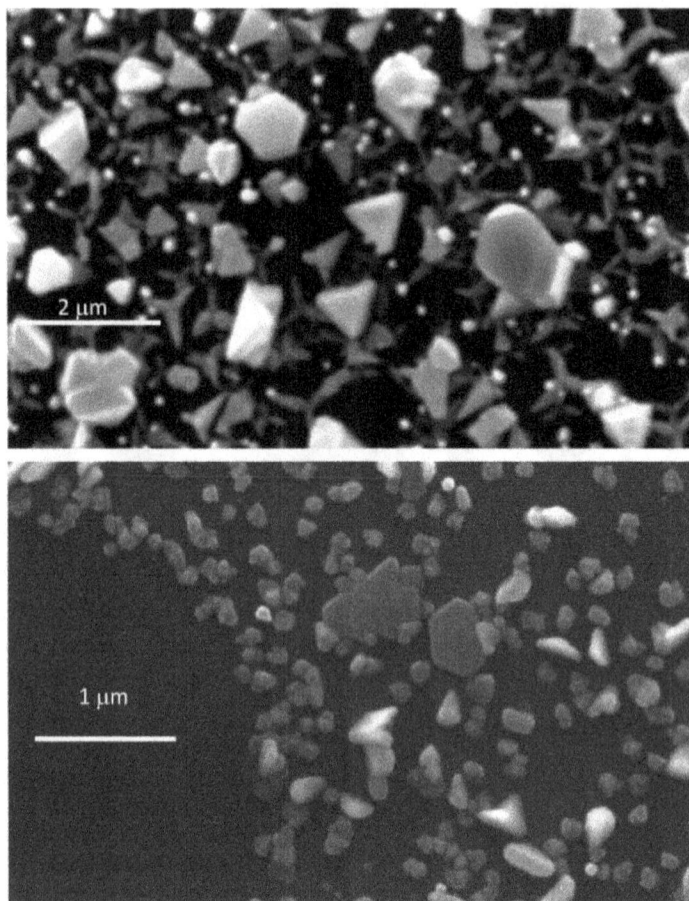

Fig. 8. SEM images comparing the effects of reaction temperature on the size of CuI crystallites. The top panel was heated in the vacuum oven at 100°C. The bottom panel was heated to 60°C. The bare area on the left of the bottom panel was initially dosed with about 50 mC/cm^2 of electron beam radiation, completely suppressing the formation of CuI.

It was also found that CuI formation was inhibited for samples which had been exfoliated for many days prior to beam exposure. Even though the samples looked almost identical to fresh surfaces in the SEM, including the presence of trapped iodine, CuI crystallites would rarely grow upon these aged samples. These results were reminiscent of an AFM study in which the copper ions could be drawn to the surface of a dichalcogenide crystal, but only for freshly exfoliated surfaces (Gunst, Klein et al. 2000).

6. Conclusions and future directions

The experimental results show that it is possible to use electron beam exposure to control the formation of CuI crystallites on the surface of a TiSe$_2$ crystal. This is a two-step process in which a fresh surface is first exposed to electron beam radiation in a controlled fashion, and then heated in the presence of iodine. Due to inconsistencies between samples, exact empirical parameters concerning the size and density of CuI formation have not been determined as yet, although qualitative relationships have been established. The size and density of CuI crystallites formed in this process are inversely proportional to the amount of electron beam radiation. The formation of crystallites can be completely eliminated in areas receiving relatively high exposure levels. It is evident this process can be used to create an array of CuI crystallites with diameters sufficiently small to be considered nanoparticles at certain processing conditions. The lateral resolution is essentially limited by the size of the crystallites, making this technique suitable for forming a pattern array of nanocrystallites with nanoscale precision.

In itself, this if of interest as CuI nanoparticles show some promise for applications in dye-sensitized solar cells (Perera and Tennakone 2003) or gas sensing applications for CO and NO (Wolpert et al. 2009). The process might also be useful for controlling the local copper dopant concentration, important for exploring competition between superconducting and charge density wave ground states in the material. However, the technique would be of far greater importance if it could be extended to other dopants and/or dichalcogenide crystals. For this to occur, one must have a better understanding of the process itself. While a complete quantitative understanding remains elusive given inconsistencies between samples, a qualitative explanation is forthcoming from our observations.

The second step of the process is relatively straightforward. Copper ions have a high mobility within dichalcogenides and can be reversibly incorporated into their structure using electrical or chemical gradients.(Gunst, Klein et al. 2000) Iodine present at the sample surface will react with surface copper ions, leading to a concentration gradient near the surface. Copper ions from deeper within the sample will be driven to the surface layers by this depletion, and thus in turn react themselves. This process will continue until the iodine is depleted, the concentration gradient is reduced below some threshold value depending on the overall percentage of copper ions in the sample, or the surface is completely coated with CuI. The increased size of the CuI crystallites at higher reaction temperatures could be due to the increased mobility of Cu ions at higher temperatures or the fact that more iodine is released from the sample. Either process could lead to the formation of more CuI at the surface before the iodine dissipates away from the sample.

The first step of the process is perhaps not as intuitive. It was not initially clear why exposure of the sample could be used to influence the reaction of copper and iodine at the sample surface. The exposure must in some way be inhibiting the copper ions from traveling to the surface during the iodine reaction. This is most likely due to the electron beam radiation inducing a localized surface chemical reaction. When TiSe$_2$ is exposed to wet or humid environments it will oxidize at the surface. Titanium diselenide is composed of molecular TiSe$_2$ layers as shown in Figure 8. Hydrogen from the water will react with Se to form gaseous H$_2$Se leaving the exposed titanium atoms to react with the remaining oxygen to form titanium oxide. Owing to the inert nature of the material and its layered structure, it can take days or even weeks for a visible oxide layer to be seen. However, it would not be surprising for the uppermost molecular layer to oxidize within even a few hours of exposure to ambient conditions. A surface layer of titanium oxide would be an effective barrier to Cu ion migration to the surface, which would explain why CuI would not form on the surface of samples that had been exfoliated for several days before electron beam exposure was initiated.

Within the SEM, the sample resides in a vacuum. However, the samples are exfoliated in air before being inserted into the microscope. This means that there will be a residual water layer of some thickness on the sample even after the measurement chamber is evacuated. In truth, such a water layer coats the entire interior surface of this and any other SEM that is not heated to at least 100°C before measurement. This minor amount of water would normally react very slowly with the $TiSe_2$ surface. However, the energy provided by the electron beam radiation could easily break up the water molecules into various radical ions that would be much more reactive. This would in turn create a titanium oxide surface layer to inhibit copper ion migration. The uniformity and thickness of such an oxide layer would depend on the radiation exposure level, making it possible to fine tune how much copper ions migrate to the surface during the time in which the sample reacts with iodine in the second step of the process. It is true that other chemical reactions could occur on the surface, such as with remnant gas molecules within the SEM, however the low pressure ($<10^{-2}$ Pa) makes this unlikely.

These results make it apparent which systems could be best utilized with this technique. A wide range of ions and molecules can be reversibly stored within various dichalcogenides (Levy 1979). This makes a wide range of surface features possible, given the proper method by which materials could be induced to leave the sample. The most straightforward extension of this technique would be to utilize silver rather than copper ions. Silver and copper both have similar mobility within a given dichalcogenide and silver also is highly reactive with halogen gasses. Thus it should not be difficult to create a patterned array of silver iodide crystallites on the surface of $TiSe_2$. AgI has been used in photographic and other optical applications for decades, and an array of AgI nanocrystallites could easily be reduced to form silver nanoparticles which are of high interest for their antibacterial properties.

Current research is being directed towards refining this technique to enable the formation of various copper and silver halide nanoparticle arrays. Titanium diselenide is used as the principle host, but Ta based dichalcogenides are also of interest for their similar chemical characteristics. It is hoped that this technique can be coupled with traditional electron beam and/or scanning probe lithographic techniques to enable to the fabrication of prototype device structures.

7. Acknowledgements

The author would like to thank Dusty Klein and Tyler Rash for assisting with the SEM/EDX measurements and Dr. Laura Strauss for assisting with crystal growth. This work was supported by Battelle and the Iowa Office of Energy Independence grant #09-IPF-11. The author was also supported by a UNI summer fellowship during portions of this research.

8. References

Balchin, A. A. (1976). Growth and the Crystal Characteristics of Dichalcogenides Having Layer Structures. Crystallography and Crystal Chemistry of Materials with Layered Structures. F. Levy. Dordrecht, D. Reidel. 2: 1-50.

Barath, H., M. Kim, J. F. Karpus, S. L. Cooper, P. Abbamonte, E. Fradkin, E. Morosan and R. J. Cava (2008). "Quantum and Classical Mode Softening Near the Charge-Density-Wave–Superconductor Transition of Cu_{x}TiSe_{2}." Physical Review Letters 100(10): 106402.

Broers, A. N., A. C. F. Hoole and J. M. Ryan (1996). "Electron beam lithography--Resolution limits." Microelectronic Engineering 32(1-4): 131-142.

Chen, J., S.-L. Li, Z.-L. Tao, Y.-T. Shen and C.-X. Cui (2003). "Titanium Disulfide Nanotubes as Hydrogen-Storage Materials." Journal of the American Chemical Society 125(18): 5284-5285.

Egerton, R. F., P. Li and M. Malac (2004). "Radiation damage in the TEM and SEM." Micron 35(6): 399-409.

Friend, R. H. and A. D. Yoffe (1987). "Electronic properties of intercalation complexes of the transition metal dichalcogenides." Advances in Physics 36(1): 1.

Gunst, S., A. Klein, W. Jaegermann, Y. Tomm, H. Crawack and H. Jungblut (2000). "Intercalation and deintercalation of transition metal dichalcogenides: Nanostructuring of intercalated phases by scanning probe microscopy." Ionics 6(3): 180-186.

Jaegermann, W., C. Pettenkofer, O. Henrion, Y. Tomm, C. Papageorgopoulos, M. Kamaratos and D. Papageorgopoulos (1996). "Surface science investigations of Cu intercalation in 1T TaSe$_2$ and TiSe$_2$ and its deintercalation by adsorbed Br$_2$." Ionics 2(3): 201-207.

Kidd, T. E., D. Klein, T. A. Rash and L. H. Strauss (2011). "Dopant based electron beam lithography in CuxTiSe2." Applied Surface Science 257(8): 3812-3816.

Kidd, T. E., T. Miller, M. Y. Chou and T. C. Chiang (2002). "Electron-Hole Coupling and the Charge Density Wave Transition in TiSe2." Physical Review Letters 88(22): 226402.

Kline, G., K. Kam, D. Canfield and B. A. Parkinson (1981). "Efficient and stable photoelectrochemical cells constructed with WSe2 and MoSe2 photoanodes." Solar Energy Materials 4(3): 301-308.

Levy, F., Ed. (1979). Intercalated Layer Materials. Physics and Chemistry of Materials with Layered Structures. Dordrecht, D. Reidel.

Liu, K., P. Avouris, J. Bucchignano, R. Martel, S. Sun and J. Michl (2002). "Simple fabrication scheme for sub-10 nm electrode gaps using electron-beam lithography." Applied Physics Letters 80(5): 865-867.

Mendes, P. M., S. Jacke, K. Critchley, J. Plaza, Y. Chen, K. Nikitin, R. E. Palmer, J. A. Preece, S. D. Evans and D. Fitzmaurice (2004). "Gold Nanoparticle Patterning of Silicon Wafers Using Chemical e-Beam Lithography." Langmuir 20(9): 3766-3768.

Morosan, E., H. W. Zandbergen, B. S. Dennis, J. W. G. Bos, Y. Onose, T. Klimczuk, A. P. Ramirez, N. P. Ong and R. J. Cava (2006). "Superconductivity in CuxTiSe2." Nature Physics 2(8): 544-550.

Perera, V. P. S. and K. Tennakone (2003). "Recombination processes in dye-sensitized solid-state solar cells with CuI as the hole collector." Solar Energy Materials and Solar Cells 79(2): 249-255.

Sipos, B., A. F. Kusmartseva, A. Akrap, H. Berger, L. Forro and E. Tutis (2008). "From Mott state to superconductivity in 1T-TaS2." Nature Materials 7(12): 960-965.

Whittingham, M. S. (1976). "Electrical Energy Storage and Intercalation Chemistry." Science 192(4244): 1126-1127.

Wilson, J. A., F. J. D. Salvo and S. Mahajan (1975). "Charge-density waves and superlattices in the metallic layered transition metal dichalcogenides." Advances in Physics 24(2): 117-201.

Wolpert, B., O. S. Wolfbeis and V. M. Mirsky (2009). "Gas sensing properties of electrically conductive Cu(I) compounds at elevated temperatures." Sensors and Actuators B: Chemical 142(2): 446-450.

6

Adhesive Properties

Anna Rudawska
Lublin University of Technology,
Poland

1. Introduction

Adhesive joints function in multiple branches of technical engineering in which the phenomenon of adhesion appears: creating adhesive joints, sealing, applying protective or decorative coating (paint or varnish), printing, decorating and many others. Among adhesive bonding techniques these are adhesive joints which are used most often in various machine structure joints.

Surface phenomena, such as adhesion, cohesion and wettability, play an exceptionally important role in creating adhesive joints, as they influence the possibility of creating such a joint and its quality. Adhesive properties are fundamental in processes in which the phenomenon of adhesion appears. These properties are referred to as the whole of physical-chemical properties heavily influencing adhesion. Adhesive properties are a crucial indicator determining, for instance, whether the surface layer is properly prepared for permanent or temporary adhesive joints to be formed. The surface layer is the external layer of the material, limited by the real surface of the object, including this surface and the outer part of the material together with its real surface. It demonstrates different physical and chemical properties or qualities when compared with the core of the material.

When analysing the issue of constituting adhesive properties, exceptional importance is ascribed to the first two groups of technological operations aimed at preparing and obtaining specific properties of the surface and the surface layer of the material, as well as a special improvement (modification) of the aforementioned. They allow, for instance, obtaining proper energy and geometric properties of the surface layer of joined materials, which positively influence adhesion.

These operations are considered crucial in terms of constituting these properties in reference to forming and the quality of hybrid adhesive joints, as they are composed of materials of different physical, mechanical and chemical properties.

Surface preparation, conducted according to the requirements, is one of the methods of constituting adhesive properties of a surface. Depending on the characteristics and required properties of adhesive joints it is possible to increase or decrease adhesion, i.e. to improve or lower adhesive properties.

The selection of a surface preparation method (including appropriate technological operations allowing to achieve desired structure and energy properties) depends on many factors, among which the most important ones is the type of materials creating the adhesive joint.

2. Surface layer

In geometry, the surface is a two-dimensional geometric figure that limits the space filled with matter, i.e. surface in a theoretical sense. In mechanics, the surface is defined as the edge of a material body, which may be analysed in a molecular scale, micro- and macrosize, at the same time distinguishing different surfaces: material, nominal, real, observed, under machining, machined (Burakowski & Wierzchoń, 1995; Sikora,1997). The real surface may be defined as the surface separating the object from the surrounding environment. However, from the point of view of adhesion, the most important concept of surface is presented in the physical-chemical sense, as it involves the phases. In physical chemistry, surface is a boundary of two touching phases, i.e. interfacial surface or an interface, where an abrupt change of properties occurs together with the phase transition. Interfacial surfaces are surfaces between bodies of different states of aggregation (Hebda & Wachal, 1980). Surface in physical-chemical sense is analysed in three dimensions, despite the difficulties in determining the thickness/depth of the interface due to its small dimensions.

The physical space is not a homogenous area between two phases. Atomically clean surface is extremely active physically and chemically, therefore, each contact with another body results in the adsorption of the substances. Newly adsorbed substances may initiate formation a new phase. Another aspect is that under a physically clean surface there may be various deformations and defects resulting from surface formation. Consequently, different properties may be observed in the physically clean surface compared to the core of the object. As a result, different layers constituting the surface layer may occur: below the surface, surface and above the surface.

2.1 The surface layer structure

At present a number of definitions of the surface layer exist (Sikora, 1996,1997; Rożniatowski, Kurzydłowski, & Wierzchnoń,1994). One of the alternatives states that the surface layer is the external layer of the material, limited by the real surface of the object, including this surface and the outer part of the material under its real surface, which demonstrates different physical and, occasionally, chemical properties when compared with the core of the material. The articles (Kuczmaszewski, 2006; Sikora, 1996,1997) contain the description and the model of the surface layer of the material resulting from the adhesive failure. The surface layer has zonal structure. The proportions and the thickness of different zones vary, in addition the zones may interpenetrate, changing into one another or occupying the same space.

The structure and properties of the surface layer depend on the type and course of multiple phenomena and processes, including physical-chemical phenomena, such as adhesion.

2.2 The non-saturated surface force field

The surface of any body consists of atoms, particles or ions, which are in different conditions than the ones inside the body. In the volume phase the particles are subject to equal forces of interaction. In the interface, however, the particles come into contact with their own phase as well as another one, which leads to the occurrence of asymmetric forces of interaction (Fig. 1, Burakowski & Wierzchnoń, 1995). The particles on the surface are more forcefully drawn into the volume phase, and as a result the surface has higher energy than the inside of the body. Such a surface is active and is able to adsorb other atoms or particles in its vicinity (Burakowski & Wierzchnoń, 1995; Dutkiewicz 1998).

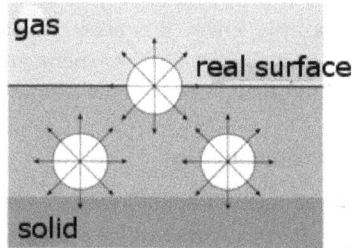

Fig. 1. A model of forces interacting with particles inside the solid and on its surface

What is equally important is the degree to which the surface particles are surrounded by other particles, i.e. whether the surface is flat or porous (Fig. 2, Dutkiewicz, 1998). The degree of non-saturation of forces is higher for a porous surface than for a flat one, therefore, the former is more active physically and chemically.

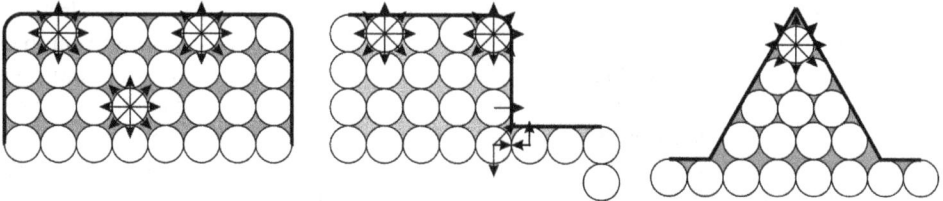

Fig. 2. The influence of porosity on the force field of various surfaces

It is the surface free energy, characteristic of solids, which is the measure of the interactions between the particles on the surface and inside the body.

3. Adhesion

3.1 Adhesion and adhesive properties definitions

The literature on the analysed subject is to some extent inconsistent in terms of contradicting terminology defining adhesion. Etymologically, 'adhesion' is derived from Latin *adhaesio* and stands for clinging or linkage. The adhesion is defined as a surface phenomenon, consisting in binding bodies in close contact as a result of force field interactions (Harding & Berg, 1997; Kuczmaszewski, 2006; Mittal, 1978, 1980; Żenkiewicz, 2000). The force field, induced by the charges of atoms constituting the surface layer (particles, ions), decreases exponentially with the distance to the surface (van der Waals interaction forces are negligible for the gap over 1-2 nm). Therefore for the adhesion to take place, the close contact of surfaces is required.

Knowledge of the adhesive propriety plays important role in processes in which appears the occurrence of the adhesion. To such processes we can number the bonding, the painting, the decoration, the printing, the lacquer finish, etc. The adhesive properties characterise the surface of the materials taking into account their applicability in the adhesive processes. Good adhesive properties have a positive influence on the strength of the adhesive joint obtained, low properties significantly lower this strength or even prevent the bonding. Knowing the properties allows as well to constitute them properly by means of required surface preparation treatment of the analysed materials (Rudawska, 2010).

3.2 Geometric structure and SEM technique

Ggeometric structure and adhesive properties are extremely important in the technology of creating adhesive joints. The geometrical structure of the material surface has an influence on the adhesive joints strength obtained, and that is the reason why it should be carefully analysed before bonding. Surface roughness is important in view of the part the mechanical adhesion plays in general adhesion; consequently, it is beneficial to know the structure of the material surface layer that will be used in the adhesion process.

A scanning electron microscope (SEM) is a type of electron microscope that images a sample by scanning it with a high-energy beam of electrons in a raster scan pattern. The electrons interact with the atoms that make up the sample producing signals that contain information about the sample's surface topography, composition, and other properties such as electrical conductivity. Due to the very narrow electron beam, SEM micrographs have a large depth of field yielding a characteristic three-dimensional appearance useful for understanding the surface structure of a sample (http://en.wikipedia.org/wiki/Scanning_electron_ microscope). Scanning electron microscopy (SEM) is generally considered micro-analytical techniques which are able to *image* or *analyze* materials we can not generally observe with the resolution offered by visible techniques. By *image* we mean photograph an object much smaller than we can see, even with the aid of an optical microscope (http://epmalab.uoregon.edu/epmatext.htm). SEM technique is very useful to analysis geometric structure of material for which is described adhesive properties (for example wettability or surface free energy).

Below there are some of example of materials for which it was determined the geometric structure (Rudawska, 2009 b, 2010).

The tests were conducted on aramide-epoxy composite samples. The composite consisted of two layers (2 x 0.3 mm) of aramide material marked KV-EP 285 199-46-003. The materials were arranged at 90 degree angle and subjected to the polymerisation process.

The geometric structure of the analysed composite was defined by means of SEM images. The results are shown in Fig. 3.

SEM images of the surface of analysed composite, show distinct differences in the surface structure, that are the result of specific character of the surface of the measures composites.

The pleat and the direction of the materials arranged at 90 degrees angle can be easily noticed.

The next tests were conducted on CP1 and CP3 titanium sheets samples. The samples of titanium sheets are made from:

1. CP1 (Grade 1- ASTM B265) and thickness 0.4 mm,
2. CP3 (Grade 3 - ASTM B265) and thickness 0.8 mm.

The results of SEM images of titanium sheets geometric structure are shown in Fig. 4 and Fig. 5.

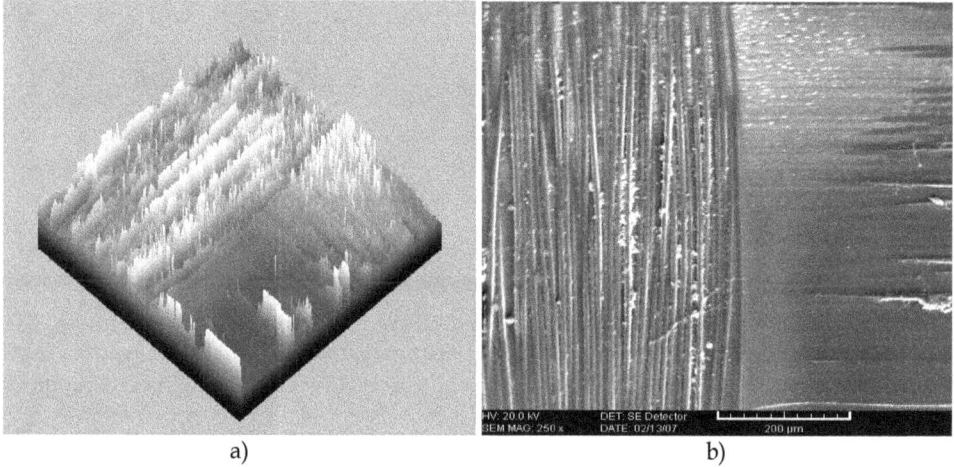

a)   b)

Fig. 3. Example of a surface topography SEM of the aramid/epoxy composite, magnification x250, a) spatial view, b) surface view (Rudawska, 2010)

a)   b)

Fig. 4. Example of a surface topography SEM of CP1 titanium sheets surface, magnification x500, a) spatial view, b) surface view (Rudawska, 2009 b)

a)   b)

Fig. 5. Example of a surface topography SEM of CP3 titanium sheets surface, magnification x500, a) spatial view, b) surface view (Rudawska, 2009 b)

SEM images of the titanium sheets surface show differences in the surface structure of analysed titanium sheets.

The following samples are concern the SEM images of aluminium sheets surface. The samples used were aluminium clad (plated) sheets type 2024-T3 (sheet thickness: 0.64 mm) The results of SEM images of aluminium sheets geometric structure are shown in Fig. 6 (own research).

a)   b)

Fig. 6. Example of a surface topography SEM of aluminium 2024-T3 sheets surface, magnification x750, a) spatial view, b) surface view

The analysis of geometric structure of the analysed sheets considered in relation to adhesion technology is extremely important since these factors influence the obtained the adhesive joints strength.

4. Wetting phenomenon and contact angle

4.1 Wettability

The wetting phenomenon is a significant issue in various technological processes (Birdi & Vu, 1993; Norton, 1992; Parsons, Buckton & Chacham, 1993; Sommers & Jacobi, 2008; Qin &Chang, 1996) . Wetting is a surface phenomenon consisting in substituting the surface of the solid and the liquid with a boundary surface, characterised by certain tension (σ), which results from the difference in the surface tension between the solid and liquid in the gaseous medium (Fig. 7), (Hay, Dragila & Liburdy 2008; Żenkiewicz, 2000).

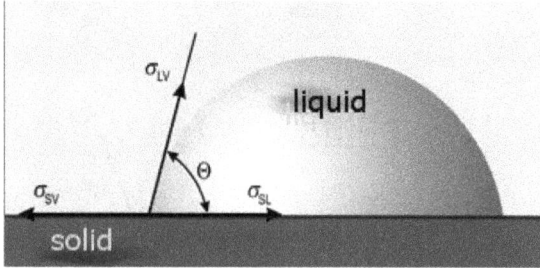

Fig. 7. Wetting a solid by a liquid

Wetting is a procedure that determines the diffusion of a liquid (adhesive) over a solid surface (substrate), creating an intimate contact between them. The air displacement caused by this physical attraction minimises the interfacial flaws. Good wettability of a surface is a prerequisite for a good adhesive bonding. Wettability is a crucial issue in the case of forming adhesive joints, because it directly affects the phenomenon of adhesion, increasing or decreasing adhesion forces.

4.2 Contact angle

The contact angle Θ provides the measure of wettability. This is the angle formed between the wetted solid surface and the tangent to the wetting liquid surface (to the meniscus of the wetting liquid), at the contact point of the liquid and the solid surface (Comyn, 1992; Hebda & Wachal 1980; Lee, 1993; McCarthy, 1998; Żenkiewicz, 2000).

When wetting the surface of a solid, the contact angle value will be lower than 90^0 (Fig. 8). The case when the contact angle $\Theta = 0^0$, indicates that the liquid spreads over the surface evenly and, furthermore, represents complete wetting of a solid surface by a liquid. If the contact angle $\Theta = 180^0$, then the result is absolute non-wetting (McCarthy, 1998).

The literature offers various tips on surface wetting, which account for the differences in size and interdependencies (as for the contact angle). In order for the liquid to wet the surface of the solid favourably, its surface tension should be lower than the surface tension of the liquid.

The contact angle can provide the measure of wettability of solids by liquids, it can determine critical surface tension, moreover, it can be used for determining surface free energy, as well as for the analysis of surface layer changes occurring when the surface is modified (Żenkiewicz, 2000).

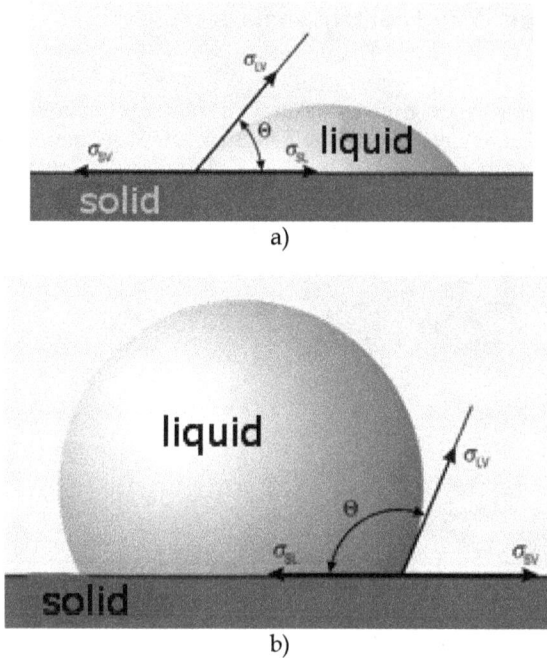

Fig. 8. Wetting of a solid surface by the liquid in the case of: a) favourable wettability $\Theta < 90^0$ and insufficient wettability $\Theta > 90^0$

There are a number of factors significantly influencing the value of contact angle and the correctness of the angle measuring process, which include: the longitudinal modulus of elasticity (surface rigidity), surface porosity, chemical and physical homogeneity of the surface (and the surface layer), surface contaminants, the type of a measuring liquid, drop volume or humidity.

The aforementioned factors contribute to disturbing the measuring of the contact angle, hinder the interpretation of results and are the cause of various metastable states of the drop itself. What is more, these phenomena result in the contact angle hysteresis (Chibowski & González-Caballero, 1993; Diaz, M. Fuentes, Cerro & Savage, 2010; Żenkiewicz, 2000).

The hysteresis is assumed to consist of two basic components: thermodynamic and dynamic. The sources of the former can be found in porosity and heterogeneity of the surface and the surface layer of the analysed material. This component of the hysteresis is independent of the surface age of the drop, provided the volume of the drop remains unchanged throughout. The other component, the dynamic hysteresis, results from the wetting liquid – test material chemical interaction, as well as from penetration of the gaps in the material by the measuring liquid. The dynamic hysteresis depends on the surface age of the drop (Żenkiewicz, 2000).

There are a number of methods for measuring the contact angle, and the most common include such techniques as: the bubble measure method, geometric method (where the contact angle is measured from the dimensions of the drop), the capillary rise method (such as Wilhelmy plate method) or the direct measurement method (Ahadian, Mohseni & Morawian, 2009; Shang, Flury, Harsh & Zollars, 2008; Mangipudi, Tirrell &Pocius, 1994; Volpe& Siboni, 1998; Żenkiewicz, 2000).

At present, this is the direct measurement of the contact angle which is a commonly applied method, and the measurement is conducted by means of specialised instruments called goniometers or contact angle analysers (Żenkiewicz, 2000).

4.3 Factors influencing the contact angle

There is a number of factors substantially affecting the contact angle and the correctness of its measurements, which include: the longitudinal modulus of elasticity (surface rigidity), surface porosity, surface (and surface layer) physical and chemical homogeneity, surface contamination, the type of measuring liquid, drop volume, humidity, etc (Ajaev, Gambaryan-Roisman & Stephan 2010; Brown, 1994; Chibowski & González-Caballero, 1993; Extrand, 1998; Thompson, Brinckerhoff & Robbins, 1993; Żenkiewicz, 2000).

One of the factors influencing the contact angle is the *drop volume*. The impact of this factor is by no means certain, since there are no prevailing conclusions, due to the fact that the contact angle measurement methods and calculating models applied in tests were different. In his work (Żenkiewicz, 2000) M. Żenkiewicz included a lot of information both on the measuring drop volumes as well as contact angle measuring methods. M. Zielecka (Zielecka, 2004) observed the influence of the size of the drop on the contact angle measurement, and arrived at a drop volume range of 2-6 mm, within which the size of the drop bears no influence on the measurement of the contact angle. X. Tang, J. Dong, X. Li (Tang, Dong & Li, 2008). conducted contact angle measurements for distilled water drops in the volume range of 3-6 µl. In their tests, K. B. Borisenko and others (Borisenko, Evangelou, Zhao & Abel, 2008). used the diiodomethane drop volume of 5 µl. Although, in the tests conducted by M. Żenkiewicz (Żenkiewicz, 2005), Q. Bénard, M. Fois and M. Grisel (Bernard, Fois, & Grisel, 2005). the measuring liquids applied were different (distilled water, glycerol, formamide, diiodomethane, α-bromonaphthalene), the volume of the drop was identical – 3 µl. In the case of many works (Hołysz, 2000; Serro, Colaço & Saramago, 2008; Żenkiewicz, 2000) the measuring drop volume ranges from 2-5 µl (2 µl, 4 µl), e.g. J. Shang and others (Shang, Flury, Harsh & Zollars, 2008). apply a 2 µl drop for static contact angle measurements and larger 5 µl in the case of dynamic contact angle measurements. According to the data collected from the literature (Żenkiewicz, 2000), the size of the drop should range between 28mm^3 and 0.5 mm^3.

The surface age of the drop, i.e. the time between the application of a drop and the measurement, is one another contact angle affecting factor. M. Żenkiewicz (Żenkiewicz, 2000) notes that the time between the application and the measurement should be as short as possible, and moreover, identical for all the drops of the test series. Following this procedure should ensure a small influence of the drop-surface interaction and reduction of the drop volume as a result of evaporation.

X. Tang, J. Dong and X. Li (Tang, Dong & Li, 2008) deal with the phenomena of wetting and contact angle and additionally present test results of the influence of the surface age of the drop on the contact angle volume for different (wet and dry) surface states.

Another factor taken into consideration is *temperature*. M. Żenkiewicz (Żenkiewicz, 2000) mentions in his paper that within the range of 80ºC, any changes in temperature only to a small degree trigger changes in the surface free energy, and natural temperature fluctuations, possible during laboratory tests, have a negligible impact on the samples contact angle measurements results. N. Zouvelou, X. Mantzouris, P. Nikolopoulos (Zouvelou, Mantzouris & Nikolopoulos, 2007) compared their tests observations with the literature data and drew a linear dependence of the surface free energy and the contact angle of certain materials on the temperature (nevertheless for high temperatures of approx. 800º C – 1173 K, 1500º C- 1773 K).

The longitudinal modulus of elasticity (surface rigidity) is yet another factor which should be considered when measuring the contact angle. M. Żenkiewicz, J. Gołębiowski and S. Lutomirski (Żenkiewicz, Gołębiewski & Lutomirski, 1999). stress that the surface of the test material where measuring drops are placed should be appropriately rigid. Therefore, the longitudinal modulus of elasticity of the material should be higher than 10 kPa, as it would prevent any drop deformations, resulting from the weight of the measuring drop.

One of the components of the thermodynamic hysteresis, *surface porosity*, is the next factor in question. R.D. Hazlett (Hazlett, 1992) describes and presents opinions of other researchers on the influence of surface porosity on the hysteresis of the contact angle, to conclude that the influence of porosity is beyond a shadow of a doubt. It can be, however, assumed that if Ra < 0,5 μm, then the impact of porosity on the contact angle is insignificant.

A.P. Serro, R. Colaço and B. Saramago (Serro, Colaço & Saramago, 2008) present test results for two samples made of UHMWPE (ultra-high-molecular-weight polyethylene) of different porosity, characterised by the Ra parameter of 3.9 and 1.0 nm, and the distance between the micropores of 23 and 6 nm respectively. They note that the wettability for given cases is irrespective of surface porosity, and that the contact angles measured for water and hexadecane are identical. However, J. Xian (Xian, 2008) points out that the wettability and the contact angle for a porous surface, e.g. analysed steel and polymers, is different for a smooth surface, adding that the change of the contact angle on a porous surface depends on the contact angle of a smooth surface of the analysed materials.

The physical and chemical homogeneity of the surface (and the surface layer) – i.e. physio-chemical homogeneity, which is the second source of the thermodynamic hysteresis, is another aspect taken into consideration when measuring the contact angle. Moreover, a considerable influence on the contact angle value may be observed on the part of the following: additive migration, diverse supermolecular structure, along with surface inhomogeneity – the result of different functional groups of different size and character formed on that surface.

What cannot be disregarded when measuring the contact angle is the analysis of *the type of the measuring liquid*. The measuring liquid penetration of the gaps in the surface layer of the material as well as of the intermolecular spaces is one of the causes of the dynamic hysteresis. The molar volume of the liquid plays an important role in the process as well – the rate of water penetration processes becomes slower and limited when the volume rises. Owing to its low molar volume water easily penetrates the structure of certain materials, therefore the importance of proper measuring liquid selection.

Other factors significantly disturbing the measurement of the contact angle are the *surface contaminants* and *air humidity* at the time of a test. Furthermore, *the sample should be firmly fixed* in order to prevent any measuring drop deformations as a result of vibrations.

Publications include plenty of information on the aspects of drop dispersion, along with the model of phenomena occurring when the contact angle measurement is taken for different liquid contact models, not to mention the characteristics of static and dynamic contact angle measurements. Some articles highlight the practical importance of wetting and wettability of different liquids in various processes, such as impregnation.

The factors mentioned in the preceding paragraphs hinder the measurement of the contact angle and the analysis of tests results, in addition they lead to different metastable states of the drop itself. These phenomena result in the contact angle hysteresis.

4.4 The contact angle hysteresis

Among many issues connected with the contact angle (the type of angle, measurements and values used in calculations) special importance is attributed to *the contact angle hysteresis,* which is the result of phenomena associated with metastable states of the measuring drop placed on the analysed surface of a solid (Bayer, Megaridis, Hang, Gamota & Biswas, 2007; Vedantam & Panchagnula, 2008; Zielecka, 2004; Żenkiewicz, 2000).

The first significant research on the contact angle hysteresis began in the middle of the 1970s and was conducted for example by R.J. Good (Good,1979). E. Chibowski and F. González-Caballero (Chibowski & González-Caballero, 1993). presented theoretical information on the contact angle hysteresis, factors causing it and the description of the observed contact angle hysteresis connected with chemical interactions. C.W. Extrand (Extrand, 1998). characterised some of the contact angle hysteresis theoretical models and presented the study of the contact angle hysteresis thermodynamic model based on the research on polymers.

The Young equation constitutes the basis for theories related to the phenomenon of wettability. This equation comprises a measurable geometric parameter – the contact angle with three thermodynamic indices, which allow explaining the properties of interactions in the interface. The Young equation (also called Young-Laplace equation) was formed in 1805 and since then its principles and description have been used in multiple publications (Diaz, Fuentes, Cerro & Savage 2010; Faibish, Yoshida & Cohen, 2002; Żenkiewicz, 2006,2000).

The Young equation describes an ideal system, which meets specific requirements of the contact angle measurement, geometric properties and qualities of the analysed surfaces (e.g. porosity, rigidity, physical and chemical homogeneity or the lack of surface contaminants). These requirements have been described in subsection 4.3.

If the surface meets the Young equation principles, the drop placed on it remains in equilibrium, which is accompanied by the lowest energy state. In such a situation, the contact angle is referred to as an equilibrium contact angle and its value does not depend on the changes of the drop volume. If the surface fails to meet the principles of Young equation, the measuring drop placed on it is in a metastable state, and then the contact angle of this drop may be higher or lower than the equilibrium angle. Initially, the gradual increase of the drop volume causes the increase of the contact angle until it reaches the limit, called *the advancing contact angle* Θ_A (Chibowski & González-Caballero, 1993; Żenkiewicz, 2000).

After this volume has been exceeded, an abrupt change of the drop position occurs - an abrupt change of the drop contour (decrease in height, increase in the contact area) and decrease in the volume of the contact angle. If the volume of the drop is gradually decreased, the value of the contact angle will initially decrease until it reaches the value called the *receding angle* Θ_R. After this value has been exceeded, the contour of the drop abruptly recedes (the height increases, the contact area decreases) and the value of the contact angle increases. A new metastable state of the drop location is, characterised by the contact angle is higher than the receding angle. Therefore, the contact angle hysteresis is defined as the difference between the advancing angle Θ_A and the receding angle Θ_R on the tree-phase contact line (Chibowski & González-Caballero, 1993; Faibish, Yoshida & Cohen, 2002; Vedantam & Panchagnula, 2008; Volpe & Siboni 1998; Żenkiewicz, 2000).

The contact angles Θ_A and Θ_R, along with their corresponding drop volumes: maximum (for Θ_A) and minimum (for Θ_R) with a constant diameter (D) of the circle created by the drop lying on the tested material, are shown in the Fig. 9 (Żenkiewicz, 2000).

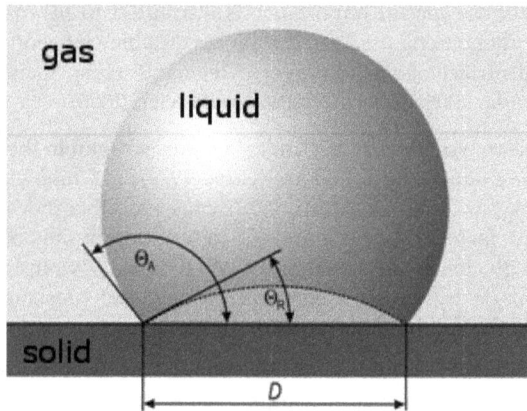

Fig. 9. The contact angles with a constant diameter D (D=const): Θ_A – the advancing angle, Θ_R – the receding angle, 1 – the maximum volume drop, 2 – the minimum volume drop

The hysteresis is assumed to consist of two basic components. First is the so called *thermodynamic hysteresis*, which results from porosity and heterogeneity of the surface and the surface layer of the tested material. This component is independent of the surface age of the drop (provided that the drop volume remains unchanged while measured). The other component is *the dynamic hysteresis*. It results from, among other things, chemical interactions of the measuring liquid with the tested material, and the measuring liquid penetration of the gaps in the material. The dynamic hysteresis depends on the surface age of the drop (Żenkiewicz, 2000).

The research on the hysteresis is extremely important from the practical point of view of, for instance, the surface free energy calculations. It is mostly connected with the question of which contact angle should be adopted in the simplified equation (3) in order to obtain the correct result. The contact angle used in calculations is the angle θ_A called the advancing angle.

5. Surface free energy

Surface free energy (SFE) is one of the thermodynamic quantities describing the state of atom equilibrium in the surface layer of materials (Hołysz, 2000; Żenkiewicz, 2000,2005). This quantity is characteristic for each substance. It reflects the specific state of unbalance in intermolecular interaction which is present at the phase boundary of two mediums.

Surface free energy is of equal number to the work necessary for creating a new surface unit while separating two phases in equilibrium, in a reversible isothermal process. It is measured in $[mJ/m^2]$ (Żenkiewicz, 2000).

5.1 Young equation

The basis for methods of calculating surface free energy from the measurements of the contact angle is the Young equation (Fig. 7) (Chibowski & González-Caballero, 1993; Lee, 1993; Thompson, Brinckerhoff & Robbins, 1993; Żenkiewicz, 2000).

It was derived from the condition of equilibrium of forces which represent surface tensions at the contact point of three phases – solid, liquid and gas.

$$\sigma_{SV} = \sigma_{SL} + \sigma_{LV}\cos\Theta_V \qquad (1)$$

where Θ_V is the equilibrium contact angle, and σ_{LV}, σ_{SV}, and σ_{SL} are the surface free energies of liquid–vapour, solid–vapour and solid–liquid interfaces, respectively.

The Young equation may also be derived from the energy balance for the triple point (Chibowski & González-Caballero, 1993; Michalski, Hardy & Saramago, 1998; Zouvelou, Mantzouris & Nikolopoulos, 2007). In this case, the equation is of the following form (Żenkiewicz, 2000):

$$\gamma_{SV} = \gamma_{SL} + \gamma_{LV}\cos\Theta_V \qquad (2)$$

where: γ denotes surface free energy and the other symbols have the same meaning as in the equation (1).

It is impossible to determine surface free energy directly from the equation (2) because of the two unknowns: γ_{SV} and γ_{SL}. For calculation purposes, the following form of the equation (2) is commonly used to determine the surface free energy of solids (Chibowski & González-Caballero, 1993; Żenkiewicz, Gołębiewski & Lutomirski, 1999):

$$\gamma_S = \gamma_{SL} + \gamma_L\cos\Theta \qquad (3)$$

where: γ_S –surface free energy of solids in a vacuum,
γ_{SL} – surface tension on the solid – liquid phase boundary,
γ_L –surface free energy of the measured liquid,
Θ_V – contact angle measured on the examined true surface.

The main drawback of the equation (1) is that it refers to an ideal system because it has been based on theoretical considerations, to a large extent not confirmed empirically. Still, this is the contact angle measurement which is the most often used method to determine energy properties of solids.

5.2 Surface free energy determination methods

The various SFE determination methods are based on specific relations, and involve the measurement of contact angles of various liquids. A number of factors have a substantial influence on the correctness of the contact angle measurement (subchapter 3.3). Some issues related to contact angle measurements and wettability have been highlighted shown in the literature.

Determination of surface free energy of solid objects involves indirect methods – direct methods can only be used in the case of liquids. Among the various indirect methods are the approaches due to Fowkes, Owens-Wendt, van Oss-Chaudhury-Good, Zisman, Wu, and Neumann (Ahadian, Mohseni & Morawian, 2009; González-Martín, Labajos-Broncano, Jańczuk & Bruque, 1999; Greiveldinger & Shanahan,1999; Hołysz, 2000; Jańczuk, Białopiotrowicz & Zdziennicka, 1999; Lee, 1993; Lugscheider & Bobzin, 2001; Żenkiewicz, 2000, 2006).

5.2.1 The Owens-Wendt (Kaelble-Owens-Wendt) method (OW)

The Owens-Wendt method (sometimes referred to as Kaelble-Owens-Wendt method) is a frequently applied method for determining the surface free energy of, e.g. polymers (Jańczuk & Białopiotrowicz, 1987; Rudawska & Kuczmaszewski, 2006; Rudawska, 2008). This method consists in determining dispersive and polar components of SFE based on Berthelot principle (Żenkiewicz, 2000) , which assumes that interaction between molecules of two bodies in their surface layers equals the geometric mean of the cohesion work between the molecules of each body.

This method assumes that the surface free energy (γ_S) is a sum of two components: polar (γ_S^p) and dispersive (γ_S^d), and that there is a relation between the three quantities:

$$\gamma_S = \gamma_S^d + \gamma_S^p \tag{4}$$

The dispersive element is the sum of components derived from such intermolecular interactions as: polar, hydrogen, induction and acid-base, with the exception of dispersive interactions. Dispersive interactions constitute the dispersive component of the surface free energy.

The work of adhesion between the solid and the liquid can be described by means of the Dupré equation:

$$W_a = \gamma_{SV} + \gamma_{LV} - \gamma_{SL} \tag{5}$$

By combining the equations 2 with 3, the Young–Dupré equation is obtained:

$$W_a = \gamma_{LV}(1 + \cos\Theta) \tag{6}$$

However, Owens and Wendt propose the following form of the work of adhesion between interacting solid and liquid.

$$W_a = 2(\gamma_S^d \, \gamma_{LV}^d)^{0,5} + 2(\gamma_S^p \, \gamma_{LV}^p)^{0,5} \tag{7}$$

If we compare and combine equations (6) and (7), the following equation is obtained:

$$\gamma_{LV}(1+\cos\Theta) \; = \; 2(\gamma_S{}^d\,\gamma_{LV}{}^d)^{0,5} + 2(\gamma_S{}^p\,\gamma_{LV}{}^p)^{0,5} \tag{8}$$

This equation allows determining the surface free energy of a solid and its SFE components.

In order to determine the polar and the dispersive components of the surface free energy, the measurements of the contact angle of the analysed samples need to be conducted with two measuring liquids. The surface free energy of the measuring liquids used in test is known, including its polar and dispersive components. One of the liquids is non-polar and the other is bipolar. Most frequently, the tests include distilled water as the polar liquid and diiodomethane as the non-polar one.

The SFE γ_S is calculated using the adjusted dependence describing the dispersive component of the surface free energy (Jańczuk& Białopiotrowicz, 1987; Rudawska & Kuczmaszewski, 2005; Rudawska, 2008; Rudawska & Jacniacka 2009).

$$\left(\gamma_s^d\right)^{1/2} = \frac{\gamma_d\left(\cos\Theta_d+1\right)-\sqrt{\dfrac{\gamma_d^p}{\gamma_w^p}}\gamma_w\left(\cos\Theta_w+1\right)}{2\left(\sqrt{\gamma_d^d}-\sqrt{\gamma_d^p\,\dfrac{\gamma_w^d}{\gamma_w^p}}\right)} \tag{9}$$

and the polar component of the surface free energy

$$\left(\gamma_S^p\right)^{0,5} = \frac{\gamma_w\left(\cos\Theta_w+1\right)-2\sqrt{\gamma_S^d\gamma_w^d}}{2\sqrt{\gamma_w^p}} \tag{10}$$

where: $\gamma_S{}^d$ – the dispersive component of the test material surface free energy, $\gamma_S{}^p$ – the polar component of the test material surface free energy, γ_d – the surface free energy of diiodomethane, $\gamma_d{}^d$ – the dispersive component of the surface free energy of diiodomethane, $\gamma_d{}^p$ – the polar component of the surface free energy of diiodomethane, γ_w – the surface free energy of water, $\gamma_w{}^d$ – the dispersive component of the surface free energy of water, $\gamma_w{}^p$ – the polar component of the surface free energy of diiodomethane, Θ_d – the contact angle of diiodomethane, Θ_w – the contact angle of water.

There is one of example of materials for which it was determined the surface free energy after various surface treatment (Rudawska, 2008, 2009).

The surface free energy of the material presented below was calculated with the Owens-Wendt method. This is a structural material applied in e.g. aircraft industry. The tests were to determine the influence of a surface preparation method on the SFE of the sample material.

The tests were conducted on glass-epoxy composite samples consisting of two layers (2x0.30 mm) of glass fibre 3200-7781. The fabric layers were arranged at a right angle and cured conforming to the technology standards.

The composite samples were tested for four surface preparation variants:

1. variant I – no surface preparation;
2. variant II – degreasing with Loctite 7036 (a detailed description of this method can be found in e.g. (Rudawska & Kuczmaszewski, 2005));

3. variant III – mechanical surface preparation with P320 abrasive tool;
4. variant IV – mechanical surface preparation with P320 abrasive tool, followed by degreasing with Loctite 7036.

The surface free energy values as well as the components of the SFE for four tested glass-epoxy composite surface preparation variants are presented in Fig. 10-13 (Rudawska, 2008, 2009).

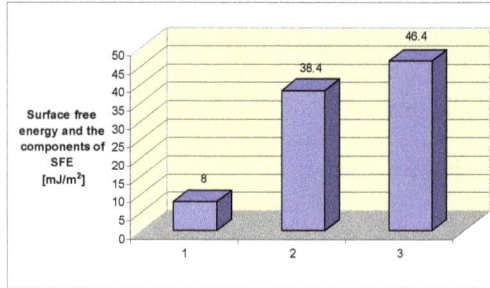

Fig. 10. Surface free energy and the components of SFE - the surface of glass/epoxy composite without surface treatment (variant I): 1 – polar component of SFE, 2 – dispersive component of SFE, 3 – surface free energy (SFE)

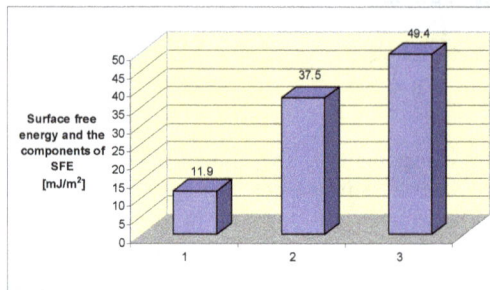

Fig. 11. Surface free energy and the components of SFE - the surface of glass/epoxy composite after degreasing (variant II): 1 – polar component of SFE, 2 – dispersive component of SFE, 3 – surface free energy (SFE)

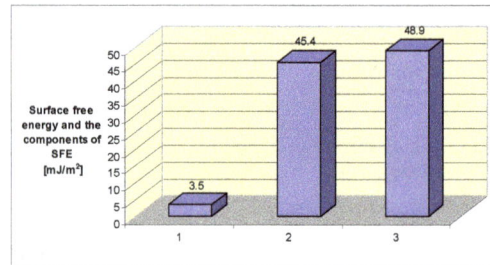

Fig. 12. Surface free energy and the components of SFE - the surface of glass/epoxy composite after the P320 grinding tool processing (variant III): 1 – polar component of SFE, 2 – dispersive component of SFE, 3 – surface free energy (SFE)

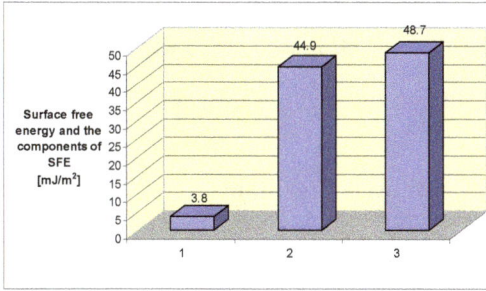

Fig. 13. Surface free energy and the components of SFE - the surface of glass/epoxy composite after the P320 grinding tool processing and degreasing (variant IV): 1 – polar component of SFE, 2 – dispersive component of SFE, 3 – surface free energy (SFE)

The results demonstrate that the highest values of the surface free energy were obtained in the case of degreasing, while the lowest were observed for variant I, with no surface preparation. Consequently, it appears that mechanical surface preparation and mechanical surface preparation followed by degreasing both increase the surface free energy. Additionally, no statistically relevant difference in the γ_S values of the two variants in question was observed.

Taking into consideration the polar component of the SFE, its highest value was noted in the case of surface preparation variant II, in which the surface free energy value was the highest as well. Additionally, the polar component constituted 24% of the total SFE. In the case of variant I, with the lowest γ_S value in the tests, the polar component constitutes 17% of the SFE. For the other two variants, III and IV, the polar component represented scant 7% and 8% respectively.

The analysis of the SFE values leads to the observation that degreasing the surface of the glass-epoxy composite has beneficial effect on the surface free energy value. It results in the increase of the SFE as compared to the surfaces with no prior surface preparation.

To conclude, it must be mentioned that, firstly, forming an adhesive joint should be preceded by certain surface preparation methods, and secondly, that this is degreasing which produces the best results in terms of adhesive properties of the analysed glass-epoxy composite.

5.2.2 The van Oss-Chaudhury-Good method (OCG)

In the case of the van Oss-Chaudhury-Good method the surface free energy is a sum of two components (Adão, Saramago & Fernandes, 1999; Żenkiewicz, 2000). While the first component γ_i^{LW} is connected with long-range interactions (dispersive, polar and inductive, referred to as Lifshitz-van der Waals electrodynamic interactions), the second component γ_i^{AB} describes the acid-base interactions (Hołysz 2000; Jansen, 1991):

$$\gamma_i = \gamma_i^{LW} + \gamma_i^{AB} \tag{11}$$

Good R.J. and van Oss C.J. (Good & van Oss, 1992) separate the acid component (electron-acceptor: γ_L^+, γ_S^+) and the base component (electron-donor: γ_L^-, γ_S^-) of the surface free energy.

Moreover, the γ_i^{AB} component can be described by means of equation for bipolar compounds (showing properties of both Lewis acids and bases), (Elftonson, Ström, Holmberg & Olsson, 1996):

$$\gamma_i^{AB} = 2(\gamma_i^+ \gamma_i^-)^{0,5} \qquad (12)$$

where: γ^+ – Lewis acid surface free energy component, γ^- – Lewis base surface free energy component, index i – subsequent measuring solids or liquids.

Determining the SFE of test materials will consist in measuring their surfaces contact angle with three different measuring liquids and calculating the γ_S of the system of three equations:

$$(\gamma_S^{LW} \gamma_{Li}^{LW})^{0,5} + (\gamma_S^+ \gamma_{Li}^-)^{0,5} + (\gamma_S^- \gamma_S^+)^{0,5} = \gamma_{Li}(1+\cos\Theta_i)/2 \qquad (13)$$

where: i=1,2,3.

Measuring the contact angle requires the application of two polar and one non-polar liquids; nevertheless, solving the equation (3) requires additional information – particular values for the applied measuring liquids. Polar liquids applied in tests are water, glycerol, formamide or ethylene glycol, and non-polar liquids (not showing properties of either Lewis acids or bases) diiodomethane or α-bromonaphthalene.

A detailed description of this method is provided in the publications (Shen, Sheng, & Parker, 1999, Żenkiewicz, Gołębiewski & Lutomirski, 1999; Żenkiewicz 2000).

Determining the SFE with the van Oss-Chaudhury-Good method is uncomplicated, nevertheless, the test results should be carefully analysed. This method is burdened with a few problems, including e.g. the fact that the test results depend heavily on the applied measuring liquids configuration. This issue has been described by e.g. C. Della Volpe and S. Siboni (Volpe & Siboni, 1998). who in addition present the Drago theory, concerning, among other issues, the properties of Lewis acids and bases.

5.2.3 The comparison OW and OCG methods

Due to the fact that the methods of calculating the surface free energy presented in the previous chapters are most frequently applied, a comparison of selected structural materials SFE values calculated with the Owens-Wendt and the van Oss-Chaudhury-Good methods should be conducted (Kuczmaszewski & Rudawska, 2002).

The structural material under analysis was electrolytic zinc coated and hot dip zinc coated sheets, which find application in such industries as automotive, construction or machine-building. The zinc coated sheets were 0.7 mm thick, the hot dip zinc coating equalled 18 μm and electrolytic zinc coating equalled 7.5 μm (following the PN-89/H-92125 and PN–EN 10152 standards).

The sample material surface was degreased with degreasing agents: Loctite 7061 and acetone. Degreasing was conducted in ambient temperature between 18 and 20 °C with relative humidity oscillating between 38% and 40%.

The method applied for measuring the contact angle was the direct measurement of the angle between the measuring drop and the tested surface.

For calculating the surface free energy with the Owens-Wendt method relationships (9) and (10) were applied. The values of both the surface free energy and its components for the applied measuring liquids are presented in Table 1 (Jańczuk& Białopiotrowicz, 1987).

No.	Measuring liquid	Surface free energy and its components [mJ/m²]		
		γ_L	$\gamma_L{}^P$	$\gamma_L{}^d$
1	Distilled water	72.8	21.8	51.0
2	Diiodomethane	50.8	2.3	48.5

Table 1. The values of the surface free energy and its components for the applied measuring liquids

The SFE components values used in the van Oss-Chaudhury-Good method are presented in Table 2 (Żenkiewicz, Gołębiewski & Lutomirski, 1999).

No.	Measuring liquid	Surface free energy and its components [mJ/m²]				
		γ_L	$\gamma_L{}^{LW}$	$\gamma_L{}^{AB}$	$\gamma_L{}^+$	$\gamma_L{}^-$
1	Distilled water	72.8	21.8	51.0	34.2	19.0
2	Glycerol	64.0	34.0	30.0	5.3	42.5
3	Diiodomethane	50.8	50.8	0	0	0

Table 2. The values of the surface free energy and its components for the applied measuring liquids

The values of the surface free energy and its components were calculated with the van Oss-Chaudhury-Good method using the data presented in Table 2 as well as relationships (3) and (4).

The surface free energy values of the electrolytic zinc coated and hot dip zinc coated sheets calculated with the Owens-Wendt method are presented in Table 3 (Kuczmaszewski & Rudawska, 2002).

No.	The type of zinc coated sheets	The type of the degreasing agent	Surface free energy and its components [mJ/m²]		
			γ_S	$\gamma_S{}^P$	$\gamma_S{}^d$
1	Electrolytic zinc coated sheets	Loctite 7061	42.0	17.8	24.2
		Acetone	35.4	10.7	24.7
2	Hot dip zinc coated sheets	Loctite 7061	44.7	9.7	35.0
		Acetone	43.8	11.6	32.2

Table 3. The zinc coated sheets surface free energy calculated with the Owens-Wendt method

The results demonstrate that the dispersive component of the surface free energy for hot dip zinc coated sheets is higher (even three times) than its polar component. In the case of electrolytic zinc coated sheets degreased with Loctite7061, this difference is less significant.

The values of the surface free energy of the electrolytic zinc coated and hot dip zinc coated sheets calculated with the van Oss-Chaudhury-Good method are presented in Table 4 (Kuczmaszewski & Rudawska, 2002). The results were obtained from the tested sheet surface layer contact angle measurement taken with distilled water, glycerol and diiodomethane as measuring liquids.

No.	The type of zinc coated sheets	The type of the degreasing agent	Surface free energy and its components [mJ/m²]				
			γ_s	γ_s^{LW}	γ_s^{AB}	γ_s^+	γ_s^-
1	Electrolytic zinc coated sheets	Loctite 7061	43.7	32.5	11.2	13.8	2.3
		Acetone	38.6	30.5	8.1	4.3	3.9
2	Hot dip zinc coated sheets	Loctite 7061	45.0	41.5	3.5	0.3	11.6
		Acetone	41.4	39.2	2.2	0.1	15.8

Table 4. The zinc coated sheets surface free energy calculated with the van Oss-Chaudhury-Good method

It can be observed that the component of the surface free energy connected with long range interactions γ_s^{LW} (polar, dispersive and inductive) is higher than the component describing acid-base interactions γ_s^{AB}. The γ_s^{AB} component is scant in hot dip zinc coated sheets. Drawn from the analysis of the acid-base interactions component γ_s^{AB}, certain regularity may be observed. Lewis acid (γ_s^+) surface free energy component is higher than Lewis base (γ_s^-) surface free energy component for electrolytic zinc coated sheets, whereas for hot dip zinc coated sheets the γ_s^+ value was negligible when compared with the γ_s^- component. Owing to the insignificant γ_s^{AB} value it may be presumed that these surfaces will show properties of monopolar or non-polar substances.

A comparison of the surface free energy calculated with both the Owens-Wendt method and the van Oss-Chaudhury-Good method for hot dip zinc coated and electrolytic zinc coated sheets degreased with Loctite 7061 is presented in Fig.14 (Kuczmaszewski & Rudawska, 2002).

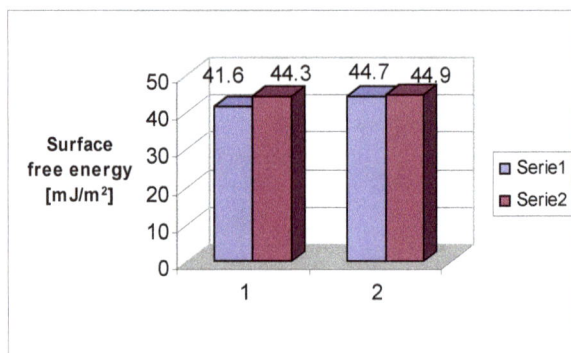

Fig. 14. The surface free energy values calculated with the Owens-Wendt method (series 1) and the van Oss-Chaudhury-Good method (series 2) for: 1- electrolytic zinc coated sheets, 2- hot dip zinc coated sheets after degreasing with Loctite 7061

A comparison of the surface free energy calculated with the Owens-Wendt method and the van Oss-Good method for hot dip zinc coated and electrolytic zinc coated sheets degreased with acetone is presented in Fig. 15 (Kuczmaszewski & Rudawska, 2002).

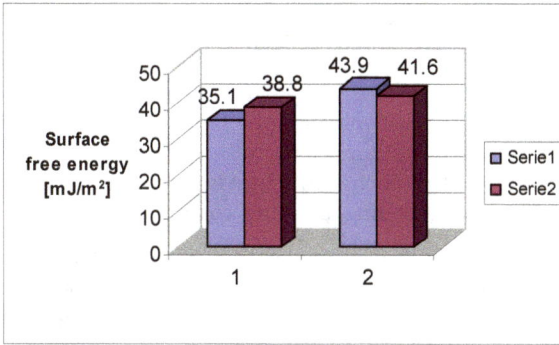

Fig. 15. The surface free energy values calculated with the Owens-Wendt method (series 1) and the van Oss-Chaudhury-Good method (series 2) for: 1- electrolytic zinc coated sheets, 2- hot dip zinc coated sheets after degreasing with acetone

The research results were subsequently analysed statistically using statistical models used for statistical verification (Krysicki et al., 1999). The statistical model – Student's t-test consisted in comparing means when the test variances were equal (Fisher - Snedecor distribution), with a predetermined level of significance $\alpha = 0.05$. The statistical analysis provided basis for formulating the following conclusions.

The analysis proved that there are no statistically significant differences in the values of the SFE calculated with either the Owens-Wendt or the van Oss-Good method when the sheets are degreased with Loctite 7061. This holds true for both electrolytic zinc coated sheets and hot dip zinc coated sheets.

When degreasing with acetone operation was applied on the sheet surface, statistically significant differences in the SFE calculated for the hot dip zinc coated sheets were observed. The γ_s value calculated with the van Oss-Chaudhury-Good method was higher. However, this difference is not too significant (lower than 10%). Still, there were no statistically significant differences in the SFE calculated with the van Oss-Good method for electrolytic zinc coated sheets.

When analysing the SFE values calculated with the Owens-Wendt method, it may be assumed that the surface layer of the electrolytic zinc coating would most likely demonstrate higher affinity with the polar substance than the hot dip zinc coating would.

6. Conclusion

Adhesion and concurrent phenomena, e.g. wettability, are present in numerous fields of engineering and life in general. Determining the factors influencing the quality of adhesion and finding technology that can increase or decrease it is of utmost importance when it comes to constituting adhesive joints. What cannot be disregarded is the structure of the surface layer of analysed materials or methods of determining adhesive properties, which assess materials suitability for adhesive processes. The existence of many methods for measuring the surface free energy stems from the fact that certain methods are suitable in particular circumstances. Existing methods describe the thermodynamic state of the surface layer differently yet all, through subsequent analyses of the surface free energy and its components, expand our knowledge of the phenomenon of adhesion.

SEM technique is very useful to analysis geometric structure of material for which is described adhesive properties. SEM micrographs have a large depth of field yielding a characteristic three-dimensional appearance useful for understanding the surface structure of a sample. The information of geometric structure is extremely important for the progress of adhesive processes like gluing, sealing, painting, coating.

In the subchapter devoted to a comparative analysis of the surface free energy measuring methods, the selection of the OW and the OCG methods was dictated by the fact that, on the one hand, these are the most frequently applied methods for measuring the surface free energy, on the other hand, due to relatively uncomplicated measurement of the contact angle with standard measuring liquids. The statistical analysis of the results evidences that, in most of the analysed cases, there are no statistically relevant differences between the values of surface free energy measured with either the Owens-Wendt or van Oss-Chaudhury-Good method.

Based on the statistical analysis it may be concluded that the choice of the surface free energy measurement method in the case of the analysed zinc coated sheets is basically of no relevance. Nevertheless, in ordinary working conditions it is the Owens-Wendt method which should be selected as a more efficient and less complicated tool for measuring the surface free energy of materials. The van Oss-Chaudhury-Good method, however, could be applied when a more detailed evaluation of the thermodynamic state of a surface (or a surface layer) is required.

Recent developments in the field of materials engineering contribute to creating structural materials or coatings, which are increasingly modern and specific – designated for particular applications. This creates the demand for continuous research into determining and describing their adhesive properties when adhesively bonding or joining such materials.

7. Acknowledgment

The some scientific study was funded from education finance for 2006-2009 as research project no. 3T10C02730, Poland

The some scientific study was funded from education finance for 2010-2013 as research project no. N N507 592538 The Ministry of Science and Higher Education, Poland

8. References

Adão, M.H.V.C.; Saramago, B.J. & Fernandes, A.C. (1999). Estimation of the Surface Properties of Styrene-Acrylonitrile Random Copolymers From Contact Angle Measurements. *Journal of Colloid Interface Science*, Vol. 217, No.1, pp. 319-328, ISSN 0021-9797

Ahadian, S.; Mohseni, M. & Morawian, S. (2009). Ranking proposed models for attaining surface free energy of powders using contact angle measurements. *International Journal of Adhesion and Adhesives* Vol.29, No. 4, pp. 458-469, ISSN 0143-7496

Ajaev, V.S.; Gambaryan-Roisman, T. & Stephan. P. (2010). Static and dynamic contact angles of evaporating liquids on heated surfaces. *Journal of Colloid Interface Science*, Vol.342, No.2, pp.550-558, ISSN 0021-9797

Bayer, I.S.; Megaridis, C.M.; Hang, J.; Gamota, D. & Biswas, A. (2007). Analysis and surface energy estimation of various model polymeric surfaces using contact angle hysteresis. *Journal of Adhesion Science and Technology*, Vol.21, No.15, pp. 1439-1467 (29), ISSN 0169-4243

Birdi, K.S. & Vu, D.T. (1993). Wettability and Evaporating Rates of Fluids from Solid Surfaces. *Journal of Adhesion Science and Technology*, Vol.7, No.6, pp. 485-493 (9), ISSN 0169-4243

Bernard, Q.; Fois, M. & Grisel M. (2005). Influence of Fibre Reinforcement and Peel Ply Surface Treatment Towards Adhesion of Composite Surface. *International Journal of Adhesion and Adhesives*, Vol. 25, No.5, pp.404-409, ISSN 0143-7496

Borisenko, K.B.; Evangelou, E.A.; Zhao, Q. & Abel, E.W. (2008). Contact Angle of Diiodomethane on Silicon-Doped Diamond-Like Carbon Coatings in Electrolyte Solutions. *Journal of Colloid Interface Science*, Vol.326, No.2, pp. 329-332. ISSN 0021-9797

Brown, S.D. (1994). Adherence Failure and Measurement: Some Troubling Question. *Journal of Adhesion Science and Technology*, Vol.8, No.6., pp. 687-711(25), ISSN 0169-4243

Burakowski, T. & Wierzchnoń, T. (1995). *Inżynieria powierzchni metali*. WNT, ISBN 83-204-1812-7, Warsaw, Poland

Comyn, J. (1992). Contact Angels and Adhesive Bonding. *International Journal of Adhesion and Adhesives*, Vol.12, No.3, pp.145–149, ISSN 0143-7496

Chibowski, E. & González-Caballero, F. (1993). Interpretation of Contact Angle Hysteresis. *Journal of Adhesion Science and Technology*, Vol.7, No. 11, pp. 1195-1209(15), ISSN 0169-4243

Diaz, M.E.; Fuentes, J.; Cerro, R.L. & Savage, M.D (2010). Hysteresis During Contact Angles Measurement. *Journal of Colloid Interface Science*, 343(2), pp.574-583, ISSN 0021-9797

Dutkiewicz, E.T. (1998). *Fizykochemia powierzchni*. WNT, ISBN 83-204-22-66-3, Warsaw, Poland

Elftonson, J.E.; Ström G.; Holmberg K. & Olsson J. (1996). Ashesion of Streptococcus Sanguis to porous and non-porous substrates with well-defined surface energies. *Journal of Adhesion Science and Technology*, Vol. 10, No.8, pp. 761-770, ISSN 0169-4243

Extrand, C.W. (1998). A Thermodynamic Model for Contact Angle Hysteresis. *Journal of Colloid Interface Science*, Vol.207, No.1, pp.11-19, ISSN 0021-9797

Faibish, R.S.; Yoshida, W. & Cohen, Y. (2002). Contact Angle Study on Polymer-Grafted Silicon Wafers. *Journal of Colloid Interface Science*, 256, pp.341-350, ISSN 0021-9797

Good, R.J. (1979). in: Surface and Colloid Science, R.J. Good, R.R. Stromberg (Eds.), vol.11. Plenyum Press, New York.

Good, R. J. & van Oss, C.J. (1992). in: Modern Approaches to Wettability, M.E. Schrader and G. Loeb (Eds), pp. 1-27, Plenum Press, New York.

González-Martín, M.L.; Labajos-Broncano, L.; Jańczuk, B. & Bruque, J.M. (1999). *Wettability and Surface Free Energy of Zirconia Ceramics and Their Constituents. Journal of Materials Science,* Vol.34, pp.5923-5926, ISSN 0022-2461

Greiveldinger, M. & Shanahan E.R. (1999). A Critique of the Mathematical Coherence of Acid/Base Interfacial Free Energy Theory. *Journal of Colloid Interface Science,* Vol.215, pp. 170-178, ISSN

Harding, H. & Berg, J.C. (1997). The role of adhesion in the mechanical properties of filled polymer composites. *Journal of Adhesion Science and Techno*logy, Vol.12, No.4, pp.471-493, ISSN 0169-4243

Hazlett, R.D. (1992). On Surface Roughness Effects in Wetting Phenomena. *Journal of Adhesion Science and Technology,* Vol.6, No.6, pp.625-633, ISSN 0169-4243

Hay, K.M.; Dragila, M.I. & Liburdy, J. (2008). Theoretical Model for the Wetting of a Rough Surface. *Journal of Colloid Interface Science,* Vol.325, No.2, pp.472-477, ISSN 0021-9797

Hebda, M. & Wachal, A. (1980). *Trybologia.* WNT, ISBN 83-204-0043-0, Warsaw, Poland

Hołysz, L. (2000). Investigation of the effect of substrata on the surface free energy components of silica gel determined by layer wicking method. *Journal of Material Science,* 35, pp.6081-6091, ISSN 0022-2461

Jansen, W.B. (1991). Overview Lecture. The Lewis Acid-Base Concepts: Recent Results and Prospects for the Future. *Journal of Adhesion Science and Technology,* Vol.5, No.1, pp.1-21, ISSN 0169-4243

Jańczuk, B.& Białopiotrowicz, T. (1987). Swobodna Energia Powierzchniowa Niektórych Polimerów. *Polimery,* pp.269-271, ISSN 0032-2725

Jańczuk, B.; Białopiotrowicz, T. & Zdziennicka A. (1999) Some Remarks on the Components of the Liquid Surface Free Energy. *Journal of Colloid Interface Science,* Vol.211, pp.96-103, ISSN 0021-9797

Krysicki, W.; Bartos. J.; Dyczka W.; Królikowska K. & Wasilewski, M. (1999). *Rachunek prawdopodobieństwa i statystyka matematyczna w zadaniach.* WNT, ISBN 83-204-2442-9, Warsaw, Poland

Kuczmaszewski, J. & Rudawska A. (2002). Porównanie wybranych metod określania swobodnej energii powierzchniowej na przykładzie blach ocynkowanych. *Farby i Lakiery,* No.6, pp.22–26, ISSN 1230-3321

Kuczmaszewski, J. (2006). Fundamentals of metal-metal adhesive joint design. Lublin University of Technology, Polish Academy of Sciences, Lublin Branch, ISBN 83-89293-11-0, Lublin, Poland

Lee, L.H.. (1993). Roles of molecular interactions in adhesion, adsorption, contact angle and wettability. *Journal of Adhesion Science and Technology,* Vol.7, No.6, pp.538–634, ISSN

Lugscheider, E. & Bobzin, K. (2001). The Influence on Surface Free Energy of PVD-Coatings. *Surface and Coatings Technology,* Vol.142-144, pp.755-760, ISSN 0257-8972

Mangipudi, V.; Tirrell, M. & Pocius, A.V. (1994). Direct Measurement of Molecular Level Adhesion Between Poly(ethylene terephthalate) and Polyethylene Films: Determination of Surface and Interfacial Energies. *Journal of Adhesion Science and Technology,* Vol.8, No.11, pp.1251–1270, ISSN 0169-4243

McCarthy, S.A. (1998). Dynamic Contact Angle Analysis and Its Application to Paste PVC Product. *Polimery* Vol.43, pp.314 – 319, ISSN 0032-2725

Michalski, M.-C.; Hardy, J. & Saramago, J.V. (1998). On the surface free energy of PVC/EVA polymer blends: comparison of different calculation methods. *Journal of Colloid Interface Science*, Vol.208, pp.319-328, ISSN 0021-9797

Mittal, K.L. (1978). (Ed.) in Adhesion Measurement of Thin Films, Thick Films, and Bulk Coatings, ASTM STP640, pp.5-17, American Society for Testing and Materials., Philadelphia, PA.

Mittal, K.L. (1980). Interfacial Chemistry and Adhesion: Recent Developments and Prospects. *Pure and Applied Chemistry*, Vol.52, pp.1295-1305, ISSN 0033-4545

Norton, M.G. (1992). The influence of contact angle, wettability, and reactivity on the development of indirect-bonded metallizations for aluminum nitride. *Journal of Adhesion Science and Technology* ,Vol.6, No.6, pp.635-651, ISSN 0169-4243

Parsons, G.E.; Buckton, G. & Chacham, S.M. (1993). Comparison of measured wetting behaviour of material with identical surface energies, presented as particles and plater. *Journal of Adhesion Science and Technology*, Vol.7, No.2, pp.95-104, ISSN 0169-4243

Qin, X. & Chang, W.V. (1996). The Role of Interfacial Free Energy in Wettability, Solubility, and Solvent Crazing of Some Polymeric Solids. *Journal of Adhesion Science and Technology*, Vol.10, No.10, pp. 963–987, ISSN 0169-4243

Rożniatowski, K.; Kurzydłowski, K.J. & Wierzchnoń T. (1994). Geometryczny opis cech mikrostrukturalnych warstwy powierzchniowej. *Inżynieria Materiałowa*, No.5, pp.141-149, ISSN 0208-6247

Rudawska, A. & Jacniacka E. (2009). Analysis of Determining Surface Free Energy Uncertainty with the Owens-Wendt method. *International Journal of Adhesion and Adhesives*, Vol.29, pp. 451-457, ISSN 0143-7496

Rudawska, A. & Kuczmaszewski, J. (2005). *Klejenie blach ocynkowanych*. Wyd. Uczelniane PL, ISBN 83-89246-43-0, Lublin, Poland

Rudawska, A. & Kuczmaszewski, J. (2006). Surface Free Energy of Zinc Coating After Finishing Treatment. *Material Science-Poland*, Vol.24, No.4, pp. 975 – 981, ISSN 2083-134X

Rudawska, A. (2008). Swobodna Energia Powierzchniowa i Struktura Geometryczna Powierzchni Wybranych Kompozytów Epoksydowych. *Polimery*, 53, No.6, pp.452-456, ISSN 0032-2725

Rudawska, A. (2009). Właściwości adhezyjne kompozytów szklano-epoksydowych po różnych sposobach przygotowania powierzchni, In: *Polimery i kompozyty konstrukcyjne*, G. Wróbel, (Ed.), pp. 97-105, ISBN 978-83-60917-40-4, Cieszyn, Poland

Rudawska, A. (2009). Wytrzymałość połączeń klejowych blach tytanowych po różnych sposobach przygotowania powierzchni. *Inżyniera Materiałowa*, No. 5, pp. 341 – 345, ISSN 0208-6247

Rudawska, A. (2010). *Geometric structure and surface layer adhesive properties of aramid-epoxy composite*. Conferencing mat. Polymer Processing Society PPS-26. Regional Meeting Istanbul, Turkey, October 20-24. 2010, pp.159.

Sekulic, A. & Curnier, A. (2010). Experimetation on Adhesion of Epoxy. *International Journal of Adhesion and Adhesives*, Vol.30 pp. 89-104, ISSN 0143-7496

Serro, A.P.; Colaço, R. & Saramago, B. (2008). Adhesion Forces in Liquid Media: Effect of Surface Topography and Wettability. *Journal of Colloid Interface Science*, Vol.325, pp.573-579, ISSN 0021-9797

Shang, J.; Flury, M.; Harsh, J.B. & Zollars, R.L. (2008). Comparison of Different Methods to Measure Contact Angles of Soil Colloids. *Journal of Colloid Interface Science*, Vol.328, pp.299-307, ISSN 0021-9797

Shen, W.; Sheng, Y.J. & Parker, I.H. (1999). Comparison of the Surface Energetics Data of Eucalypt Fibers and Some Polymers Obtained by Contact Angle and Inverse Gas Chromatography Methods. *Journal of Adhesion Science and Technology*, Vol.13, No.8, pp.887-901, ISSN 0169-4243

Sikora, R. (1996). Warstwa wierzchnia tworzyw wielkocząsteczkowych. *Polimery*, pp.96-113, ISSN 0032-2725

Sikora, R. (1997). Konstytuowanie warstwy wierzchniej tworzyw wielkocząsteczkowych. *Inżynieria Materiałowa*, No. 4, pp.160-164, ISSN

Sommers, A.D. & Jacobi, A.M. (2008). Wetting Phenomena on Micro-Grooved Aluminium Surfaces and Modeling of the Critical Droplet Size. *Journal of Colloid Interface Science*, Vol.328, pp.402-411, ISSN 0021-9797

Tang, X., Dong, J.& Li, X.(2008). A Comparison of Spreading Behaviors Silwet L-77 on Dry and Wet Lotus Leaves. *Journal of Colloid Interface Science*, Vol.325, pp.223- 227, ISSN 0021-9797

Thompson, P.A., Brinckerhoff, W.B. & Robbins, M.O.J. (1993). Microscopic Studies of Static and Dynamic Contact Angles. *Journal of Adhesion Science and Technology*, Vol.7, No.6, pp. 535-554, ISSN 0169-4243

Vedantam, S. & Panchagnula, M.V. (2008). Constitutive modeling of contact angle hysteresis. *Journal of Colloid Interface Science*, Vol.321, pp. 393-400, ISSN 0021-9797

Volpe, C.D. & Siboni, S. (1998). Analysis of dynamic contact angle on discoidal samples measured by the Wilhelmy method. *Journal of Adhesion Science and Technology*. Vol.12, pp.197–224, ISSN 0169-4243

Xian, J. (2008). Wettability of rough polymer, metal and oxide surfaces as well as of composite surface. *Journal of Adhesion Science and Technology*, Vol.22, No.15, pp.1893-1905, ISSN 0169-4243

Zielecka, M. (2004). Methods of Contact Angle Measurement as a Tool for Characterization of Wettability of Polymers. *Polimery* Vol.49, pp. 327-332, ISSN 0032-2725

Zouvelou, N.; Mantzouris, X. & Nikolopoulos, P. (2007). Interfacial energies in oxide/liquid metal systems with limited solubility. *International Journal of Adhesion and Adhesives* 27, pp.380-386, ISSN 0143-7496

Żenkiewicz, M.; Gołębiewski, J. & Lutomirski, S. (1999). Doświadczalna Weryfikacja Niektórych Elementów Metody van Ossa-Gooda. *Polimery*, Vol.44, No.3 ,pp. 212–217, ISSN 0032-2725

Żenkiewicz, M. (2000). *Adhezja i modyfikowanie warstwy wierzchniej tworzyw wielkocząsteczkowych*, WNT, ISBN 83-204-2547-6, Warsaw, Poland

Żenkiewicz, M. (2005). Wettability and Surface Free Energy of a Radiation-Modified Polyethylene Film. *Polimery*, Vol.50, No.5, pp.365-370, ISSN 0032-2725

Żenkiewicz, M. (2006). New Method of Analysis of the Surface Free Energy of Polymeric Materials Calculated with Owens-Wendt and Neumann Method. *Polimery*, Vol.51, http://en.wikipedia.org/wiki/Scanning_electron_microscope http://epmalab.uoregon.edu/epmatext.htm

7

Palmtop EPMA

Jun Kawai, Yasukazu Nakaye and Susumu Imashuku

Department of Materials Science and Engineering,
Kyoto University, Sakyo-ku, Kyoto,
Japan

1. Introduction

We have been developping palmtop electron probe X-ray microanalyzers (EPMA) for these several years [1-4] and succeeded to make such an instrument recently [3], and in the present chapter, we describe how to make an instrument in detail. The EPMA is an instrument to perform microanalysis (micrometer area elemental analysis) of any kinds of samples, such as metals, alloys, minerals, environmental and biological samples, by using an electron beam. Usually 10-30 keV kinetic energy electron beam is focused less than 1 µm in order to irradiate a sample, and consequently to excite characteristic X-rays such as Kα (2p → 1s) or Lα (3d → 2p) lines. From the energy and intensity of the characteristic X-ray lines, the kind (qualitative analysis) and concentration (quantitative analysis) of the elements in the specimen can be analyzed. Commercially available EPMA instruments are usually large (need a room of at least 3 m × 5 m to install) and expensive instruments (a few 10^5 USD). The palmtop EPMA we describe in the present chapter has features as follows and different from the conventional EPMA.

1. Small size. The size of the main part (sample holder, X-ray emission part, and the electron gun) is palmtop size. Typically less than 3 cm diameter and 5 cm length (**Fig.1**), but can be smaller than this size. The limitation of the size is due to the high voltage discharge distance.
2. Electric battery driven. The electron gun is driven by two 1.5 V electric D-batteries (**Fig. 2**), i.e. 3 V is enough for high energy (>10 keV) electron beam in order to excite characteristic X-rays.
3. The X-ray detector is Amptek Si-PIN detector (**Fig. 3**). Thus the size of the detector is also small. We use analog type X-ray detector amplifier. The size of the detector pre-amplifier is typically around 7 cm × 4 cm × 3 cm. The temperature control of the detector unit and bias power supply from the Amptek Co. is needed (**Fig.3**). Thus we need power supply for the Si-PIN controller to cool down the detector and bias voltage. However usually low voltage (5-12 V) DC is enough.
4. The pulse height analyzer (PHA) commercially available is not used, but we use a musician's amplifier and Windows computer as an alternative to the commercially available PHA [5-8]. This part is usually called DSP (digital signal processor). Usually a DSP is an expensive device which costs between 5000 and 10000 USD. However the musician's amplifier (**Fig. 4**) is typically less than 500 USD.

5. Vacuum in the sample chamber is of the order of 10^{-2} Torr (or ~1 Pa). This vacuum level is equivalent to the vacuum reached by a small rotary pump. Higher the vacuum than 10^{-2} Torr, in other words, too good or too high vacuum is not effective for high intensity X-ray emission. Too high vacuum down to 10^{-6} Torr needs expensive turbo molecular pump or oil diffusion pump, however, the residual gas is an electron source for the electron gun, thus such a high vacuum reduces the intensity of X-rays emitted. Too low vacuum up to 10^{-1} Torr will make electric discharge and 10^{-2} Torr is suitable for the present instrument. This vacuum level is suitable for oil sealed rotary pump (**Fig. 5**).
6. The sample exchange is easy because the vacuum seal is usually rubber O-ring (**Fig. 6**) and can be easily opened after air leaked.
7. The electron beam size is not micrometer, but as wide as the vacuum vessel diameter. It irradiates whole part of the sample and thus sample holder as well as vacuum vessel materials are excited. However the sample size can be as small as possible depending on the signal intensity. A typical size of the specimen is 50 µm diameter × 5 mm length metal wire. Thus single sand particle can be measured.
8. The palmtop EPMA can be made in laboratory, even by students without special experience.

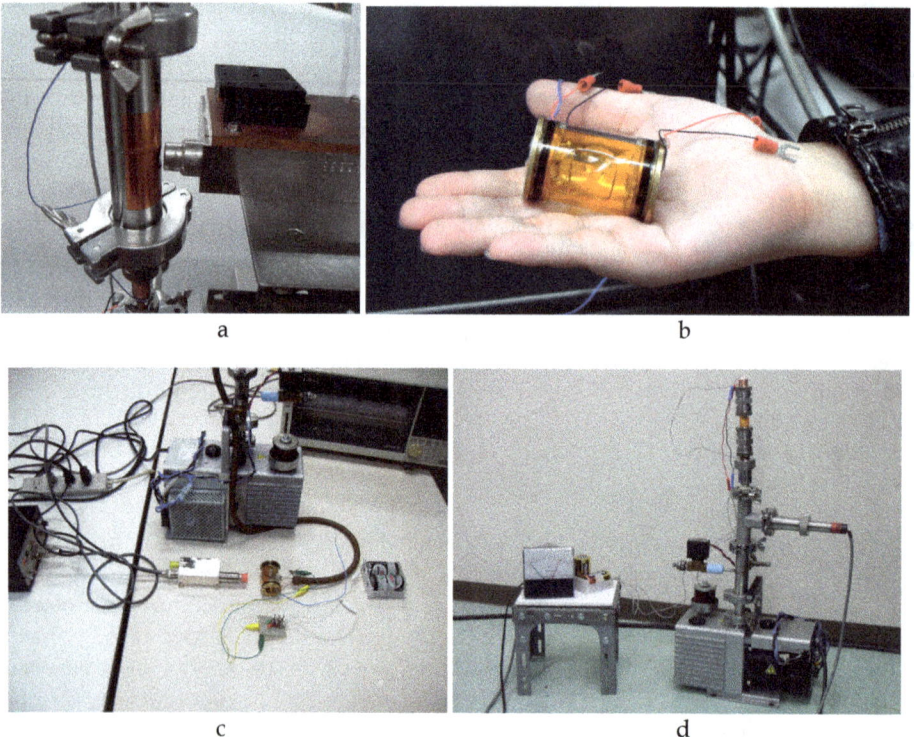

a

b

c

d

Fig. 1. Various kinds of palmtop EPMA made in our laboratory. (a) NW25 nipple is used for vacuum chamber, a hole is sealed by Kapton tape. (b) Glass type vessel is on the palm. (c) The whole system of the palmtop EPMA. (d) Another type of palmtop EPMA, with Pirani vacuum gauge.

Fig. 2. Electric D-batteries.

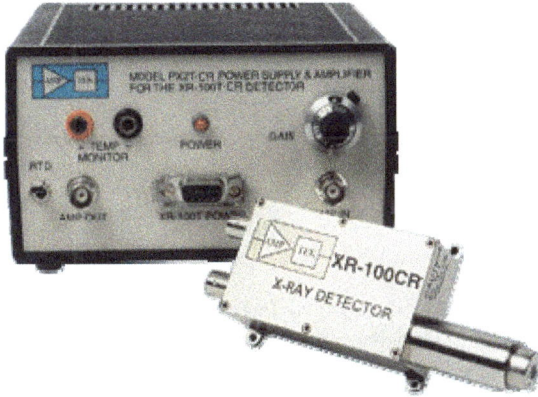

Fig. 3. Amptek Si-PIN X-ray detector (taken from Amptek Web page)

Fig. 4. Musician's amplifier for DSP.

2. The parts prepared before to build a palm-top EPMA

The following mechanical or electronic parts should be prepared to build up the instrument.

1.   Rotary pump. Oil sealed rotary pump to reach to the vacuum level of the order of 10^{-2} Torr or around 1 Pa is needed. The smaller size rotary pump is preferable. Single phase 100 V power supply is enough. A typical rotary pump used in the present work is shown in **Fig. 5**.

Fig. 5. A small rotary vacuum pump.

2.   Si-PIN X-ray detector. Amptek X-ray detector (**Fig. 3**) is needed. It costs usually less than 4000 USD including the preamplifier. The controller of Peltier device inside the Si-PIN detector to cool the Si-PIN device is needed. Usually the power supply associated with the Si-PIN detector is preferable. If Si-PIN X-ray detector is not available, Geiger-Müller counter is alternatively used for check the X-ray emission. The Geiger-Müller counter is available less than 500 USD (**Fig. 7**).
3.   Notebook size computer. Windows computer to control the Amptek Si-PIN detector, to display the X-ray spectra, and to control and analyze the digital X-ray signal is needed as is shown in **Fig. 4**. Several USB devices are connectable at the same time.
4. $LiTaO_3$ single crystal (one piece). The size is around 3 mm (x) × 3mm (y) × 5 mm (z) (**Fig. 8**). The z direction should be known. From Shin-Etsu Chemical Co. Ltd., Japan, this single crystal is available by less than 200 USD. Several other companies treat $LiTaO_3$ single crystal of the size of 10 mm length and 3 mm diameter with similar price. Other alternative is $LiNbO_3$. These materials are called pyroelectric crystals. If these single crystals are not available, at the first stage, you can use PZT stone used in the cigarette lighter to ignite.
5.   Peltier device. The size of Peltier device in **Fig. 9** is 8 mm × 8 mm. To drive this Peltier device, 3 V electric D-batteries are needed. (**Fig. 2**)
6.   Glass vessel or steel nipple. Single crystal, peltier device, and sample should be inside of the vessel or nipple to evacuate by the vacuum pump (**Fig. 1**).
7.   If possible, Pirani vacuum gauge is helpful to build up the instrument.

Fig. 6. O-ring, with brass specimen holder. Graphite is better than brass.

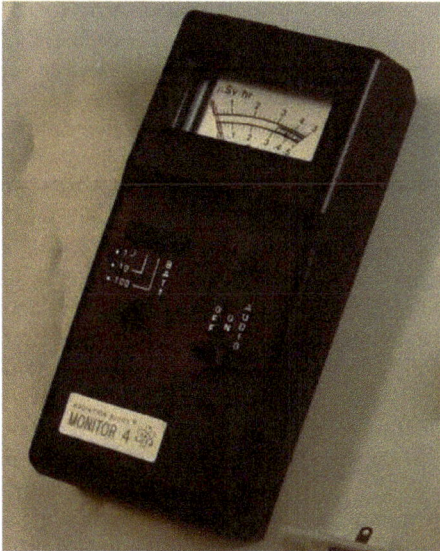

Fig. 7. Geiger-Müller counter

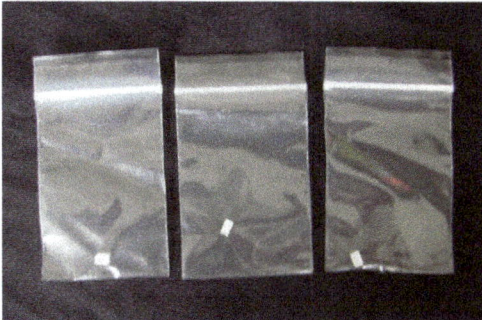

Fig. 8. LiTaO$_3$ single crystals in plastic bags.

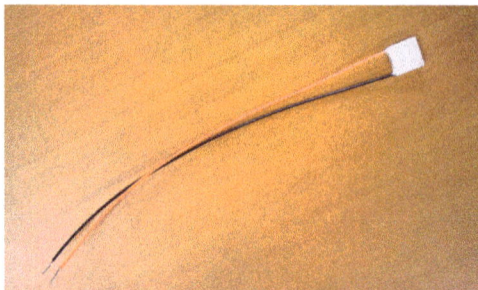

Fig. 9. Peltier device of size 8 mm × 8 mm.

3. Principle

The pyroelectric crystal is a material usually called as ferroelectric material. A mineral named tourmaline has similar characteristics. This is because the center of gravity of the plus charge and minus charge ions are not the same place in the crystal, but has a distance along z direction. Thus the extension or compression of these materials will produce electric high voltage. If a single crystal of the thickness of 1 mm in z direction of $LiTaO_3$ changes the temperature (usually the heating will expand, and cooling will compression) from room temperature (25 °C) to say 50 or 100 °C, then 10 kV high voltage will be produced. Thus when the crystal with thickness of 5 mm will produce 50 kV and 10 mm will produce 100 kV. When this single crystal is put into low vacuum such as 10^{-2} Torr, the electrons in the residual gas will be accelerated by this high voltage to hit the surface of the pyroelectric crystal or counter electrode of the pyroelectric crystal [9-19], as shown in **Fig. 12** below. If a specimen is attached on the counter electrode, the electrons accelerated by the high voltage between the pyroelectric (-HV) and counter electrode (0 V, grounded to earth) will hit the specimen. Consequently the X-rays are excited by the ionization of electrons by the bombardment of the accelerated electrons. Then the ionized electrons will contribute next instance to be accelerated to hit the specimen again and again, until the surface charge is neutralized. The vacuum should be not too good. Usually better side of 10^{-2} Torr is the suitable vacuum for this experiment. The heating and cooling of the pyroelectric crystal should be performed by a Peltier device. The temperature control of 50 °C from the room temperature (25 °C) is possible. The polarity of the surface changes when heated and cooled. Thus the electrons are moved inversely when heated and cooled. Usually it is heated for a few minutes, then next few minutes the pyroelectric crystal should be cooled. When the electron hits the pyroelectric surface, the X-rays are not from the specimen but Ta X-rays are observable from $LiTaO_3$.

4. How to build the main part of the palmtop EPMA

The Peltier device and the pyroelectric crystal are on the copper rod as shown in **Fig. 10**. They are glued by silver paste. The reason using the copper rod is its good heat conductivity as well as the electric conductivity. The one side of pyroelectric crystal glued to the Peltier device should be grounded to the earth. The heat created by the Peltier device should be diffused through the copper rod. Thus both silver paste and copper rod are suitable for this purpose.

Fig. 10. Pyroelectric crystal, Peltier device, and copper rod, glued by silver paste.

The anode is graphite rod as shown in **Fig. 11**. The specimen will be put on the graphite rod by double sided carbon adhesive tape. This is also because of the electric conductivity. The graphite rod and copper rod should be electrically connected to make the same ground potential. The X-ray intensity becomes weak if the electric connection between the copper and graphite rods is removed. Here the graphite is used as the specimen holder. The reason we use graphite and carbon adhesive tape is because the specimen and holder are hit by the electron beam from the pyroelectric crystal, and the carbon Kα X-rays from graphite as well as carbon adhesive tape are negligible because the C Kα energy is about 300 eV and thus such a low energy soft X-rays are strongly absorbed by the air and window of the vessel (Kapton), resulting the X-rays from graphite rod and carbon adhesive tape are not detectable.

Fig. 11. Graphite anode as specimen holder.

The lead wire from the Peltier device is inside of the vacuum vessel, and should go through the vacuum boundary to the outside of the vessel without the leak of the vacuum by using epoxy glue. If possible addition of thermocouple is helpful to measure the temperature of the Pettier device.

Both of the electrodes are sealed inside the glass or stainless pipe by rubber O-ring. Viton rubber O-ring is preferable. If NW-25 type quick coupling flange is directly used, this is quite easy to seal the vacuum. The stainless steel nipple of NW-25 or glass pipe should have ca 5 mm diameter through-hole to go through the X-rays, and this through-hole should be sealed by thin Kapton film. Adhesive type Kapton tape is commercially available. Polyester, PET (polyethylene terephthalate), or Mylar films glued by epoxy resin adhesives are alternative to the Kapton tape. The structure of the palmtop part is illustrated in **Fig. 12**.

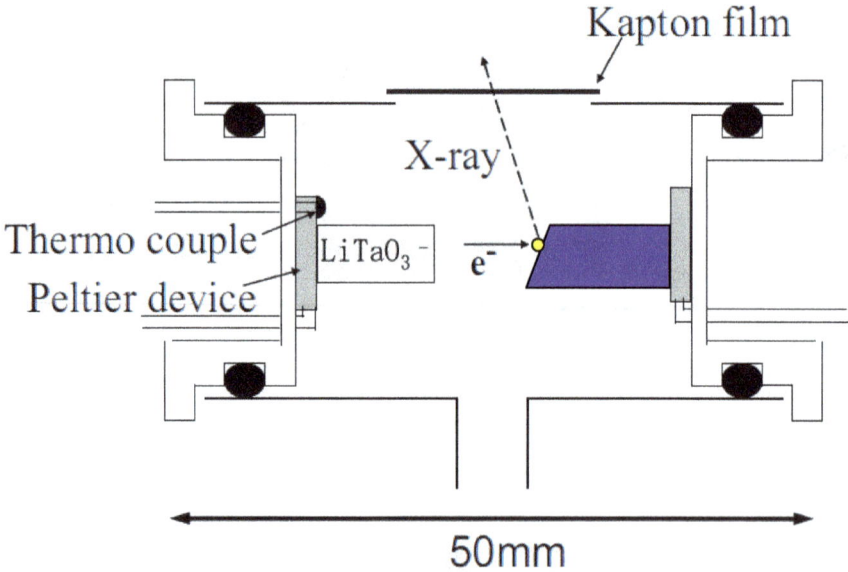

Fig. 12. Illustration of the palmtop EPMA.

5. How to operate

Putting stainless steel small plate, the size of which is typically 5 mm × 5 mm, on the graphite rod, by double sided carbon adhesive tape, then the vacuum vessel should be evacuated by the rotary pump. The connection between the vessel and the pump should be thick rubber tube. At the first experiment, the stainless steel as the specimen should be as large as possible to get enough X-ray intensity. The stainless steel is usually composed of 18 % Cr, 8 % Ni, and the rest Fe. Thus we can observe Cr, Fe, and Ni Kα and Kβ lines. After one or two minutes evacuation by the rotary pump, the vacuum reached to better than 2×10^{-2} Torr. Then the Peltier lead wires should be connected to 3 V D-batteries. Waiting for 10-20 seconds, the Geiger-Müller counter reacts the X-rays, and the intensity is not very strong but we can hear the X-ray counts sounds from the Geiger-Müller counter as almost continuous sounds from the counter. It is more than a few tens of cps (counts per seconds). The polarity of D-battery is no problem. If the polarity is for the pyroelectric crystal being +HV, the end of crystal is hit by the electron and X-rays of Ta are emitted.

If you confirm the emission of X-rays from the vessel, the next step is to measure the spectra by the Si-PIN detector. It is better to become accustomed to use the Si-PIN detector to measure the X-ray spectra by using other method, such as weak radio isotopes. Then the Si-PIN detector should be close to the Kapton window of the palmtop EPMA to measure the spectra. Usually the time interval to change the polarity of the D-battery from heat to cool the crystal is 1 or 2 minutes. A typical temperature change and X-ray intensity decay are shown in **Fig. 13**.

The spectra measured by the Si-PIN detector is shown in **Fig. 14**, where the peak intensity is several thousands of counts for a few minutes one cycle. The X-ray intensity is not very strong, but it is better to protect the exposure to the X-rays. Usually lead containing acrylic plate is good for protecting from the X-rays, but steel plate of the thickness of 1 or 2 mm is enough. The Geiger-Müller counter is not saturated during this experiment, but if the X-ray intensity is too strong when Geiger-Müller counter is directly irradiated by an 1 watt X-ray tube, then the Geiger-Müller counter will be saturated, and we cannot see the difference between no X-rays. Thus the present experiment should be performed with an expert of X-ray experiments.

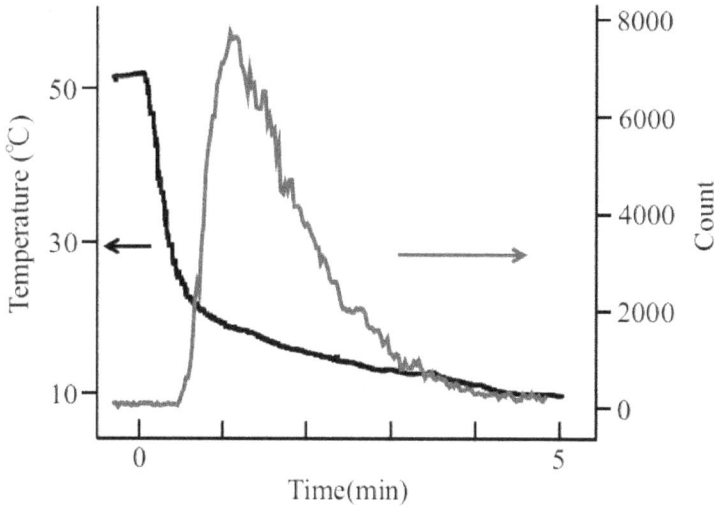

Fig. 13. Time dependent temperature and X-ray intensity after the crystal became to be cooled.

6. Digital signal processor

The X-ray signal from the Si-PIN X-ray detector is voltage signal. If the energy of X-rays is high, the voltage increase step-like according to the energy of the X-rays. Thus when the X-ray energy is high, the step height is larger. If the X-ray intensity is strong, then the step frequency in unit time increases. The step height increases again and again, and finally the voltage is larger than the voltage of the power source, then the step signal is reset and starts again. Thus if the time dependent voltage increase is recorded as voice signal in the memory of a computer as the voice recorder, we can plot the X-ray spectra by differentiating the step-like voltage signal and plotting the frequency of the pulses against the height of the peak. This is the X-tray spectra. One of the X-ray spectra obtained in such a way is shown in **Fig. 14**. The X-ray intensity of our palmtop EPMA is not very strong (less than 10000 counts per minute), thus the response of analog/digital (A/D) converter for music purpose is enough (**Fig. 15**). The important points to use the music A/D convertor, or musician's amplifier are as follows.

1. The input of music amplifier is usually biased by a DC voltage to drive a microphone, and thus direct DC connection between X-ray pre-amplifier and music A/D converter will cause to destroy the pre-amplifier, and thus AC coupling should be used.
2. Impedance should be matched.
3. Any kind of notebook computer has microphone input, and can be used for the similar purpose, but the notebook computer inside is full of digital noises. Consequently too high level of digital noises makes it impossible to measure the X-ray spectra. This is the reason we use separate musician's amplifier.

The above function is identical to the digital oscilloscope and if you have a digital oscilloscope, you can connect your oscilloscope to the computer by a USB and import the X-ray signal into your computer. After recording or during the recording, you can differentiate the X-ray signal numerically, and plot the X-ray spectra on the computer display. When you record the X-ray spectra, the peaks are not assigned at all, and the peaks should be assigned using linear relation between the voltage and X-ray energy, and the energy of the spectra of elements contained in the specimen should be assigned by an X-ray database. The resulted X-ray spectrum is shown in **Fig. 14**. An example of the software is provided from X-ray Precision Co. Ltd. Kyoto by the price of around 200 USD by CD-ROM, but you can make such a program by yourself.

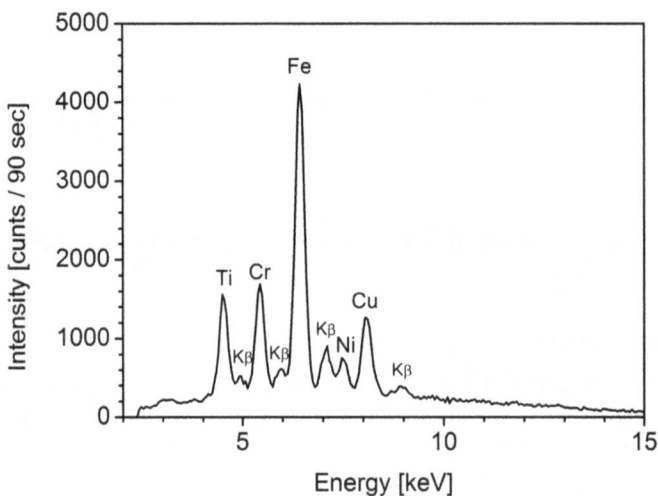

Fig. 14. X-ray spectra of steel and Ti (about 1 minute measurement).

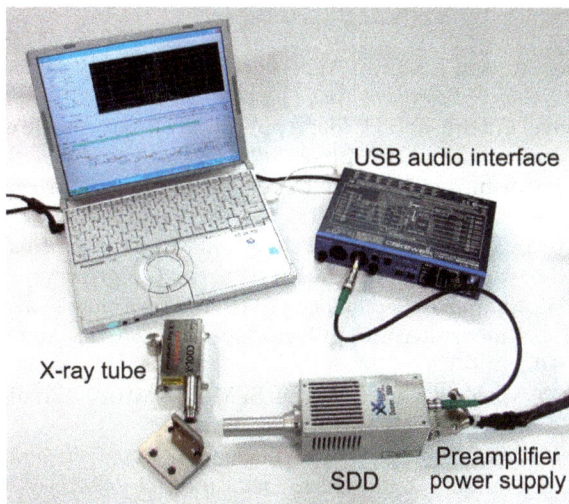

Fig. 15. Musician's amplifier connected to X-ray detector and computer.

7. Maintenance and safety

The surface of the pyroelectric crystal becomes dark because of the sputtered particles by the high voltage discharge will be deposited on the surface. These black fine particles are electric conductive, and thus the high voltage is not accumulated in the pyroelectric crystal. In such a case, the pyroelectric surface should be cleaned by cotton stick with ethanol to remove the dark sputtered particles. Then the X-rays will come back again.

The earth electric line is important to avoid the electric shock. Also the electric connection between the two electrodes is important to emit strong X-rays, as mentioned in the text. The X-ray is dangerous to be exposed even it is very weak. Thus monitoring of the X-ray intensity and shielding the X-rays are important for experiments.

After the experiment, the rotary pump should be leaked to the atmospheric pressure. If one forgets to leak the air into the vacuum system, the oil in the rotary pump will rise up to the vacuum vessel of the palmtop EPMA to fill the vessel, and consequently the experiment is not possible any more because of the contamination by the oil. To avoid this, auto-leak valve is preferable to attach just above the rotary pump. Since the auto-leak valve, manual air leak valve, and Pirani gauge are attached to the vacuum system, the instrument shown in **Fig. 1d** is not a simple one but complicated by vacuum parts. All these vacuum parts are connected to the vacuum system by the NW-25 flanges.

8. Acknowledgements

E. Hiro and S. Terada, who contributed at the early stage of the palmtop EPMA development, are acknowledged. The authors would like to thank Asahi Glass Foundation for financial support. The computer program for music amplifier has been developed by the JST support. The first version of the palmtop EPMA was a by-product of JST Sentan project.

9. References

[1] E. Hiro, T. Yamamoto, and J. Kawai: Applying pyroelectric crystal to small high energy X-ray source, *Adv. X-Ray Chem. Anal., Japan*, 41, 195-200 (2010).

[2] J. Kawai, Y. Nakaye, E. Hiro, and H. Ida: Application of pyroelectric crystal---a safety X-ray source, *Radioisotopes*, 60 (6), 249-263 (2011).

[3] S. Imashuku, A. Imanishi, and J. Kawai: Development of miniaturized electron probe X-ray microanalyzer, *Anal. Chem.* 83 (15), 6011-6017 (2011).

[4] S. Imashuku, A. Imanishi, and J. Kawai: Palmtop EPMA by electric battery, submitted to the Proceedings of the 21st ICXOM (Intern. Congress on X-ray Optics and Microanalysis, Campinas, September, 2011), *AIP Conference Proceedings* (submitted).

[5] Y. Nakaye, and J. Kawai: Recording X-ray spectra with an audio digitizer, *X-Ray Spectrom.*, 39 (5), 318–320 (2010).

[6] L. Ze, Y. Nakaye, Y. Morikawa and J. Kawai: SEM-EDX with audio digitizer, *Adv. X-Ray Chem. Anal. Japan*, 42, 111-114 (2011).

[7] Y. Nakaye and J. Kawai: Observation of pulsed electron field emission driven by a pyroelectric crystal, *Adv. X-Ray Chem. Anal. Japan*, 42, 249-253 (2011).

[8] Y. Nakaye and J. Kawai: X-Ray measurement using an audio A/D converter, *Adv. X-Ray Chem. Anal. Japan*, 42, 255-259 (2011).

[9] J. D. Brownridge: Pyroelectric X-ray generator, *Nature*, 358, 287-288 (1992).

[10] J. D. Brownridge, and S. Raboy: Investigations of pyroelectric generation of x rays, *J. Appl. Phys.*, 86, 640-647 (1999).

[11] S. M. Shafroth, W. Kruger, and J. D. Brownridge: Time dependence of X-ray yield for two crystal X-ray generators, *Nucl. Instrum. Meth. Phys. Res.*, A422, 1-4 (1999).

[12] J. D. Brownridge, and S. M. Shafroth: Electron and positive ion beams and X-rays produced by heated and cooled pyroelectric crystals such as $LiNbO_3$ and $LiTaO_3$ in dilute gases: Phenomenology and applications, in "Trends in Laser and Electro-Optics Research", Ed. W. T. Arkin, Nova Science Pub. (2006) pp. 59-95.

[13] B. Naranjo, J. K. Gimzewski, and S. Putterman: Observation of nuclear fusion driven by a pyroelectric crystal, *Nature*, 434, 1115-1117 (2005).

[14] J. Geuther, Y. Danon, F. Saglime, and B. Sones: Electron acceleration for X-ray producing using paired pyroelectric crystals, Abstracts of the 6th Intern. Meeting on Nuclear Applications of Accelerator Technology (AccApp'03), San Diego, June 1-5 (2003) pp.124-128. (http://www.rpi.edu/~danony/Publications.htm).

[15] J. A. Geuther and Y. Danon: Electron and positive ion acceleration with pyroelectric crystals, *J. Appl. Phys.*, 97, 074109 (2005).

[16] J. A. Geuther and Y. Danon: High-energy X ray production with pyroelectric crystals, *J. Appl. Phys.*, 97, 104916 (2005).

[17] J. A. Geuther, Y. Danon, and F. Saglime: Nuclear reaction induced by a pyroelectric accelerator, *Phys. Rev. Lett.*, 96, 054803 (2006).

[18] J. A. Geuther and Y. Danon: Application of pyroelectric particle accelerators, *Nucl. Instrum. Meth. Phys. Res.*, B261, 110-113 (2007).

[19] J. A. Guther and Y. Danon: Enhanced neutron production from pyroelectric fusion, *Appl. Phys. Lett.*, 90, 174103 (2007).

Part 2

Biology, Medicine

8

Contribution of Scanning Electron Microscope to the Study of Morphology, Biology, Reproduction, and Phylogeny of the Family Syllidae (Polychaeta)

Guillermo San Martín and María Teresa Aguado

Departamento de Biología (Zoología), Facultad de Ciencias, Calle Darwin 2, Universidad Autónoma de Madrid, Canto Blanco, Madrid, Spain

1. Introduction

Syllidae is a highly diverse family of the polychaetes (Annelida, Phyllodocida), with 72 described genera and almost 700 species (San Martín, 2003; Aguado & San Martín, 2009; Aguado et al., in press), and continuously new taxa are being described. They are small marine worms, usually of few mm long, although some species can reach up to 90 mm. Contrariwise to their small size, they are very complex, with a body exhibiting numerous structures, external and internal, some of them difficult to examine properly under light microscope, even using higher magnifications and Nomarsky system of polarized light. Description of most species before around the year 2000 was based only on examinations and drawings made with camera lucida under light microscopes. The use of SEM to study syllids is relatively recent, but produced the discover and descriptions of a number of new, unknown structures, or only incompletely known before, whose physiology and significance open a new field of research. The oldest SEM picture of a syllid is from 1980 in which Heacox showed the head of a *Chaetosyllis* stolon and a larval compound chaeta in a study of the life cycle of *Syllis pulchra* Berkeley & Berkeley, 1948 (Heacox, 1980); somewhat later, Pocklington & Hutchenson (1983) reported the viviparity of the interstitial species *Parexogone hebes* (Webster & Benedict, 1884) showing excellent and surprising SEM photographs of juveniles emerging through segmental apertures (probably nephridial pores) and also some characteristic crenulations of the ventral surface of female's body after releasing juveniles; in the same year, Pawlick published the first SEM photos of general aspect of body, chaetae, and details of ciliation on the basis of dorsal cirri on the species *Branchiosyllis oculata* Ehlers, 1887. This is the first paper, in our knowledge, in which some previously overlooked structures were showed and described thanks to SEM; finally, Sardá & San Martín (1992) redescribed one species of syllid from East coast of USA, with some SEM pictures. However, few descriptions of new taxa with SEM and only few papers with SEM pictures of syllids were published during next years. Most of examinations came from our own relatively recent papers; one is the book "Fauna Ibérica. Syllidae", published in 2003, in which the 161 recorded species of the Iberian Peninsula are described and figured, with 122 SEM plates, showing morphology, and also details of the reproduction; the SEM study revealed some new, tiny structures, unknown before (San Martín, 2003). However, since that book was published in Spanish, several of these discoveries could be overlooked by some non-Spanish speakers. Same kind of study and descriptions, with both light microscope and SEM photos was followed since then for descriptions of syllids from other parts of the world (Aguado et al., 2006 and others), especially in the still unfinished series on Australian Syllidae (San Martín, 2005; San Martín & Hutchings, 2006; San Martín et al., 2008a, b; 2010, and others) and revisions of genera (Aguado & San Martín, 2008). In these papers, and others published by other authors profusely cited below, more unknown structures and new details on chaetae, ciliation pattern, pharyngeal armature, etc., were described, as well as important observations on the reproductive biology of many species. A review of these discoveries and their implications on some aspects of the knowledge of the family Syllidae, with especial relevance in taxonomy and systematics, will be analyzed in detail in this chapter.

Detailed morphological and reproductive observations under SEM had important relevance to the knowledge of the family Syllidae for two reasons.

- The first one is related with the phylogeny of the family; discoverment and descriptions of these structures and details of the reproductive modes provided more morphological features to analyze and, consequently, more robust hypothesis about the relationships among the different genera of the family.
- The second one is the great help to differentiate sibling complexes of species; species apparently identical morphologically can be differentiated by minute details, only perceptible under SEM.

Two books about polychaetes also includes excellent SEM pictures of Syllidae (as well as many others of different families of polychaetes): Rouse & Pleijel (2001), which shows 6 photos with details of morphology and reproduction, and Beesley et al. (2000) in which a couple of SEM photos about syllid reproduction are shown.

2. Material and methods

Syllids are generally of small size and hence, the process of preparation is usually complex and sometimes difficult, so, it is desirable to prepare several specimens for examination. At least, one to be examined dorsally and another one ventrally; specimen with exerted pharynx is also strongly recommended to examine and take photos of details of the anterior end of the pharynx and their armature, which are important characters for identification to genus level.

Minute specimens are very easily lost during the process of preparation, so it is important to be extremely careful, especially having a short number of specimens.

Fixation of the specimens is a very important process for taking good and sharp pictures. Specimens of syllids are usually dirty, especially those of the genera with dorsal papillae, which produce a sticky secretion which agglutinate debris. However, details of papillation is an important taxonomic trait for identification of species and therefore, it is necessary to clean up them, using a brush of a single, slender hair.

Techniques could be different depending upon the authors and the type of SEM used. The specimens for taxonomic studies are usually fixed in formalin, examined in pure water and finally stored in 70% ethanol. To be prepared for SEM, the selected specimens experience a series of progressive baths in more concentrated alcohol (80%, 90%, and pure ethanol), then dried on critical point and covered of a coat of gold. Rouse & Pleijel (2001) recommend Osmium tetroxide (OsO$_4$) as preferred fixative, although its high toxicity (see Rouse & Pleijel, 2001, p. 7).

3. Results

3.1 Morphology

3.1.1 Ciliation

Many syllids are provided with numerous cilia, whose arrangement has taxonomic importance, but they are difficult to see under light microscope; sometimes, it is possible to observe the presence of tufts of cilia in some appropriate areas, but a detailed description of the arrangement of these cilia is almost impossible or extremely difficult. However, well prepared specimens show, under SEM, a system of transversal rows of cilia, which can be single (fig. 1A) or double (fig. 1B), sometimes some anterior segments with single and from one segment backwards being double (fig. 1B), sometimes only tufts of cilia on some areas; in some species only the peristomium is dorsally provided with a single band of cilia and remaining segments lack them. These details were not included in the descriptions until recently, and they are certainly useful for segregation of species and even similar genera. Descriptions arrangement of ciliary bands, based on SEM examinations are in San Martín (2003) for the species *Paraehlersia ferrugina* (Langerhans, 1881), *Odontosyllis fulgurans* (Audouin & Milne Edwards, 1834), *Myrianida benazzi* (Cognetti, 1953), *Myrianida convoluta* (Cognetti, 1953), *Myrianida edwarsi* (Saint-Joseph, 1887), and *Myrianida dentalia* (Imajima, 1966); San Martín & Hutchings (2006) for the species *Eusyllis kupfferi* Langerhans, 1879, *Odontosyllis polycera* Schmarda, 1863, *O. australiensis* Hartmann-Schröder, 1979; *Paraehlersia weissmannioides* (Augener, 1913), *P. ehlersiaeformis* (Augener, 1913), *Perkinsyllis koolalya* (San Martín & Hutchings, 2006), and *P. serrata* (Hartmann-Schröder, 1984); and Nogueira & Fukuda (2008) for *Trypanosyllis zebra* (Grube, 1860) and *T. aurantiacus* Nogueira & Fukuda, 2008. Also some ciliated areas on bases of dorsal cirri, lateral of segments, palps, ventral surface or in ventral cirri were described and figured by Pawlik (1983) for *B. oculata*; Licher & Kuper (1998) for *Syllis tyrrhena* (Licher & Kuper, 1998); López et al. (2001) for *Pionosyllis magnifica* Moore, 1906; San Martín (2003) for *Xenosyllis scabra* (Ehlers, 1864), and *Trypanosyllis aeolis* Langerhans, 1879; San Martín & Hutchings (2006) for *O. australiensis*; Aguado & San Martín (2008) for *Brachysyllis infuscata* (Ehlers, 1901); Ramos et al. (2010) for *Streptodonta exsulis* Ramos, San Martín & Sikorski, 2010; and Salcedo-Oropeza et al. (2011) for *Trypanosyllis microdenticulata* Salcedo-Oropeza, San Martín & Solís-Weiss, 2011.

3.1.2 Nuchal organs

Nuchal organs are only present in polychaetes and are thought to be a synapomorphy for the group (Rose & Fauchald, 1997; Rouse & Pleijel, 2001), although they show different chemoreceptor structures and are different in shape among the families. The most typical shape appears as two semicircular, densely ciliated pits between prostomium and peristomium; these are the kind of nuchal organs most common in the Syllidae, in fact most genera have the typical nuchal organs (fig. 2A), sometimes extending laterally to prostomium, forming two semicircular ciliated areas (fig. 2B) as in the genera *Eusyllis* Malmgren, 1867, *Odontosyllis* Claparède, 1863, and *Trypanosyllis* Claparède, 1864; however, some genera show other kind of nuchal organs, called nuchal lappets, forming two dorsal, longitudinal evaginations over a number of segments, sometimes only one, more or less spherical or elongate, straight (fig. 2C) or sinuous (fig. 2D), ciliated or lacking of cilia. Numerous descriptions of species include SEM photos of the nuchal organs, as those in Lanera et al. (1994); Lewbart & Riser (1996); Licher & Kuper (1998); Martín et al. (2002); Nogueira & San Martín (2002); San Martín (2003; 2005); San Martín & Hutchings (2006), Aguado & San Martín (2007, 2008); Lattig et al. (2007; 2010 a, b); Nogueira & Fukuda (2008); Lattig & Martin (2009); Ramos et al. (2010); San Martín et al. (2008 a, b; 2010); Salcedo-Oropeza et al. (2011); Lattig & Martín (in press a, b). San Martín & López (2003) described a new genus and species form Australia with nuchal organs laterally located, protected by two lips.

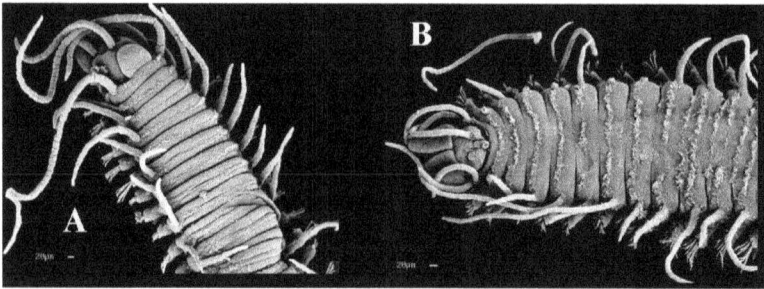

Fig. 1. A, Anterior end, dorsal view of *Odontosyllis australiensis*, showing single row of cilia on each segment. B, Anterior end, dorsal view of *Perkinsyllis serrata* showing single row of cilia on anterior segments and double row form chaetiger 9. San Martín & Hutchings (2006).

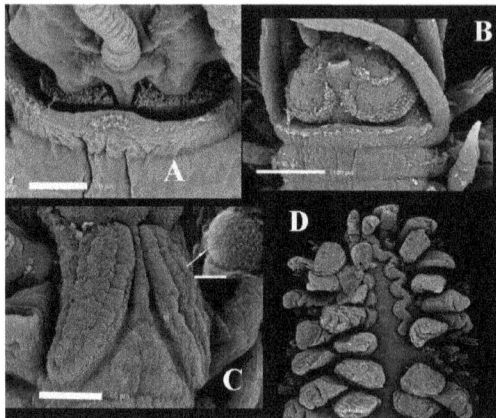

Fig. 2. Nuchal organs of: A *Syllis corallicola*, B, *Eusyllis kupfferi*, showing the ciliation extending to lateral sides of prostomium; C, *Proceraea aurantiaca*, and detail (arrow, right) of the rugose bulks; D, *Clavisyllis alternata* Knox, 1957. A, C, San Martín (2003); B, San Martín & Hutchings (2006); D, Aguado & San Martín (2008).

In the species *Proceraea aurantiaca* Claparède, 1868, San Martín (2003) found and described an enigmatic structure on the nuchal lappets, as semicircular, rugose bulks (fig. 2C, right, up, arrow) provided with pores and tiny hairs, whose function is totally unknown.

3.1.3 Papillation

Some genera of the family Syllidae, as *Sphaerosyllis* Claparède, 1863, *Prosphaerosyllis* San Martín, 1984, *Erinaceusyllis* San Martín, 2005, *Rhopalosyllis* Augener, 1913; and *Paraopisthosyllis* Hartmann-Schröder, 1991, as well as some species of other genera, as *Opisthosyllis* Langerhans, 1879, *Branchiosyllis* Ehlers, 1887, and *Trypanosyllis* Claparède, 1864, have the dorsal, and also sometimes the ventral, surface covered by conspicuous papillae (figure 3), sometimes also extending through parapodia, antennae, cirri and palps. These papillae can be scarce and scattered or densely distributed, in different sizes or all similar, hazardous distributed or arranged in rows. Information on size, shape and distribution of papillae has a very important taxonomic meaning, and it is very useful to differentiate species; although presence of papillae is easily perceptible under light microscope, the arrangement and sizes of papillae are much better evaluable under SEM. In *Sphaerosyllis*, *Prosphaerosyllis* and *Erinaceusyllis*, the papillae produce an adhesive secretion which sticks detritus for mask the individuals; so, it is sometimes difficult to see them because they are covered by detritus and the specimens requires a previous cleaning by a minute brush and washing of water. Numerous descriptions of species of these three genera are in San Martín (2003; 2005), Nogueira et al. (2004); Musco et al. (2005); Álvarez & San Martín (2009); and Olivier et al. (2011). Details of papillae of the species *Syllis papillosus* (Tovar-Hernández, Granados-Barba & Solís-Weiss, 2002) in Tovar-Hernández et al. (2002); *Trypanosyllis troll* Ramos, San Martín & Sikorski, 2010 are in Ramos et al. (2010), *Paraopisthosyllis alternocirra* San Martín & Hutchings, 2006 in San Martín & Hutchings (2006); *Opisthosyllis viridis* Langerhans, 1879 in San Martín et al. (2008 a); and *Branchiosyllis verruculosa* (Augener, 1913) in San Martín et al. (2008b).

3.1.4 Pharyngeal armature

The pharyngeal armature in syllids is an important diagnostic character at different taxonomic levels; for this reason, a careful examination of the pharyngeal armature is always necessary for identification. SEM photos help considerably in these observations, and also contributed greatly to the discover of some overlooked details, as ciliation and presence of pores on the pharyngeal papillae, coats of cilia on the pharyngeal opening, secondary crown of papillae, total absence of papillae, etc. The pharynx of syllids can be unarmed or provided with a single middorsal tooth (fig. 4A), usually surrounded by a crown of papillae, and sometimes with a complete or incomplete crown of teeth (trepan) (fig. 4B) sometimes there is a trepan but not a pharyngeal middorsal tooth; teeth of trepan are usually of the same size, but sometimes they are of different sizes (figs. 4C, D). Detailed observations under SEM of the pharyngeal armature, papillae and ciliation are in numerous papers: Capa et al. (2001); Tovar-Hernández et al. (2002); Nogueira & San Martín (2002); Martin et al. (2002; 2003; 2009); San Martín (2003; 2005); San Martín & Hutchings (2006); Aguado & San Martín (2007,2008); Nogueira & Fukuda (2008); San Martín et al. (2008a, b; 2010); Lattig & Martín (2009; in press a, b); Lattig et al. (2007; 2010 a, b); Ramos et al. (2010); Olivier et al. (2011); Salcedo-Oropeza et al. (2011).

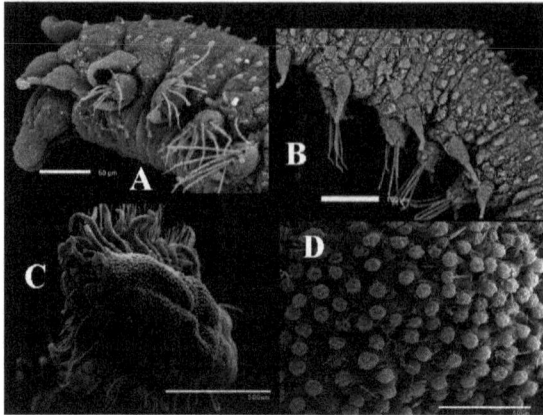

Fig. 3. Papillae: A, anterior end, lateral view,*Sphaerosyllis hirsuta*. B, midbody of the same. C, anterior end, dorsal view, *Opisthosyllis papillosa*. D, details of the papillae of the same. A, B, San Martín (2005). C, D, San Martín et al. (2008 a).

3.1.5 Chaetae

Chaetae have a great value for taxonomic identifications; small details of chaetae can be crucial to assign one specimen to one species. Therefore, a careful examination of chaetae is decisive. They are usually the most photographied structures of syllids; together with the pharyngeal armature. The chaetae are the unique hard structures in syllids, so they are relatively easy to be examined and photographied by SEM; however, they are not always clean enough. Additionaly, length and shape can vary from anterior to posterior part of body and from dorsal to ventral. In our knowledge, the oldest publications showing chaetae of any syllid are those of Heacox (1980) and Pawlik (1983) cited above. Later, Westheide (1990) showed the chaetae and also a special genital chaeta of a new species (*Sphaerosyllis hermaphrodita*) from Thailand. There are many variations on the chaetal types in syllids, and since then, numerous papers showing details of chaetae, many of them profusely cited in this chapter, and also others, like Licher et al (1995) or Martínez et al (2002), not mentioned in any other sections of this chapter. Relative length and orientation of spines on margin of blades of compound chaetae are difficult to see properly under light microscope but they are absolutely precise under SEM. In the figure 5, there is a selection of photos of different kinds of chaetae in syllids (see also figure 11). One interesting is the Australian species *Odontosyllis freycinetensis* Augener, 1913; compound chaetae of that species are apparently unidentate under light microscope, even examined with Normarsky system and high magnifications, but in fact they are bidentate, with a minute, spine-like distal tooth and a big, curved proximal tooth. The use of SEM for study of chaetae in syllids also clarify some other details; for instance, hyaline hoods on the margin of blades appear in few some species, but SEM examinations showed that in fact they are several rows of minute spines, instead of one row, with more or less well defined spines (fig. 6).

Images of the same chaeta under light microscope or SEM can be remarkably different, being the latter much more precise and useful for descriptions and phylogenetic inferences. Also study under SEM of special chaetae for reproduction and brooding of eggs produce interesting discoveries (see below).

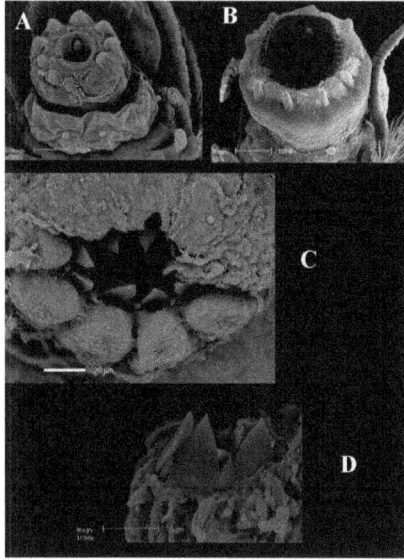

Fig. 4. Everted pharynx of: A, *Syllis gerundensis*, showing the pharyngeal tooth, and two
crowns of papillae. B, *Eusyllis assimilis*, showing the middorsal tooth, incomplete trepan, and
two crowns of papillae. C, D, frontal and lateral view of the trepan of *Proceraea picta*,
showing the different sizes of teeth. San Martín (2003).

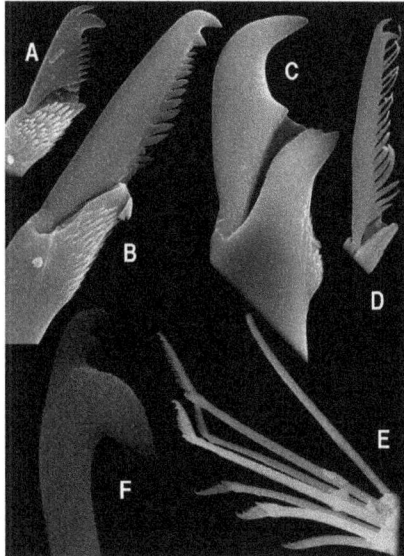

Fig. 5. Some examples of Syllid chaetae. A, *Paraehlersia elersiaeformis*; B, C, *Branchiosyllis
maculata*; D, *Perkinsyllis heterochaetosa*; E, chaetal fascicle of *Syllis rosea*; F, *Parahaplosyllis
brevicirra*. A, D, San Martín & Hutchings (2006); B, C, San Martín et al.(2008a); E, San Martín
(2003); F, San Martín, et al. (2010).

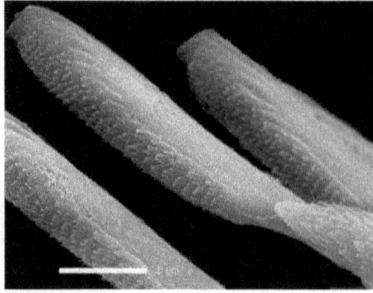

Fig. 6. Compound chaetae of *Streptodonta pterochaeta*, showing the rows of minute spines. San Martín (2003).

3.1.6 Glands

Syllids are provided with numerous kinds of glands, whose function is mostly unknown; some of them were known from long time ago but others were discovered and described recently using SEM. Some of the previously known glands were studied under SEM, and some details of their structures were showed after SEM observations. Most of the different kinds of glands in Syllidae are present in the subfamilies Exogoninae Langerhans, 1879 and Syllinae Grube, 1850, especially in interstitial, minute species, but also in some large syllids. Members of the genus *Syllis* Lamarck, 1818 (including *Typosyllis* Langerhans, 1879) have convoluted, refringent glands within the articles of cirri. They are easily visible under light microscope, even under compound microscope. These glands are usually opened by means of minute pores, only perceptible under SEM (fig. 7A). The first picture of these pores are in Licher & Kuper (1998), for *Syllis thyrrena*, and later San Martín (2003) for *Trypanosyllis zebra*, *Trypanosyllis aeolis*, and *Syllis amica* Quatrefages, 1865, but they are probably present in many other species. San Martín et al. (2008b) showed these pores in the species *Parasphaerosyllis indica*. San Martín (2003) also discovered and published some photos of pores in some anterior ventral cirri of some species, as *Eurysyllis tuberculata*, *Plakosyllis brevipes*, *Xenosyllis scabra*, and *Trypanosyllis zebra*; especially interesting are those of *P. brevipes*, since they are half-moon like, with an arranged alveolar organization (fig. 7B). Since that species is strictly interstitial in sands, these glands could be adhesive; however, they are also present, although less developed, in other non-interstitial species, as *T. zebra*. Pores on dorsum of several species have been also described, as in *P. brevipes* (San Martín, 2003). San Martín et al. (2008b) described a species, *Branchiosyllis carmenroldanae*, from Australia, with the granular dorsum covered by numerous pores. Pores on the dorsal tubercles have been reported in *E. tuberculata* by San Martín (2003) and in the tips of dorsal crests (see below) in *Xenosyllis moloch* San Martín, Hutchings & Aguado, 2008 (San Martín et al., 2008b) (fig. 7C). Also dorsum densely provided of granules, opened by pores was described in the Anoplosyllinae *Syllides fulvus* (Marion & Bobretzky, 1875) by San Martín (2003). Dorsal pores have been reported in *P. ferrugina* by the same author, *P. ehlesiaeformis*, *P. weissmannioides*, by San Martín & Hutchings (2006). Especially interesting are the pores described and figured by the same authors for the species *Brevicirrosyllis mariae* (San Martín & Hutchings, 2006), located on bases of dorsal cirri, similar to the pores of parapodial glands of some Exogoninae (see below) (fig. 7D), and also at tips of dorsal cirri. Pores in ventral cirri have been also reported for other few species, as *Amblyosyllis madeirensis* Langerhans, 1879, and *Eusyllis lamelligera* Marion & Bobretzky, 1875.

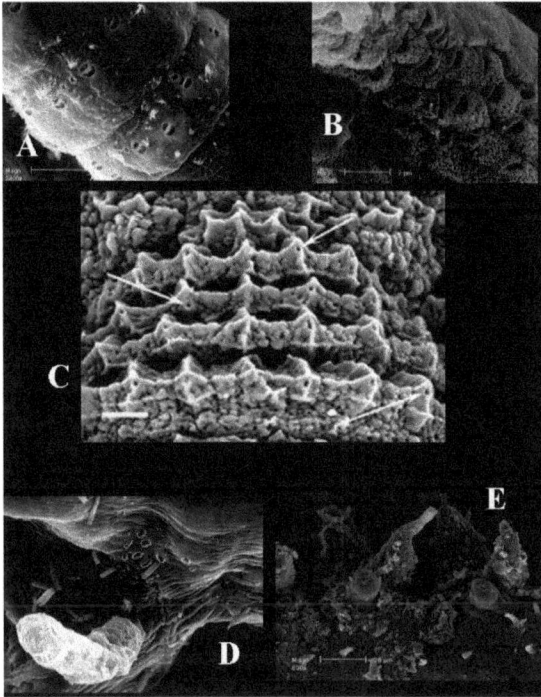

Fig. 7. Opening of different glands. A, articles of dorsal cirri, *Trypanosyllis zebra*; B, ventral
cirri, *Plakosyllis brevipes*; C, dorsal crests, *Xenosyllis moloch*; D, base of dorsal cirri,
Brevicirrosyllis mariae; E, parapodial glands, *Sphaerosyllis capensis*. A, B, San Martín (2003);
C, San Martín et al. (2008a); D, San Martín & Hutchings (2006); E, San Martín (2005).

In some genera of the subfamily Exogoninae (*Brania* Quatrefages, 1865, *Sphaerosyllis*
Claparède, 1863, *Parapionosyllis* Fauvel, 1923) there is a typical kind of glands, the
parapodial glands, which might show rod, granular, or hyaline material. In *Sphaerosyllis*, the
parapodial glands are connected to a special papilla, opened by one pore (fig. 7E) described
for *S. capensis* Day, 1953 by San Martín (2005). San Martín (2003) took some detailed SEM
pictures of pores of parapodial glands in some species of *Parapionosyllis*, as *P. labronica*
Cognetti, 1965, *P. brevicirra* Day, 1954, *P. elegans* (Pierantoni, 1903), *P. minuta* (Pierantoni,
1903), and *P. cabezali* (Parapar, Moreira & San Martín, 2000). These pores are very well
defined, with a thickened, circular area around, and a digitiform, eversible and contractile
structure inside, even one SEM image shows one of these digitiform structures with
glandular material emerging from the body (fig. 8, below).

3.1.7 Other structures

Additionally, SEM, has provided the possibility to find some enigmatic structures, such as
transversal lines of spines or papillae in some species of *Trypanosyllis* (San Martín, 2003;
Nogueira & Fukuda, 2008; San Martín et al, 2008a; Ramos et al., 2010; Salcedo-Oropeza et al.,
2011) (fig. 9A), a detailed arrangement of crests in *Xenosyllis* (San Martín, 2003; San Martín et

al., 2008b) (fig. 7C), dorsal glands in *Syllis pulvinata* (Langerhans, 1881), ciliated areas close to prostomium in *Sacconereis* stolons (San Martín, 2003) (fig. 9D), terminal papillae in palps of some species of *Syllides* and *Streptosyllis*, (San Martín, 2003; San Martín & Hutchings, 2006), subcirral papillae in some species of *Paraehlersia* (figs. 9B, C), etc. Furthermore, SEM has been used to take detailed photos of internal structures, as the proventricle, a distinct muscular and glandular structure of the gut, after dissection, formed by columns ending in hexagons (fig. 9E), which were photographed in SEM by San Martín (2003), San Martín et al., 2008a, 2010, Martín et al., 2009).

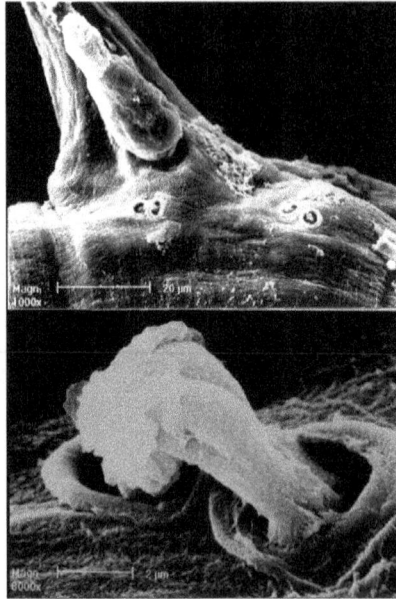

Fig. 8. Parapodial glands of *Parapionosyllis brevicirra* (above), detail of a papilla with glandular material (below). San Martín (2003).

3.2 Biology and reproduction

Some aspects of the biology has been also remarked using SEM, such is the case of specimens of the genus *Haplosyllis* crawling through the galleries inside sponges (Magnino et al., 1999 a, b), or specimens produced by asexual reproduction (Lattig et al., in press b), or in *Procerastea*. However, most important contributions of SEM to the knowledge of syllids are related with reproductive aspects, well documented by means of numerous pictures in several papers cited below. We have already mentioned the SEM photos of young *E. hebes* emerging from the mother's body (Pocklington & Hutchenson, 1983), stolons (Heacox, 1980), and genital especial chaetae (Westheide, 1990); after these papers, others also dealt with same or similar topics, as Qian & Chia (1989) for the larval development of one species of *Myrianida*.

Many interesting contributions were made on the external gestation of Exogoninae; Küper & Westheide (1998) demonstrated, including SEM photos, that in the genera *Prosphaerosyllis*

and *Salvatoria*, the eggs are attached by means of tiny, simple notochaetae which penetrate
into the eggs, also examined posteriorly in other species of the same genera by ourselves,
and in species of the genera *Erinaceusyllis* San Martín, 2005 (San Martín, 2003; 2005) (figs. 10
A, B). Especially interesting were the SEM photos taken to the new Australian genus and
species *Nooralia bulgannabooyanga* San Martín, 2002, because it is up to date, the unique syllid
which develops compound notochaetae for dorsal brooding of eggs (San Martín, 2002).
Ventral brooding of eggs and development of juveniles attached to females in some species
were recorded and photographied by San Martín (2003), Mastrodonato et al. (2003),
Nogueira et al. (2004), and Böggemann & Purschke (2005) (figs. 10C, D). SEM pictures were
also very useful to the study of reproductive stolons in the subfamily Syllinae, as in San
Martín (2003), San Martín & Nishi (2003), San Martín et al. (2008a,b, 2010), and Nogueira &
Fukuda (2008).

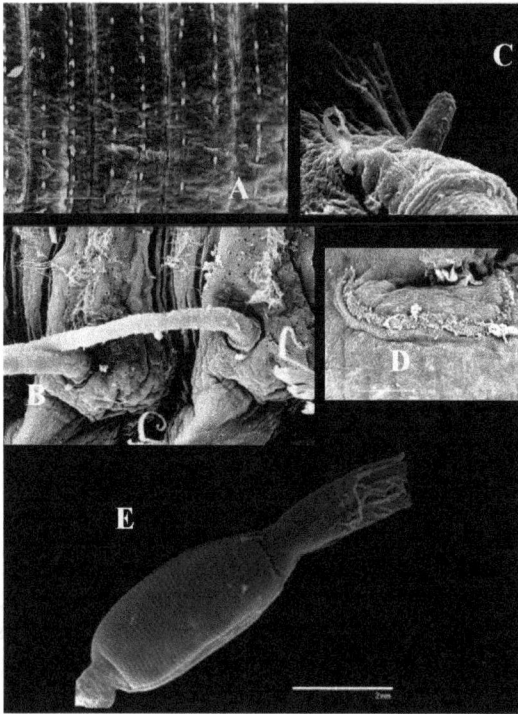

Fig. 9. A, transversal rows of spines, *Trypanosyllis aeolis*. B, subcirral papilla (under bases of
dorsal cirri), *Paraehlersia ehlersiaeformis*. C, detail of subcirral papilla, *Paraehlersia
weissmannioides*. D, ciliated area (nuchal organs?) of a *Sacconereis* stolon. E, pharynx and
proventricle, *Megasyllis inflata*. A, D, San Martín (2003); B, C, San Martín & Hutchings (2006);
E, San Martín et al. (2008a).

Especially interesting were some photos taken to stolons of the Australian species *Megasyllis
corruscans* (Haswell, 1885), showing pores on ventral bases of parapodia, never reported
before (fig. 10E). Also of great interest was the information taken on stolons in the complex
of species of the genus *Haplosyllis* (see below).

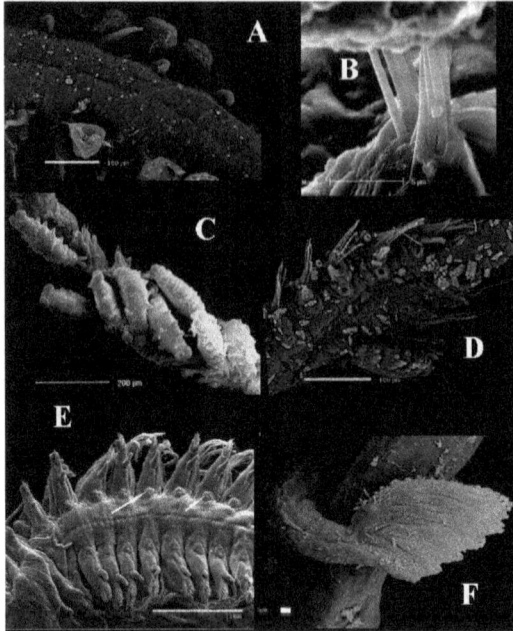

Fig. 10. A, dorsal brood of eggs by means of capillary notochaetae, *Prosphaerosyllis sexpapillata*. B, detail of capillary notochaetae inserting into an egg, *Salvatoria vieitezi*. C, juveniles attached to a female of *Exogone verugera*. D, juvenile attached to a female of *Exogone africana*, showing nephridial pores. E, ventral view of an stolon of *Megasyllis corruscans*, showing pores on parapodial bases. F, distal end of a dorsal cirri, *Murrindisyllis kooromundroola*. A, D, San Martín (2005); B, C, San Martín (2003); E, San Martín et al. (2008 a). F, San Martín et al. (2007).

3.3 Taxonomy

All these observations, discoveries, and more detailed descriptions of characters, thanks to SEM had important consequences, not only descriptions of new structures, opening new topics for future research, but also in the taxonomy and systematics of this complicate family of polychaetes. Detailed descriptions, using SEM, and comparison of different populations of supposed species of large distribution can help to differentiate cryptic species. The genus *Haplosyllis* has been recently revised, comparing specimens from all around the world, and using SEM techniques for all characters, especially the chaetae, apparently identical or very slightly different; however, the comparison of SEM photos show important differences (fig. 11). Before these studies, most of the reports all around the world of this genus were attributed to the species *Haplosyllis spongicola*, but that species is actually only present in the European seas, and with doubts in Australia. This study produced a number of papers (Martín et al., 2002, 2003; Lattig & Martín, 2009, in press a, b; Lattig et al., 2007, 2010 a, b) and the descriptions of numerous new species. Similarly, the genus *Haplosyllides* was also revised with similar results (Martín et al., 2009). Revisions of other genera of Syllidae are currently in process.

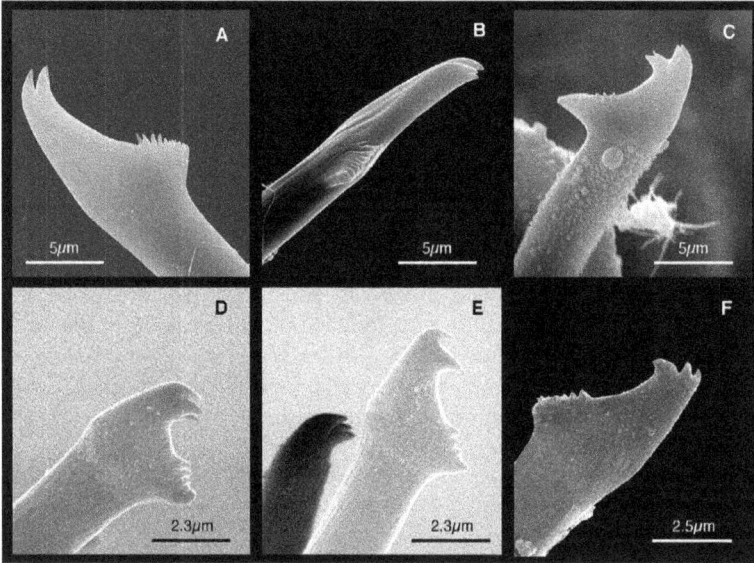

Fig. 11. Details of chaetae in different species of the genus *Haplosyllis*, showing the minute details, very difficult to see under light microscope. Martin et al. (2003).

3.4 Phylogeny

All these observations, above explained, under SEM on numerous species of syllids made possible the description of a number of new characters, previously unknown, as well as a better understanding of other features already known. This had important consequences in taxonomy, as already explained, and also on the phylogeny of the family. Sharp and detailed SEM photos are extremely useful to illustrate the characters used for phylogenetic analysis, as done by Nygren (1999), Aguado & San Martín (2009), and Aguado et al. (in press). Nygren (1999) used one plate with six SEM photos to illustrate 8 characters, and Aguado & San Martín (2009) 5 plates with a total of 33 SEM photos, in which ciliation of segments, body shape, ornamentation (papillae, tubercles, crests, among others) nuchal organs and details of nuchal lappets, details of peristomium and prostomium, size and shape of antennae and cirri, palps, parapodia, chaetae, several kinds of glands, pharyngeal armature, kinds of reproduction, etc., were profusely illustrated. The increasement on the number of characters and detailed descriptions supported more robust hypotheses about evolutionary relationships of syllids. The information of new morphological traits or new details about well known structures has been extremely useful to provide evidence for synapomorphies of some clades. For instance, the glands in anterior ventral cirri highly contributed to support a clade with the genera *Eurysyllis*, *Plakosyllis* and *Trypanosyllis*; a character that is only perceptible under SEM.

4. Perspectives and conclusions

The use of SEM in the study of syllids has been proved of high usefulness, showing characters not previously described and for giving appropriate descriptions of other already known. However, since it is also possible the discoverment of new taxa in wide unprospected areas of the world, and a detailed study under SEM still lacks for numerous syllid genera, new morphological structures may still be undescribed. For instance, San Martín et al. (2007) described a new genus and species from Australia (*Murrindisyllis kooromundroola*) with numerous autopomorphic characters. The most shocking characteristic is the tip of dorsal cirri, ending in a structure amazingly similar to a hand, with five "fingers" jointed by a membrane, giving a frog leg appearance (fig. 10F). These and other unusual characters of this genus were carefully detailed with SEM photos. At this point, the use of SEM is absolutely necessary for providing good and detailed descriptions of new taxa and numerous already known genera for which a detailed examination is still lacking. However, the use of SEM for the study of syllids is not the unique tool for improving the knowledge of the family; other techniques are absolutely necessary for phylogenetic studies in this family, such as molecular information, as shown by Aguado et al. (in press) who reorganizes the classification of the family using combined molecular and morphological data. Similarly, a worldwide catalogue for all the species of the family Syllidae is currently in process, in which SEM pictures acquire crucial importance.

5. Acknowledgements

We wish express our gratitude to the Editorial Board of the publisher,Open Access Publisher, and especially to Ms. Daria Nahtigal, for their kind offer to participate in this book and for all the given facilities. We wish to acknowledge to Fauna Ibérica, JMBA, Recods of the Australian Museum, Zootaxa, and Springer Verlag, for allow us to reproduce the SEM photos, previously published. Contribution to the project no. CGL2009-12292 BOS funded by the Ministerio de Ciencia e Innovación of the Spanish Government.

6. References

Aguado, M. T. & San Martín, G. 2007. Syllidae (Polychaeta) from Lebanon with two new reports for the Mediterranean Sea. *Cahiers of Biologie Marine*, 48: 207-224.

Aguado, M. T. & San Martín, G. 2008. Re-description of some genera of Syllidae (Phyllodocida: Polychaeta). *Journal of Marine Biological Association United Kingdom* 88(1): 35-56.

Aguado, M.T. & San Martín, G. 2009. Phylogeny of Syllidae (Annelida, Phyllodocida) based on morphological data. *Zoologica Scripta*, 38 (4): 379-402.

Aguado, M. T., San Martín, G. & Nishi, E. 2006. Two new species of Syllidae (Polychaeta) from Japan. *Scientia Marina*, 70: 9-16.

Aguado, M. T., San Martín, G. & Siddall, M. E. in press. Systematics and evolution of syllids (Syllidae, Annelida). *Cladistics*.

Álvarez, P. & San Martín, G. 2009. A new species of *Sphaerosyllis* (Annelida: Polychaeta: Syllidae) from Cuba, with a list of syllids from the Guanahacabibes Biosphere Reserve (Cuba). *Journal of Marine, Biological Association United Kingdom*, 1-9.

Beesley, P. L., Ross, G. J. B. & Glasby, C. J. 2000. *Polychaetes & Allies: the Southern Synthesis. Fauna of Australia. Vol. 4A. Polychaeta, Myzostomida, Pogonophora, Echiura, Sipuncula.* CSIRO Publishing: Melbourne XII, 465 pp.

Böggemann, M. & Purschke, G. 2005. Abyssal benthic Syllidae (Annelida: Polychaeta) from
    the Angola Basin. *Organism, Diversity & Evolution* 5: 221-226.
Capa, M., San Martín, G. & López, E. 2001. Syllinae (Syllidae: Polychaeta) del Parque
    Nacional de Coiba (Panamá). *Revista de Biología Tropical*, 49 (1): 103-115.
Heacox, A. E. 1980. Reproduction and Larval Development of *Typosyllis pulchra* (Berkeley &
    Berkeley) (Polychaeta: Syllidae). *Pacific Science*, 34 (3): 245-259.
Küper, M. & Westheide, W., 1997. Ultrastructure of the male reproductive organs in the
    interstitial annelid *Sphaerosyllis hermaphrodita* (Polychaeta, Syllidae). *Zoomorphology
    (Berlin)*, 117: 13-22.
Lanera, P., Sordino, P. & San Martín, G., 1994. *Exogone (Parexogone) gambiae*, a new species of
    Exogoninae (Polychaeta, Syllidae) from the Mediterranean Sea. *Bolletino di Zoologia*,
    61: 235-240.
Lattig, P. & Martín, D. 2009. A taxonomic revision of the genus *Haplosyllis* Langerhans, 1887
    (Polychaeta: Syllidae: Syllinae). *Zootaxa*, 2220 : 1-40.
Lattig, P. & Martín, D. in press a. Two new endosymbiotic species of *Haplosyllis* (Polychaeta :
    Syllidae) from the Indian Ocean and Red Sea, with new data on *H. djiboutienesis*
    from the Persian Gulf. *Italian Journal of Zoology*.
Lattig, P. & Martín, D. in press b. Sponge associated *Haplosyllis* (Polychaeta : Syllidae :
    Syllinae) from the Caribbean Sea, with the description of four new species. *Scientia
    Marina*.
Lattig, P., San Martin, G. & Martín, D. 2007. Taxonomic and morphometric analyses of the
    Haplosyllis spongicola complex (Polychaeta: Syllidae: Syllinae) from the Spanish seas,
    with re-description of the type species and description of two new species. *Scientia
    Marina* 71(3): 551-570.
Lattig, P., Martín, D. & San Martín, G. 2010a. Syllinae (Syllidae: Polychaeta) from Australia.
    Part 4. The genus *Haplosyllis* Langerhans, 18789. *Zootaxa* 2552: 1-36.
Lattig, P., Martín, D. & Aguado, M. T., 2010b. Four new species of *Haplosyllis* (Polychaeta :
    Syllidae : Syllinae) from Indonesia. *Journal of the Marine Biological Association of the
    United Kingdom*, 90 (4): 789-798.
Lewbart, G. A. & Riser, N. W. 1996. Nuchal organs of the polychaete *Parapionosyllis manca*
    (Syllidae). *Invertebrate Biology* 115 (4): 286-298.
Licher, F. & Küper, M., 1998. *Typosyllis tyrrhena* (Polychaeta, Syllidae, Syllinae), a new
    species from the island of Elba, Tyrrhenian Sea. *The Italian Journal of Zoology
    (Modena)*, 65: 227-233.
Licher, F., Ding, Z., Fiege, D. & Sun, R. 1995. Redescription of *Typosyllis magnipectinis*
    (Storch, 1967) from the South China Sea. *Senckenbergiana maritime*, 25 (4/6): 107-
    113.
López, E., Britayev, T., Martín, D. & San Martín, G. 2001. New symbiotic associations
    involving Syllidae (Annelida: Polychaeta), with taxonomic and biological remarks
    on *Pionosyllis magnifica* and *Syllis* cf. *armillaris*. *Journal of the Marine Biological
    Association of the United Kingdom*, 81: 399-409.
Magnino, G., Pronzato, R., Sarà, A. & Gaino, E. 1999a. Fauna associated with the horny
    sponge *Anomoianthella lamella* Pulitzer-Finali & Pronzato, 1999 (Ianthellidae,
    Demospongiae) from Papua New-Guinea. *Italian Journal of Zoology*, 66: 175-181.

Magnino, G., Sarà, A., Lancioni, T. & Gaino, E. 1999b. Endobionts of the coral reef sponge *Theonella swinhoei* (Porifera, Demospongiae). *Invertebrate Biology*, 118 (3): 213-220.

Martín, D., Núñez, J., Riera, R. & Gil, J. 2002. On the association between *Haplosyllis* (Polychaeta: Syllidae) and gorgonians (Cnidaria: Octocorallia), with the description of a new species. *Biological Journal of the Linnean Society*, 77: 455-477.

Martín, D., Britayev, T., San Martín, G. & Gil, J., 2003. Inter-population variability and character description in the sponge associated *H. spongicola* species-complex (Polychaeta: Syllidae). *Hydrobiologia*, 496:145-162.

Martin, D., Britayev, T.A. & Aguado, M.T. 2009. Review of the Symbiotic genus *Haplosyllides* (Polychaeta, Syllidae), with description of a New Species. *Zoological Science*, 26: 646-655.

Martínez, J., I. Adarraga & San Martín, G. 2002. *Exogone (Exogone) mompasensis* (Exogoninae: Syllidae: Polychaeta), a new species from the Iberian Peninsula. *Proceedings of the Biological Society of Washington*, 115 (3): 676-680.

Mastrodonato, M., Sciscioli, M., Lepore, E., Gherardi, M., Giangrande, A., Mercati, D., Dallai, R. & Lupetti, P. 2003. External gestation of *Exogone naidina* Örsted, 1845 (Polychaeta, Syllidae): ventral attachement of eggs and embryos. *Tissue & Cell*, 35: 297-305.

Musco, L., Çinar, M. E. & Giangrande, A. 2005. A new species of *Sphaerosyllis* (Polychaeta, Syllidae, Exogoninae) from the coasts of Italy and Cyprus (eastern Mediterranean Sea). *Italian Journal of Zoology*, 72: 161-166.

Nogueira, J. M. & Fukuda, M. V. 2008. A new species of *Trypanosyllis* (Polychaeta: Syllidae) from Brazil, with a redescription of Brazilian material of *Trypanosyllis zebra*. *Journal of the Marine Biological Association of the United Kingdom*, 88 (5): 913-924.

Nogueira, J. M. & San Martín, G., 2002. Species of *Syllis* Lamarck, 1818 (Polychaeta: Syllidae) living in corals in the state of Sao Paulo, southeastern Brazil. *Beaufortia*, 52 (7): 57-93.

Nogueira, J. M., G. San Martín & Fukuda, M. V. 2004. On some exogonines (Polychaeta, Syllidae, Exogoninae) from the northern coast of the State of São Paulo, southeastern Brazil Results of BIOTA/FAPESP/Bentos Marinho Project. *Meiofauna Marina*, 13: 45-77.

Nygren, A. 1999. Phylogeny and reproduction in Syllidae (Polychaeta). *Zoological Journal of the Linnean Society*, 126: 365-386.

Olivier, F., Grant, C., San Martín, G., Archambault, P. & McKindsey, C. V. 2011. Syllidae (Annelida: Polychaeta: Phyllodocida) from the Chausey Archipelago (English Channel, France), with a description of two new species of the Exogoninae *Prosphaerosyllis*. *Marine Biodiversity*, 12526: 92.101.

Pawlick, J. R., 1983. A sponge-eating worm from Bermuda: *Branchiosyllis oculata* (Polychaeta, Syllidae). *Marine Ecology*, 4: 65-79.

Pocklington, P. & Hutchenson, M. S. 1983. New record of viviparity for the dominant benthic invertebrate *Exogone hebes* (Polychaeta: Syllidae) from the Grand Banks of Newfoundland. *Marine Ecology Progress Series*, 11: 239-244.

Qian, P.Y. & Chia, F. S. 1989. Larval development of *Autolytus alexandri* Malmgren, 1867 (Polychaeta, Syllidae). *Invertebrate Reproduction and Development*, 15: 49-56.

Ramos, J., San Martín, G. & Sikorski, A. 2010. Syllidae (Polychaeta) from the Arctic and sub-Arctic regions. *Journal of the Marine Biological Association of the United Kingdom*: 1-10.

Rouse, G. & Fauchald, K. 1997. Cladistics and polychaetes. *Zoologica Scripta*, 23: 271-312.

Rouse, G. & Pleijel, F. 2001. *Polychaetes*. Oxford University Press, 354 pp.

Salcedo-Oropeza, D. L., San Martín, G. & Solís-Weiss, V. 2011. Two new species of Syllidae (Annelida: Polychaeta) from the Southern Mexican Pacific. *Zootaxa*, 2800: 41–52.

San Martín, G. 2002. A new genus and species of Syllidae (Polychaeta) from Australia brooding eggs dorsally by means of compound notochaetae. *Proceedings of the Biological Society of Washington* 115(2): 333-340.

San Martín, G. 2003. *Annelida Polychaeta II. Syllidae*. IN: Fauna Ibérica, vol. 21. Ramos, M. A. *et al.* (EDS). Museo Nacional de Ciencias Naturales. CSIC. Madrid, 554 pp.

San Martín, G. 2005. Exogoninae (Polychaeta: Syllidae) from Australia, with the description of a new genus and twenty-two new species. *Records of the Australian Museum*, 57 (1): 39-152.

San Martín, G. & Hutchings, P. 2006. Eusyllinae (Polychaeta, Syllidae) from Australia, with the description of a new genus and fifteen new species. *Records of the Australian Museum* ,58: 257-370.

San Martín, G. & López, E. 2003. A new genus of Syllidae (Polychaeta) from Western Australia. *Hydrobiologia* 496: 191-197.

San Martín, G. & Nishi, E. 2003. A New species of *Alcyonosyllis* Glasby and Watson, 2001 (Polychaeta: Syllidae: Syllinae) from Shimoda, Japan. Commensal with the Gorgonian *Melithaea flabellifera*. *Zoological Science*, 20: 371-375.

San Martín, G., Aguado, M. T. & Murray, A. 2007. A new genus and species of Syllidae (Polychaeta) from Australia with unusual morphological characters and uncertain systematic position. *Proceedings of the Biological Society of Washington*, 120 (1): 39-48.

San Martín, G., Hutchings, P. & Aguado, M. T. 2008a. Syllinae (Polychaeta, Syllidae) from Australia. Part. 2. Genera *Inermosyllis*, *Megasyllis* n. gen., *Opisthosyllis*, and *Trypanosyllis*. *Zootaxa*, 1840: 1-53.

San Martín, G., Hutchings, P. & Aguado, M. T. 2008b. Syllinae (Polychaeta, Syllidae) from Australia. Part. 1. Genera *Branchiosyllis*, *Eurysyllis*, *Karroonsyllis*,*Parasphaerosyllis*, *Plakosyllis*, *Rhopalosyllis*, *Tetrapalpia* n. gen., and *Xenosyllis*. *Records of Australian Museum*, 60(2): 119–160.

San Martín, G., Hutchings, P. & Aguado, M. T. 2010. Syllinae (Polychaeta: Syllidae) from Australia. Part 3. Genera *Alcyonosyllis*, Genus A, *Parahaplosyllis*, and *Trypanosyllis* (*Trypanobia*). *Zootaxa*, 2493: 35-48.

Sardá, R. & San Martín, G. 1992. Systematic description of *Syllides verrilli* as *Streptosyllis verrilli* (Moore, 1907). Life cycle, population of a salt marsh in Southern England (MA, USA). *Bulletin of Marine Science*, 51 (3): 407-419.

Tovar-Hernández, M. A., Granados-Barba, A. & Solís-Weiss, V. 2002. *Typosyllis papillosus*, a new species (Annelida: Polychaeta: Syllidae) from the southwest Gulf of Mexico. *Proceedings of the Biological Society of Washington*, 115 (4): 760-768.

Westheide, W. 1990. A hermaphroditic *Sphaerosyllis* (Polychaeta: Syllidae) with epitokous genital chaetae from intertidal sands of the Island of Phuket (Thailand). *Canadian Journal of Zoology*, 68: 2360-2363.

9

The Application of Scanning Electron Microscope (SEM) to Study the Microstructure Changes in the Field of Agricultural Products Drying

Hong-Wei Xiao and Zhen-Jiang Gao
College of Engineering, China Agricultural University, Qinghua Donglu, Beijing, China

1. Introduction

The objective of this part: Highlight the significance of microstructure investigation in the field of agricultural product drying or dehydration.

It is a common sense that structure of material determines its function. The change of macroscopic properties of materials is caused by the changes of its microstructure. For example, a porous structure as a honeycomb would facilitate rapid water diffusion or promote a rapid water uptake during drying or cooking. On the contrary, a compact structure or fewer pores at the surface of the product can cause a slower moisture migration during drying or water penetration into the interior during rehydration or cooking. Therefore, microstructure investigation can help quantifying product changes during processing and may also improve the understanding of mechanisms and changes in quality factors, especially the changes in food texture (**Aguilera & Stanley, 1999; Xiao et al., 2009**). For example, the pore sizes and the number of pores can significantly influence the texture of food. Smaller number of pores and small sizes led to the dense structure. While, larger number of pores and large pore size can cause a decrease of the hardness of the product.

2. The classification of the specific research

The objective of this part: Through concrete examples to introduce the main content and conclusions of the microstructure investigation in the field of agricultural materials drying.

2.1 Observing the microstructure of the material surface before and after the processing to find out the effect of drying on the product microstructure

The scanning electron micrographs on the surface of fresh and dried samples can be used to analyze the microstructure changes during drying or dehydration process. Fresh apple tissue has a well-organized structure consisting of cells and intercellular spaces, however, the breakdown of cell walls, a decreased intercellular contact and collapse of cell structure were found in the dried apple tissues (Deng &Zhao, 2008 a and b).

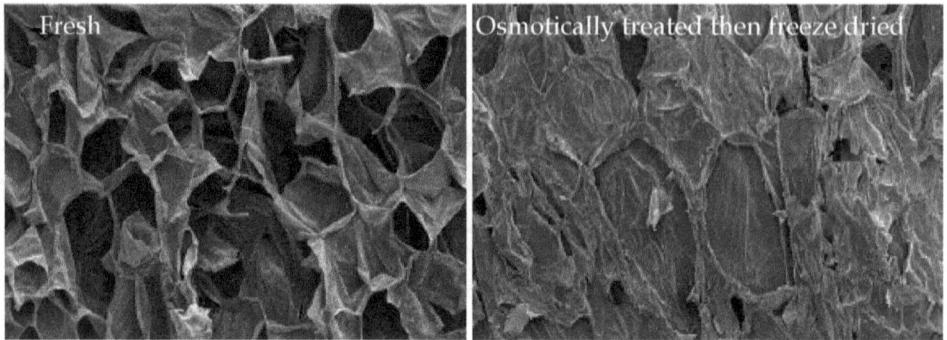

Fig. 1. Scanning electron micrograph of fresh and dried apple (Fuji)
(Deng & Zhao, 2008 a and b)

Take the pressure pulsed osmotic dehydration (PPOD) of salt eggs using NaCl solution for another example. Osmotic dehydration of salt eggs is an ancient method of egg preservation, which can be traced back to several hundreds of years. With special taste and flavour, salt eggs is one of the most popular egg food in Asian countries especially in China. Whereas, presently in China the osmotic dehydration of egg is done manually, the process of which is tedious, time consuming and labor intensive. The traditional osmotic dehydration process involves submerging eggs in NaCl solution in a static situation which last about 30 days at room temperature. During this process the NaCl transfer from salt solution through eggshell into the egg white and yolk. In order to increase the automation and decrease the processing time, the PPOD technology has been applied in salting eggs, which has been proved more efficient reducing the osmotic time from 30 days to 2 or 3 days (Chen and Gao, 2006; Wang and Gao, 2010). However, the mechanisms of PPOD of salt eggs hasn't be explored.

The authors try to find the mechanism using the microstructures of the eggshell before and after PPOD, as shown in Figure 2. From Figure 2, it can be found that the pore sizes and the number of pores in the surface of eggshell was increased after the processed of PPOD, which could significantly facilitate the NaCl immigration from the salt solution to the egg inside. As regards the cross section of eggshell, it can be observed that after PPOD the eggshell become less dense and a few pore channels were formed, which enabled the NaCl to penetrate easily from solution to egg inside and increased the moisture transport from egg to the solution, and thus accelerated the osmotic dehydration process. This phenomenon may be due to the tunneling effect of PPOD, which can create canals in the microstructure of eggshell during PPOD by stretching and enlarging the pore sizes. The membrane of eggshell is the most resistance for mass transfer during osmotic dehydration of salt egg (Chen et al., 1999). From Figure 2, it can be found that the eggshell membrane comprised of multi-layers of "fibers" as a bird's nest built of sticks. It can be also found that after PPOD the microstructure of eggshell membrane become looser. Certainly, such a loose structure can promote mass transport and improve the overall process rate compared to the dense structure.

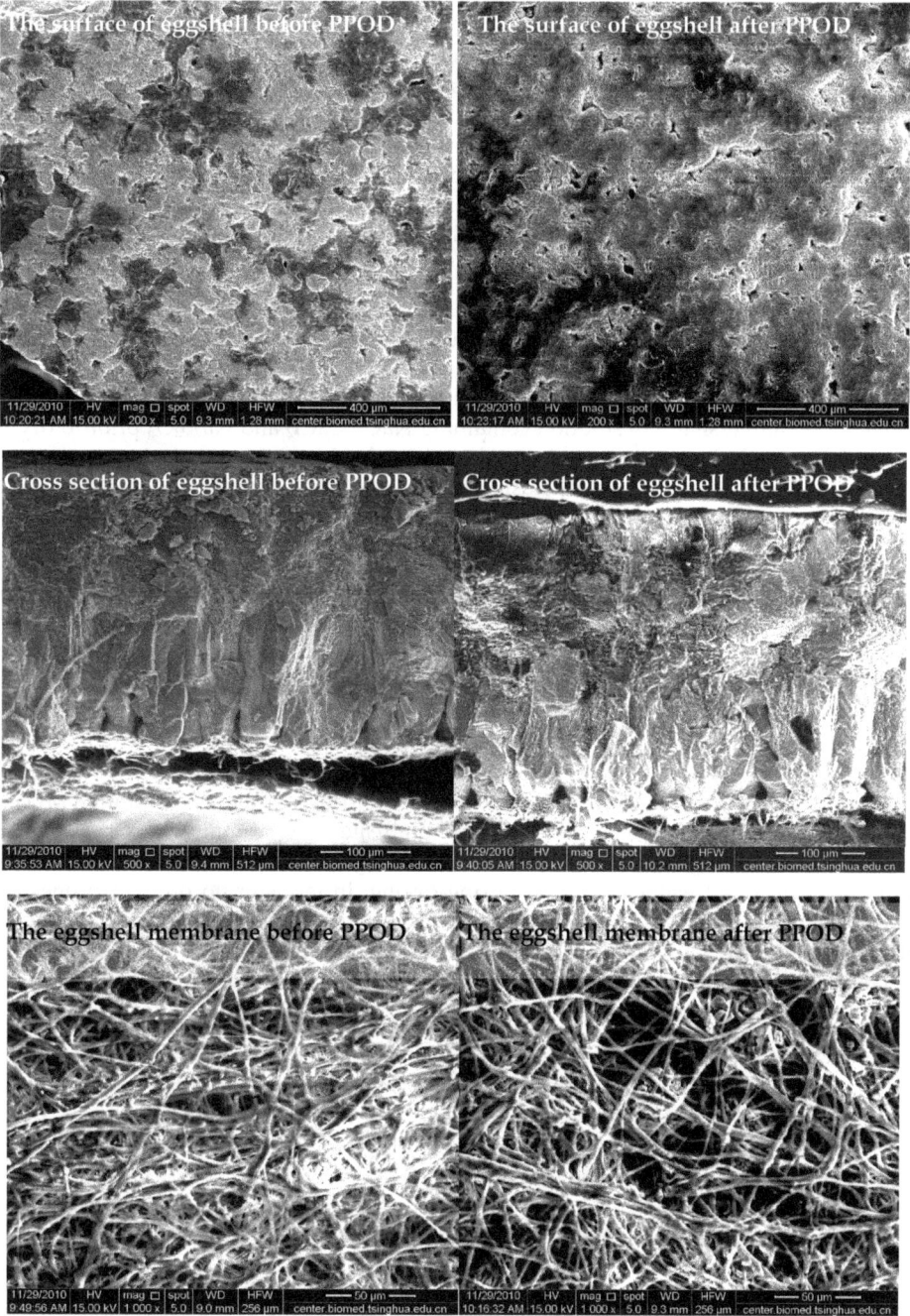

Fig. 2. Scanning electron micrograph of eggshell before and after the pressure pulsed osmotic dehydration (PPOD).

2.2 To investigate the effect of different pretreatment methods and drying conditions on the microstructure of the samples

The information on microstructure changes is essential for enabling better process control and improvement in the appearance by optimizing the pretreatment and drying parameters. The microstructure observation of the product's surface can be carried out to explore the effects of different pretreatment methods and drying temperatures on the microstructure changes of the samples. Recently, many researchers have investigated the microstructure changes of various fruits, vegetables and other food materials during pretreatment and drying process.

Vega-Gálvez et al. (2008) investigated the influence of pretreatment and air drying temperature on the quality and microstructure properties of rehydrated dried red bell pepper. Microscopic evaluation of the rehydrated pepper samples, as shown in **Figure3**, illustrated that the damage to cellular structure was minimized when the samples was pretreated by immersing in a solution containing NaCl, $CaCl_2$ and $Na_2S_2O_5$ prior to drying in comparison with the no pretreated samples. They also found that the drying temperature had an significant effect on the microstructure of the dried sample and the damage to cellular structure could be alleviated by decreasing the drying temperature.

Xiao et al. (2009) studied the effect of different pretreatments on drying kinetics and quality of sweet potato bars in terms of textural properties, microstructure, and colour undergoing air impingement drying. Microstructure observation of the surface of dried sweet potato bars, as shown in **Figure4**, was carried out to evaluate the effects of different pretreatments on the microstructure changes of the dried samples. It was found that when the sweet potato bars were subjected to hot water blanching and superheated steam blanching pretreatments, the dried samples had a homogeneous compact structure. In addition, no pores and starch granules were found on the surface of the samples. As a result, the structure would slower water transfer or penetration during drying or rehydration process. However, the samples subjected to citric acid pretreated for 30 min had large with non uniform pores and lots of starch granules on its surface. Absolutely, such porous structure would facilitate rapid water migration during drying. In terms of no pretreated ones, the dried sweet potato tissues showed more numerous starch granules and fewer pores than the citric acid pretreated samples on its surface. Therefore it is interesting to note that different pretreatments cause various changes of microstructure of the samples and lead to product properties varied differently.

Bondaruk, Markowski, Blaszczak (2007) investigated the effect of drying conditions on the quality of vacuum-microwave dried potato cubes in terms of colour, starch content, sugar content, mechanical properties and microstructure. Concerning the microstructure, it was observed that compared with the forced convection air drying the application of microwave energy led to different physical changes in the sample microstructure. It was also found that in the case of hot-air drying the intensity of structural changes depended on the drying temperature. In addition, a higher temperature causes greater damage to the microstructure of potato cubes.

Pimpaporn et al. (2007) reported that potato chips dried at 80 and 90°C had more uniform pore size and pore distribution compared with the chips dried at 70°C and more extensive surface shrinkage was found on the samples dried at 70°C. However, Fang et al. (2011)

reported that lower drying temperature led to relatively uniform size and shape with smooth particle surface, whereas higher drying temperature resulted in size variations and wrinkled particle surfaces when they carried out the milk spray drying under different drying temperatures. The SEM micrographs of milk powder under different spray drying temperatures were shown in **Figure5**.

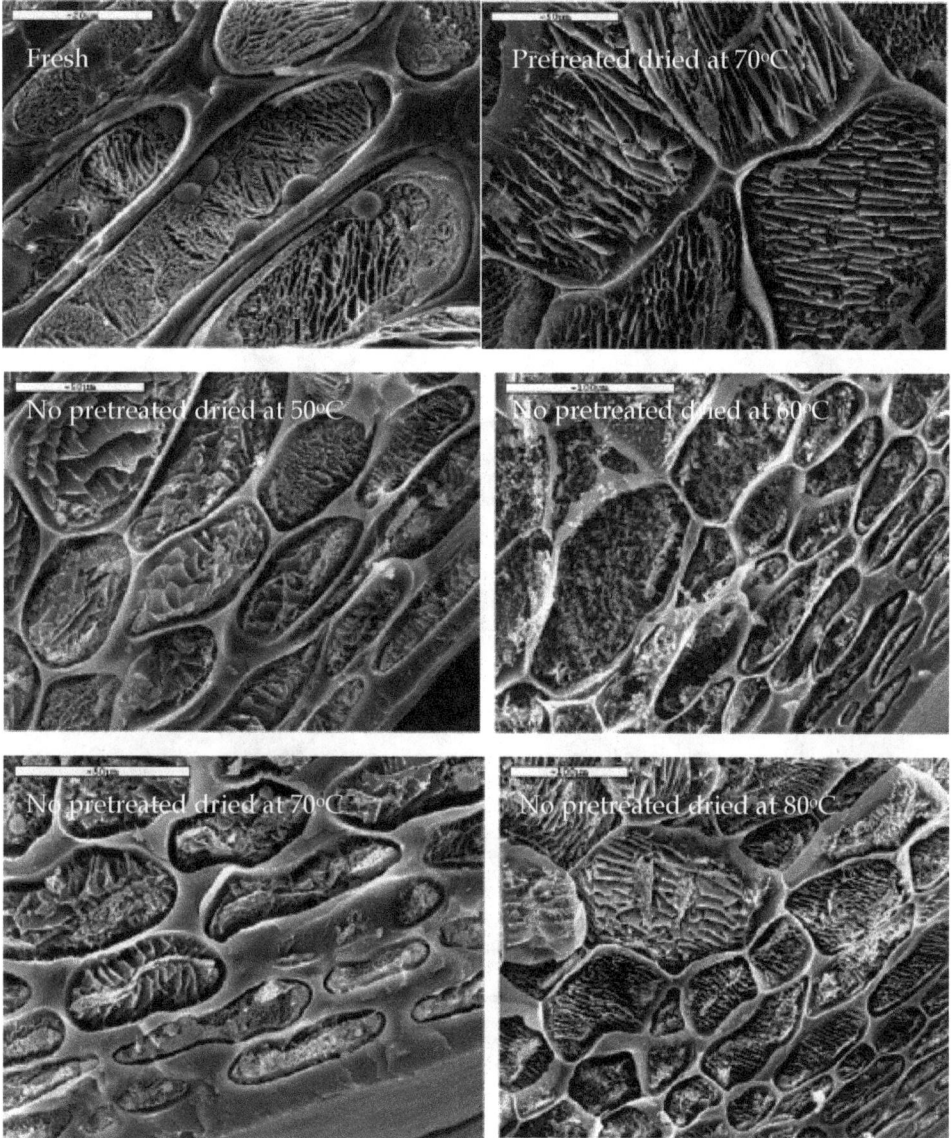

Fig. 3. SEM micrographs of fresh and rehydrated red pepper samples with and without pretreatment dried at different temperatures (Vega-Gálvez et al., 2008).

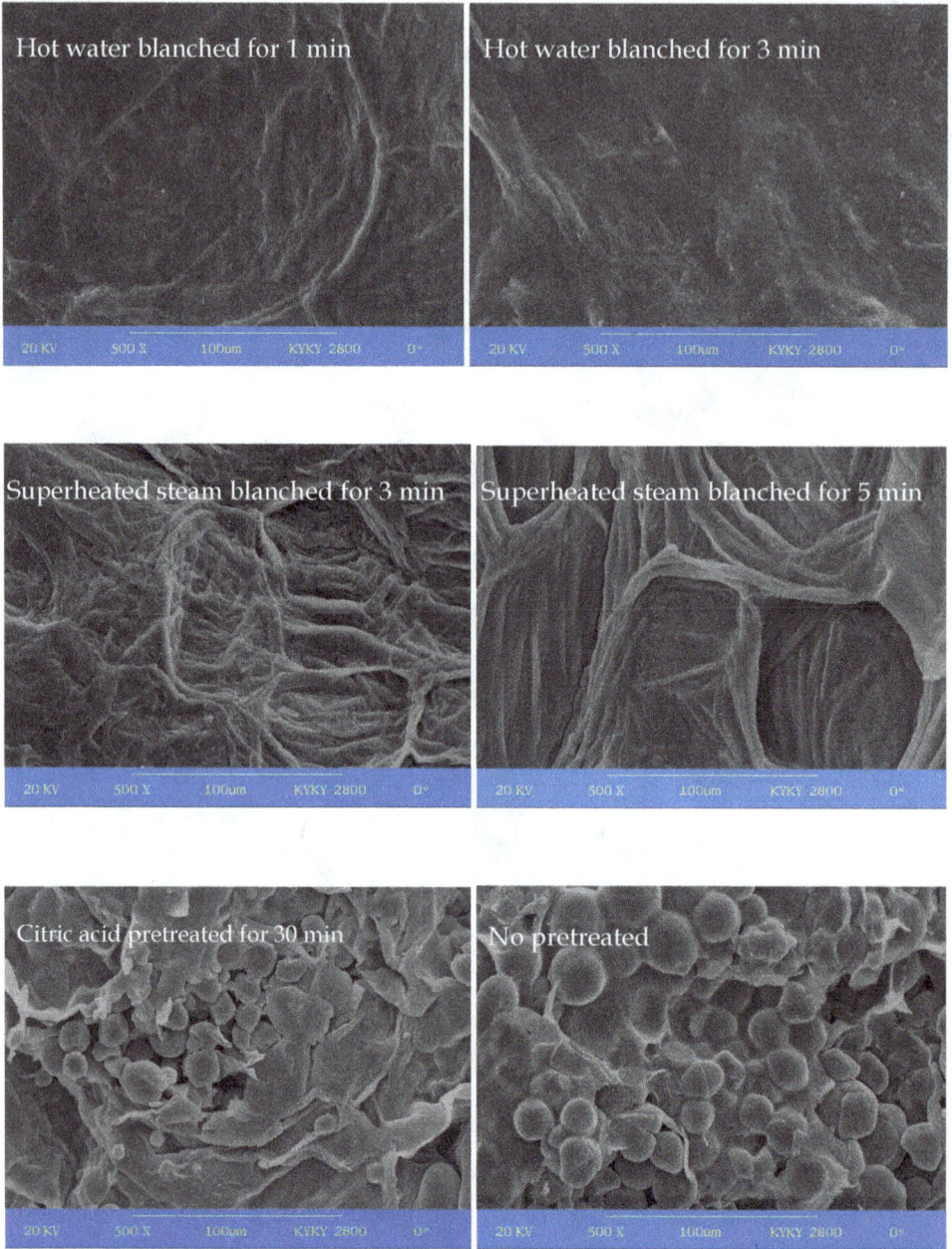

Fig. 4. SEM micrographs of the surface of dried sweet potato bars underwent different pretreatments (Xiao et al., 2009).

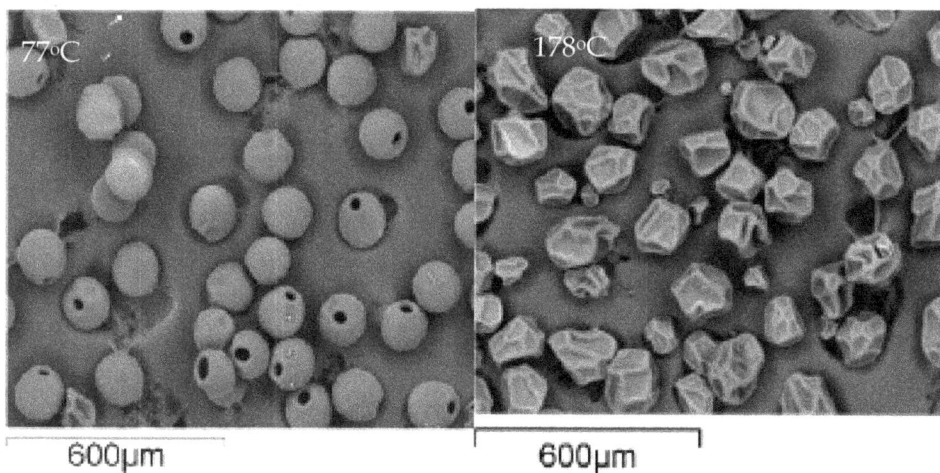

Fig. 5. SEM micrographs of milk powder under different spray drying temperatures
(Fang, e t al., 2011).

2.3 Using SEM micrographs of materials to analyze the moisture transfer mechanisms during drying process and interpret the rehydration characteristics or the texture properties of the dried products

In general, most of the drying occurs in the falling rate period and moisture migration controls the whole process. Due to the limited information on the mechanism of moisture movement during drying and the complexity of the process that may involve molecular diffusion, capillary flow, Knudsen flow, hydrodynamic flow, surface diffusion and all other factors which affect drying characteristics, the moisture transfer mechanism hasn't been described completely (Madamba et al., 1996). Knowledge of the microstructure in which moisture and heat transfer take place may assist in finding the mechanisms and their relative contributions to the transport phenomena. Yang et al. (2010) made a try when they carried out an experiment to investigate the influence of glutinous components in plant tissue on the drying characteristics of plant materials taking Chinese angelica and Astragalus slices as the samples.

From the SEM micrographs of the sample (as shown in **Figure6**), they found that the trachea with relative large pore diameter was surrounded by massive parenchyma cells. It implied that during the drying process there was two parallel ways for moisture transfer from parenchyma cells to surrounding drying media: the direct moisture diffusion through some pores or open structures on the surface layer and the moisture emigrate from the parenchyma cells inside matrix to the surrounding drying media. Further more, they also reported that the process of moisture transfer from the parenchyma cells inside matrix to the surrounding drying media included three steps: firstly from the parenchyma cell to the adjacent cell via plasmodesma; secondly from the parenchyma cell to the adhered trachea via aperture; thirdly from tranchea to the surrounding drying media.

Rehydration ratio and texture of the samples, which is the macro performance of the material microstructure, is strongly dependent on the product microstructure. Therefore, the SEM micrographs can be used to analyze, interpret or even predict the rehydration characteristics or the texture properties of the dried products.

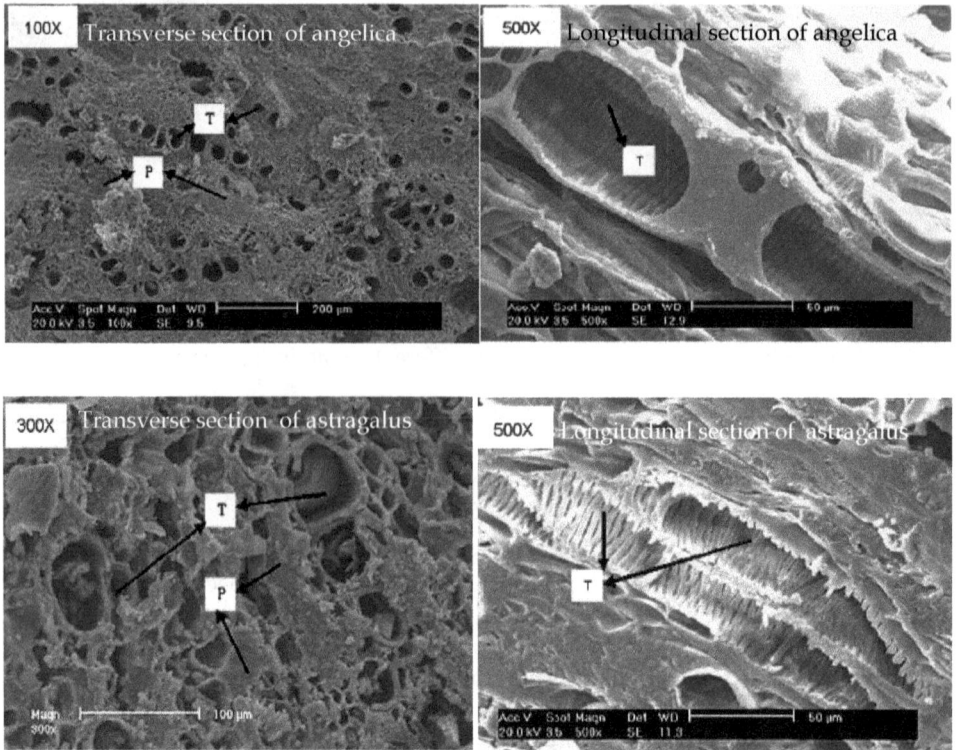

Fig. 6. The SEM micrographs of transverse section and longitudinal section of Chinese angelica slices and Astragalus sices (T part: trachea; P part: parenchyma cells). (Yang et al., 2010)

Thuwapanichayanan et al. (2011) studied the influence of drying temperatures on the moisture diffusivity and quality attributes of the dried banana slices in terms of volatile compound, shrinkage, colour, texture and microstructure. On the subject of microstructure, as shown in Figure7, they found that the drying temperature strongly affected the microstructure of dried banana and on the surface the pore sizes and the number of pores increased with increasing drying temperature, which significantly influenced the product texture in terms of hardness. The hardness of the samples dried at 90ºC was lower than that dried at 70ºC but was not significantly different from that dried at 80ºC. This might be due to the effect of puffing that occurred more at higher temperatures and probably increased the porosity and resulted in a decrease of hardness and less shrinkage of the samples.

Brown et al. (2008) evaluated the microstructure characteristics of carrot pieces that had
been dried using different techniques. They found that samples dried in ethano-modified
supercritical carbon dioxide possessed many pores which could facilitate the movement of
water into the internal structure and decrease the rehydrated and cooked time compared
with the air-dried samples. Recently, similar results has been reported by Yang et al. (2010),
who revealed that the larger porosity and total volume of the sample, the higher
rehydration ratio of the sample.

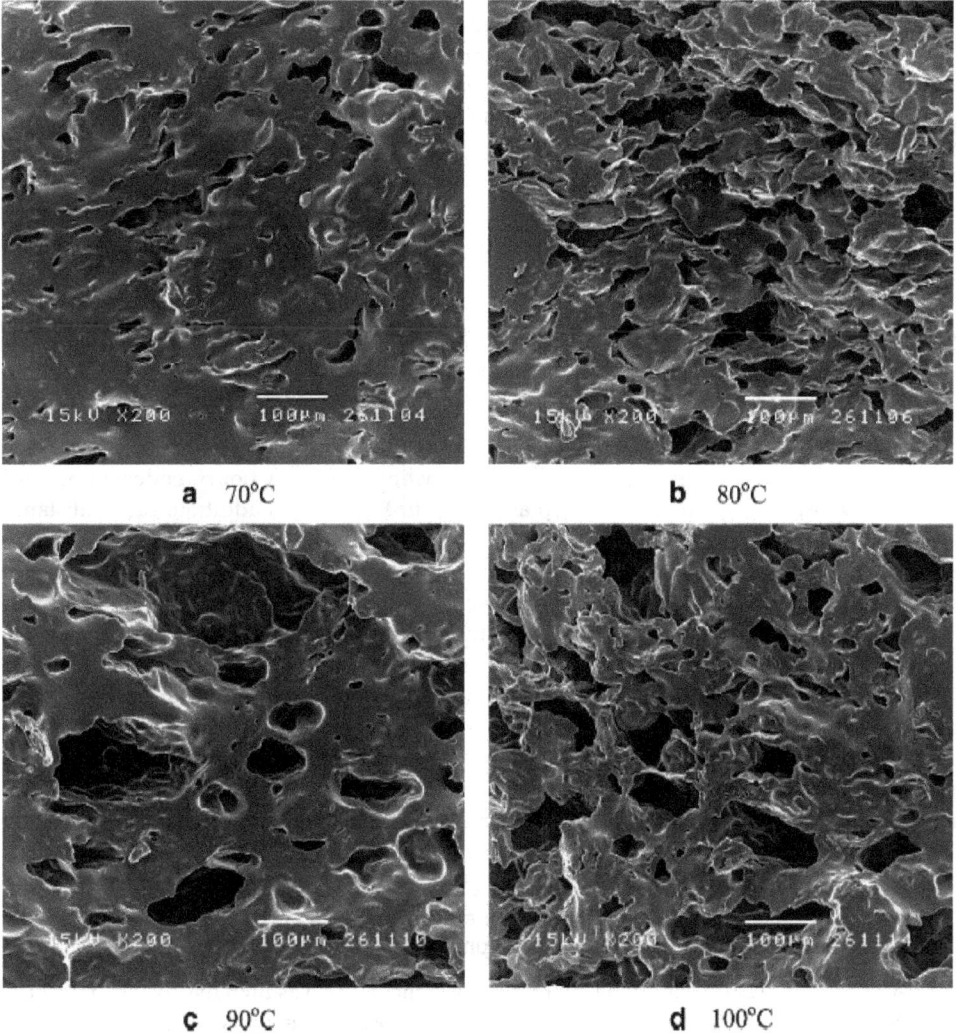

Fig. 7. SEM micrographs of banana slices dried at different temperatures
(Thuwapanichayanan et al., 2011).

3. The causes leading to microstructure changes

The objective of this part: Analysis of the reasons behind the microstructure change phenomenon through specific case

3.1 Thermal and moisture gradients can cause cell wall disruption, deformation and folding during drying process

Deng & Zhao (2008b) explored the effect of different osmoconcentration pretreatment on glass transition temperature, texture, microstructure and calcium penetration of dried apples. In terms of microstructure, which was observed using SEM, it was found that it impossible to distinguish between cells and other spaces because of the disruption of cell walls, a decreased intercellular contact and the collapse of cell structure. Furthermore, they pointed that this phenomenon might be due to the fact that during drying process the transient thermal and moisture gradients causing cell wall breakdown, deformation and folding. Additionally, they also reported that the structural deformation, folding and collapses of cell structure might also associated with surface tension, environment pressure, moisture transport mechanism, and generation of internal pressure in samples during drying.

3.2 The microstructure changes of samples is closely associated with the stress developed in the tissue, which may be set up by shrinkage

Drying is a simultaneous heat and mass transfer process, shrinkage takes place when the materials are heated and moisture is lost. Since the composition of materials is very difference, take corn as an example which is constituted of soft floury endosperm, hard vitreous endosperm, germ and pericarp as shown in **Figure 8**. In addition, each substance has different shrinkage characteristics. These fact leads to the occurrence of non-uniform shrinkage during drying process. The non-uniform shrinkage within the product results in two types of internal stresses: the thermal stress which is due to temperature gradients within the material and the hydro stress which is due to moisture gradients within the product. When the combination of thermal and hydro stress exceeds the binding force between cells of the material stress crack or burst phenomena occur, which can change the macrostructure and microstructure of the product during drying process. As Wang and Brennan (1995) pointed that during drying internal cracks are formed and shrinkage stresses pull the tissue apart. Lewicki & Pawlak (2003) also demonstrated that physical changes are mostly due to stress and are pronounced by macro and micro alterations of size, shape and internal structure.

3.3 Stress in cell walls, phase changes in membrane lipids and chemical changes can also cause structural modifications of the product

Vega-Gálvez et al. (2011) explored different pretreatments such as high hydrostatic pressure, blanching, enzymatic and microwaves on the microstructure of Aloe vera gel during convective drying at 70°C, as shown in **Figure9**. It was found that the intact cellular structure of aloe was transformed into a more separated and ruptured cellular structure with non-distinct middle lamella. They ascribed this effect to the degradation of pectinacious material and the damage of most cell wall due to excessive strain in membranes and stress in cell walls during processing.

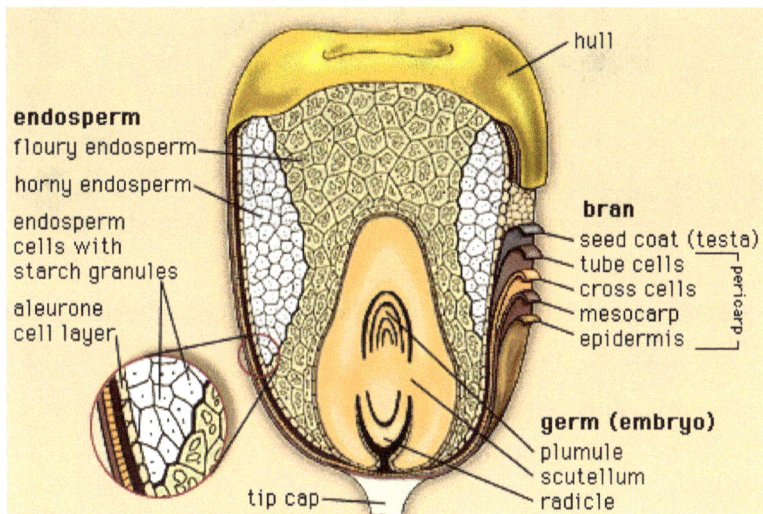

Fig. 8. The structure and compositon of corn kernel.

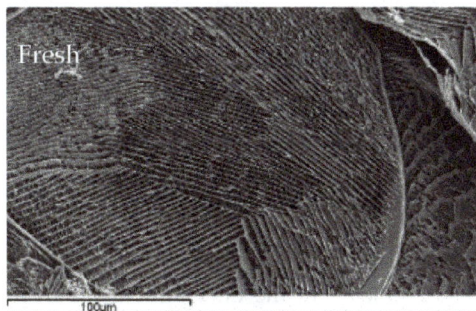

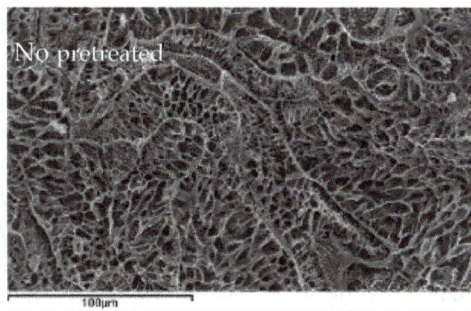

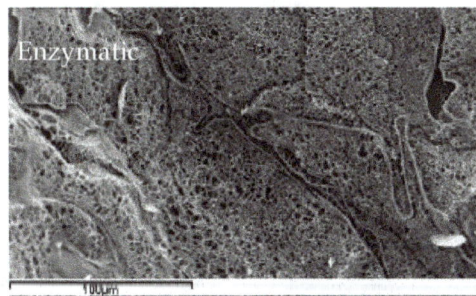

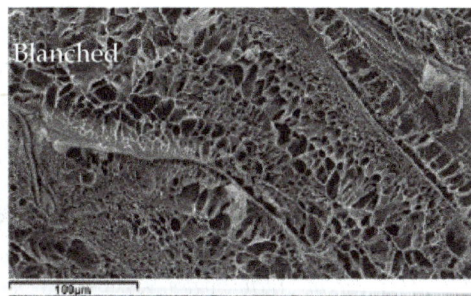

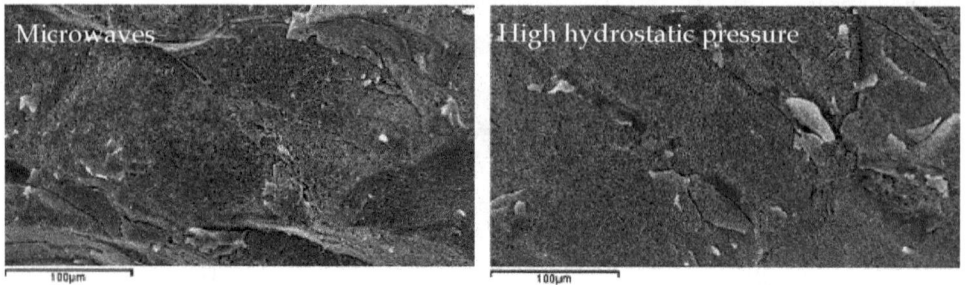

Fig. 9. The SEM micrographs of fresh and dried Aloe vera gel under different pretreatments (Vega-Gálvez et al., 2011).

4. Further research suggestion

The objective of this part: Point out the inadequacy of current research and the future research directions.

The microstructure of fresh and dried samples can provide a powerful tool and strong evidence for analyzing the properties changes of the samples during drying process. In this sense, research about the relationship between microstructure and the properties of the dried samples should be enhanced. In detail, more work needs to be performed on how the microstructure changes of dried food affects the food properties such as texture, rehydration ratio as well as its functionality, or even the availability of bioactive components.

Since information on microstructure change kinetics during products processing would be useful in predicting the quality changes during drying such as the texture and the surface shape of the product, thereby enabling better process control and improvement in the appearance by optimizing the process parameters. In addition, understanding the microstructure changes of the product is very important to clarify the change mechanism of quality during it processing. Therefore, the change kinetics of the microstructure during processing should be carried out in the further research.

5. References

Aguilera, J.M., & Stanley, D.W.(1999). Microstructural principles of food processing and engineering (second ed.). Gaithersburg: Aspen Publishers.

Bondaruk, J., Markowski, M., Blaszczak, W. (2007). Effect of drying conditions on the quality of vacuum-microwave dried potato cubes. *Journal of Food Engineering*, 81, 306-312.

Brown, Z.K., Fryer, P.J., Norton, I.T., Bakalis, S., Bridson, R.H. (2008). Drying of foods using supercritical carbon dioxide-Investigations with carrot. *Innovative Food Science and Emerging Technologies*, 9, 280-289.

Chen, S.T., Gao, Z.J. (2006). Study on the process of salted eggs under pulsed pressure. *Transaction of the CSAE*, 22, 163-166. (in Chinese with English abstract)

Chen, X.D., Freeman, Y., Guo, F., Chen, P. (1999). Diffusion of sodium chloride through chicken eggshell in relation to an ancient method of egg preservation. *Trans IchemE*, 77, Part C, 40-46.

Deng, Y., Zhao, Y.Y. (2008a). Effects of pulsed-vacuum and ultrasound on the osmodehydration kinetics and microstructure of apples *(Fuji)*. *Journal of Food Engineering*, 85, 84-93.

Deng, Y., Zhao, Y.Y. (2008b). Effect of pulsed vacuum and ultrasound osmopretreatments on glass transition temperature, texture, microstructure and calcium penetrationof dried apples (Fuji). LWT-Food Science and Technology, 41, 1575-1585.

Fang, Y., Rogers, S., Selomulya, C., Chen, X.D. (2011). Functionality of milk protein concentrate: effect of spray drying temperature. *Biochemical Engineering Journal*, doi:10.1016/j.bej.2011.05.007

Lewicki, P.P., & Pawlak, G. (2003). Effect of drying on microstructure of plant tissue. *Drying Technology*, 21, 657-683.

Madamba, P. S., Griscoll, R. H., & Buckle, K. A. (1996). The thin layer drying characteristics of garlic slices. *Journal of Food Engineering*, 29, 75 - 97.

Pimpaporn, P., Devahastin, S. and Chiewchan, N. (2007). Effects of combined pretreatments on drying kinetics and quality of potato chips undergoing low-pressure superheated steam drying. *Journal of Food Engineering*, 81, 318-329.

Thuwapanichayanan, R., Prachayawarakorn, S., Kunwisawa, J., Soponronnarit, S. (2011). Determination of effective moisture diffusivity and assessment of quality attributes of banana slices during drying. *LWT-Food Science and Technology*, 44, 1502-1510.

Vega-Gálvez, A., Lemus-Mondaca, R., Bilbao-Sáinz, C., Fito, P., Andrés, A. (2008). Effect of air drying temperature on the quality of rehydrated dried red bell pepper (var. Lamuyo). *Journal of Food Engineering*, 85, 42-50.

Vega-Gálvez, A., Uribe, E., Perez, M., Tabilo-Munizaga, G., Vergara, J., Garcia-Segovia, P., Lara, E., Scala, K.D. (2011). Effect of high hydrostatic pressure pretreatment on drying kinetics, antioxidant activity, firmness and microstructure of Aloe vera (Aloe barbadensis Miller) gel. LWT-Food Science and Technology, 44, 384-391

Wang, N., Brennan, J.G. (1995). Changes in structure density and porosity of potato during dehydration. *Journal of Food Engineering*, 24, 61 - 76.

Wang, X.T., Gao, Z.J. (2010). Technologic optimization of pickled salted eggs under pulsed pressure. *Food Science*, 31, 97-101.(in Chinese with English abstract)

Xiao, H.W., Lin, H., Yao, X.D., Du, Z.L., Lou, Z., Gao, ZJ. (2009). Effects of different pretreatments on drying kinetics and quality of sweet potato bars undergoing air impingement drying. *International Journal of Food Engineering*, 5, Article 5.

Yang, J.H., Di, Q.Q., Zhao, J. (2010). Effect of glutinous components on matrix microstructure during the drying process of plant porous materials. *Chemical Engineering and Processing: Process Intensification*, 49, 286-293.

10

Effects of Er:YAG Laser Irradiation on Dental Hard Tissues and All-Ceramic Materials: SEM Evaluation

Bülent Gökçe
Ege University, School of Dentistry,
Department of Prosthodontics,
Turkey

1. Introduction

A reliable bond to dental hard tissues and materials has always been one of the most significant contributions for restorative dentistry (Leinfelder, 2001). A durable and stable bond between resins and dental hard tissues and restorative materials which has to integrate all parts of the system into one coherent structure is fundamental for the long-term retention and clinical success of the restorations. However, micromechanical attachment is one of the key mechanisms for a reliable adhesion to dental hard tissues and restorative materials (Matinlinna & Vallitu, 2007; Van Noort, 2002b; Fabienelli, et al., 2010). Advances in adhesive dentistry have resulted in the recent introduction of modern surface conditioning methods in order to achieve high bond strengths through increased surface roughness of both dental hard tissues and the restorative materials (Matinlinna & Vallitu, 2007; Van Noort, 2002b).

The use lasers in dentistry has evolved since their development in 1962. Researches have been carried out on effects of lasers on dental hard tissues and materials and applications of different wavelengths as they become available (Roberts-Harry, 1992; Convissar & Goldstein, 2001; White, et al., 1993; Frentzen, et al., 1992; Arima & Matsumoto, 1993; Wilder-Smith, et al., 1997; Cernavin, 1995; Keller & Hibst, 1989; Burkes, et al., 1992; Wigdor, et al., 1993; Visuri, et al., 1996b). According to current literature there is no optimum wavelength for all dental applications. Each wavelength has distinct treatment advantages and offers various treatment options. Understanding the differences between laser wavelengths will help to choose the adequate wavelength for each application in the dental office (Kutsch, 1993).

Laser light has properties such as being coherent, monochromatic and collimated. Laser light travels in specific wavelengths in a predictable pattern (coherent) and parallel (collimated) and it has one color (monochromatic). Lasers and target tissues interact in four ways. When a laser light hits the target it can be reflected, absorbed, scattered throughout the target or transmitted into the target (Kutsch, 1993). During laser application light energy is converted into heat and energy absorption on the target surface causes the vaporization. This process is called ablation or photoablation by vaporization (Cardoso, et al., 2008; Esteves–Oliveira, et al., 2007; Tachibana, et al., 2008; Lee, et al., 2007). Among currently available lasers, the erbium:yttrium-aluminum-garnet (Er:YAG) and Erbium,Chromium:Yttrium-Scandium-Gallium-Garnet (Er,Cr:YSGG) lasers have been proposed for different dental applications, including carious dentin removal, cavity preparation, surface conditioning, and as a surface treatment method for indirect restorations (Trajtenberg, et al., 2004; Atsu, et al., 2006; Bottino, et al., 2005; Gökçe, et al., 2007; Harashima, et al., 2005).

2. Morphological analysis of Er:YAG laser treated enamel and dentin

Etching of enamel with phosphoric acid was first recommended by Buonocore in 1955. (Buonocore, 1955). Resin bonding to tooth ensured by acid etching of enamel and/or dentin with total etch or self-etching techniques and followed by the use of a dentin adhesive (Fusayama, et al., 1979). Phosphoric acid removes the matrix phase of enamel and increases the surface area as well as creating high-energy hydrophilic surface with honey-comb-like structure (Sharpe, 1967; Reynold, 1975). Acid etching results in dissolution of the hydroxyapatite and enhances the penetration of adhesive monomers (Van Meerbeek, et al., 2003) forming resin tags in situ after polymerization (Barkmeier & Cooley, 1992; Leinfelder, 2001).

Conversely bonding to dentin is more complex due to its hydrated biological structure. To obtain intimate association of adhesive and dentin is hard when dentin is conditioned with total etch technique (Marshall, et al., 1997; Pashley, 1992). Etching dentin results in smear-free surface, open dentinal tubules with widened orifices due to removal of peritubular dentin, increased permeability by the loss mineralized dentin within the collagen matrix and exposed collagen web (Marshall, et al., 1997; Pashley, 1992; Pashley & Carvalho, 1997; Schein, et al., 2003).

Micromechanical retention is still the key factor for bonding to dentin. Monomers containing hydrophilic radicals infiltrate through the collagen fibrils and polymerized to develop the micromechanical retention. Many efforts have been spent to promote this dentin-resin interdiffusion zone, hybrid layer, since its description in 1982 (Nakayabashi, et al., 1982).

Air abrasion has also been introduced for enamel pretreatment by Olsen et al., in 1940. It was used for cavity preparation (Olsen, et al., 1997a). In this method, alumina particles were applied under air pressure to roughen the enamel surface (Zachrisson & Buyukyılmaz, 1993).

Etching dental hard tissues with laser has recently been proposed and may enable strong bonds with the restorative materials. Pulsed Nd:YAG lasers are sometimes used to etch enamel in preparation for bonding of restorative materials but some studies suggest that Nd:YAG etching alone results weaker bonds compared with acid etching (Roberts-Harry, 1992). It was suggested that to use the Nd:YAG laser efficiently for surface roughening a topical absorber must be applied to enamel surfaces and low pulse energies (100 mj or less) should be used (Roberts-Harry, 1992). SEM evaluation of the surface of Nd:YAG laser treated dentin was partially obliterated due to resolidification of molten dentin with grooves, fissures and concavities but without smear layer (Ariyaratnam, et al., 1999). They also stated that lased dentin surfaces produced a rougher surface compared to untreated dentin. This difference was suggested to maintain the micromechanical interlocking with the dentin adhesive. It was concluded that although laser irradiation with Nd:YAG laser produced a favorable surface for bonding, the bond strength to dentin did not differ from the conventionally treated dentin.

Both enamel and carious dentine were suggested to be removed with Nd:YAG and excimer lasers without signs of thermal damage (White, et al., 1993; Frentzen, et al., 1992; Arima & Matsumoto, 1993; Wilder-Smith, et al., 1997). When compared with Nd:Yag laser, Ho:YAG laser was shown to remove dental hard tissues more effectively with less cracks (Cernavin, 1995).

Some investigations suggest that CO_2 laser etching results in bonds of comparable strength on enamel and higher bond on dentin surfaces, compared to acid etching (Cooper, et al., 1988; Liberman, et al., 1984). Therefore CO_2 lasers can be recommended for enamel etching prior to composite restorations and fissure sealants without need of an absorber (Walsh, 1994). However excessive heat generated by some lasers may cause pulpal damage (Akova, et al., 2005). Adequate laser parameters can supply limited pulpal temperature increases within safety limits (Obata, et al., 1999). Controversially CO_2 laser at high fluencies and in continuous wave mode may cause cracking, flaking, crater formation, charring, melting and recrystallization of dental hard tissues (Stern, et al., 1972; Boehm, et al., 1997; McCormack, et al., 1995; Malmström, et al., 2001).

Other pulsed lasers whose wavelengths are strongly absorbed by dental hard tissues and hydroxyapatite, e.g. erbium lasers (Er:YAG and Er,Cr:YSGG), can successfully be used for dental hard tissue procedures including conditioning or etching without any side effects. Again no absorber is required (Liberman, et al., 1984; Keller & Hibst, 1989; Burkes, et al., 1992; Wigdor, et al., 1993; Visuri, et al., 1996b).

The water and the hydrated components of dental hard tissues absorb the high energy of erbium lasers and evaporate with micro explosions resulting in particle removal (ablation). (Cardoso, et al., 2008; Esteves–Oliveira, et al., 2007; Tachibana, et al., 2008; Lee, et al., 2007). This thermomechanical effect of erbium lasers on dental hard tissues can vary according to the tissue composition and mainly the water concentration. The mechanism of ablation of dental hard tissues with erbium lasers is still unclear but it was proposed that it takes place by the expansion of subsurface water resulting in microexplosions. This microexplosion induce strong mechanical separation of the calcified tissue (Kayano, et al., 1989). This constitutes the major principle of erbium laser ablation and produce non-uniform tissue removal with ejection of both organic and inorganic tissue microparticles, creating the micro-crater like appearance typical of lased surfaces (Corona, et al., 2007)

Erbium lasers have a shallow thermal penetration depth and can ablate sound and carious enamel and dentine (Keller & Hibst, 1989; Burkes, et al., 1992; Wigdor, et al., 1993; Visuri, et al., 1996b). Besides rough and irregular surface with sharp edged craters without color changes indicative of thermal damage (burning or carbonization) of surrounding tissues and/or the pulp have been reported. Concave and convex surfaces caused by microablation have been observed (Harashima, et al., 2005; Oelgiesser, et al., 2003). Er:YAG laser with appropriate parameters proposed to can selectively remove enamel hydroxyapatite crystals resulting in irregular surface that would enhance the micromechanical retention (Hibst & Keller, 1989; Hossain, et al., 1999).

Sasaki, et al., (2008) made a structural analysis of acid and Er:YAG laser etched enamel. They stated that acid etching exhibited a more homogenous etching pattern whereas Er:YAG alone showed areas of ablation. Er:YAG laser irradiation followed by acid etching resulted in more homogenous surface pattern than the only lased surfaces.

Harashima, et al., (2005) reported that cavities prepared by Er:YAG laser showed characteristic rough surface similar to an acid etched surface with open dentinal tubules and stripped surfaces. They also stated very clean surfaces, almost free of debris when the laser tip was aligned perpendicular to the surface. Scratched appearance with interspersed open dentinal tubules at areas covered by melted surfaces was found with angulated laser application (Harashima, et al., 2005). Unlike acid etching it was shown that the collagen fibrils were not found forming a porous network responsible for the increased porosity of dentin surface and subsurface. The morphological analysis of resin-dentin interface of acid etched dentin revealed triangular hybridization with resin tags in different lengths at the transition between peri- and intertubular dentin. But little or no hybridization zones with fewer and thinner tags at the intertubular dentin areas could be observed due to scarcity and discontinuity of the interdiffusion area at the resin-dentin interface (Schein, et al., 2003).

Literature review also states crater formations, mineral meltdowns and enamel melting, cracks, fissuring in enamel and smooth edged voids (Frentzen & Koort, 1992; Olsen et al., 1997b). Parameter factors and wavelength specificity relate to the degree of change that can be induced to enamel. Varying pulse width, pulse mode and spot size can produce significant changes in enamel and dentin surface morphology (Frentzen & Koort, 1992).

Erbium lasers also denatures the organic content and reduces the solubility of hydroxyapatite (Keller & Hibst, 1989; Hibst & Keller, 1989; Bader & Krejci, 2006). The interaction of erbium lasers with dental hard tissues results in negatively effected bond between the composite resins and dentin and collagen fibrils (Moretto, et al., 2010; Ceballo, et al., 2002; Ramos, et al., 2010, Oliveira, et al., 2010). Carvalho, et al., (2011) suggested that removal of laser irradiated dentin with phosphoric acid gel and sodium hypochlorite had increased the bond strength to dentin.

In a recent study phosphoric acid etching of enamel was compared with Er:YAG laser and Er:YAG laser+acid etching, and it was concluded that Er:YAG laser+acid group exhibited the highest bond strength, followed by acid and laser groups. The lower bond strength with only laser group was attributed to the non-homogenous laser application leaving untouched areas on the surface. Laser application followed by acid etching effectively conditioned the non-lased spots remained within the irradiated area (Sasaki, et al., 2008).

On the other hand some authors reported that the microretentive pattern resulting from laser irradiation could be favorable to bonding procedures (Hossain, et al., 2001; Li, et al., 1992; Visuri, et al., 1996a). Some studies suggest that laser irradiated dentinal tissue resulted in lower bond strength than does non-irradiated dentin. Visuri, et al. (1996a) reported a significantly higher shear bond strength of composite to dentin prepared with an Er:YAG laser. In contrast, Sakakibara, et al. (1998), Ceballo, et al. (2002) and Dunn, et al. (2005) reported a decrease in bond strength to laser-irradiated dentin, and Armengol, et al. (1999) and Kataumi, et al. (1998) found no difference between laser- irradiated and non-irradiated specimens.

Treating dentin erbium lasers (Er:YAG and Er,Cr:YSGG) creates a rough, smear layer-free surface with open dentinal tubules. SEM observations of Carvalho, et al., (2011) revealed irregular and rugged dentinal surfaces, following Er,Cr:YSGG laser. Harashima, et al., (2005) observed smaller width and stripped surfaces on the cavities prepared by Er:YAG laser. They may also cause fissures and cracks that can be considered as drawbacks of using erbium lasers for surface pretreatment (Aoki, et al., 1998; Hossain, et al., 1999; De Munck, et al., 2002; De Oliveira, et al., 2007; Moretto, et al., 2010). Increase in acid resistance of dental hard tissues after laser irradiation was also been reported by some authors (Fried, et al., 1996; Hossain, et al., 2000; Apel, et al., 2002; Liu, et al., 2006).

SEM evaluation of Er:YAG laser treated enamel and dentin revealed different surface morphologies in accordance with literature reviewed depending on the laser parameters.

2.1 Morphological analysis of Er:YAG laser treated enamel

Fig. 1. Enamel. 100 mj. 10 Hz. With water cooling. Honey-comb appearance can be seen but not throughout the surface which is due to non-homogenous application of the laser.

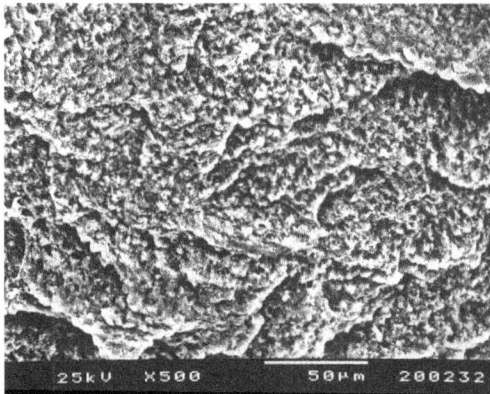

Fig. 2. Enamel. 100 mj. 10 Hz. With water cooling. Honey-comb appearance can be seen on the surface similar to acid etching.

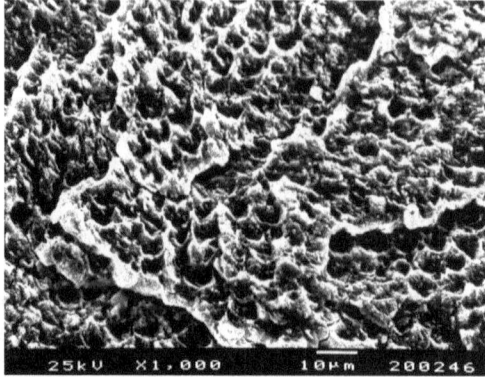

Fig. 3. Enamel. 100 mj. 10 Hz. With water cooling. Higher magnification of the surface in Fig. 2. No signs of thermal damage. Honey-comb appearance.

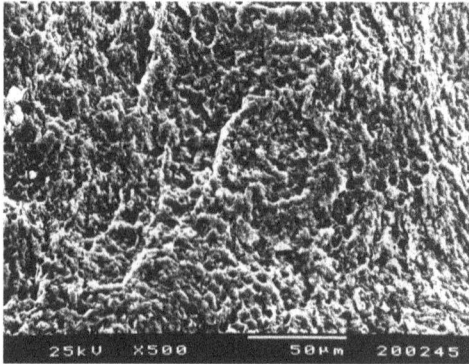

Fig. 4. Enamel. 250 mj. 10 Hz. With water cooling. Serrated surface with honey-comb appearance.

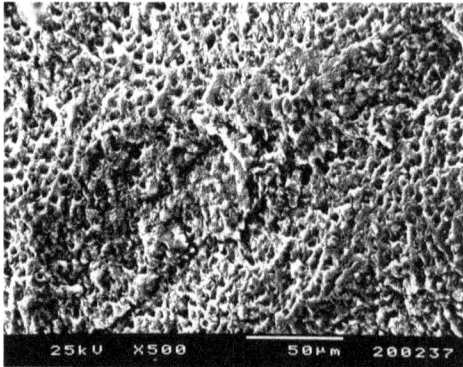

Fig. 5. Enamel. 500 mj. 10 Hz. With water cooling. Interprismatic matrix has been removed. Similar to acid etching but some melting points probably due to repeated shots at the same point can be observed.

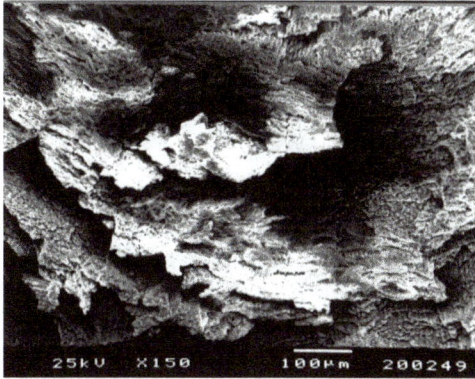

Fig. 6. Enamel. 600 mj. 10 Hz. Without water cooling. Layered enamel surface possibly due to dehydration of enamel during laser application.

Fig. 7. Enamel. 750 mj. 10 Hz. Without water cooling. Higher magnification of the previous Fig. Layered enamel surface.

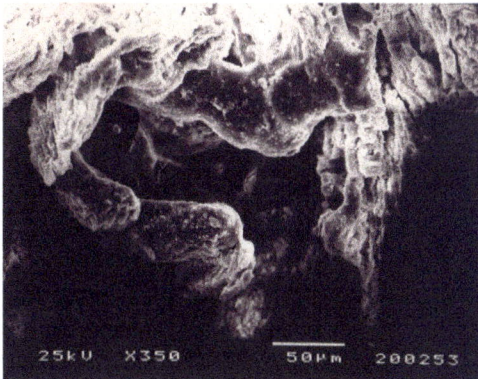

Fig. 8. Enamel. 800 mj. 5 Hz. Without water cooling. Melted and resolidified enamel. This texture is highly acid resistant.

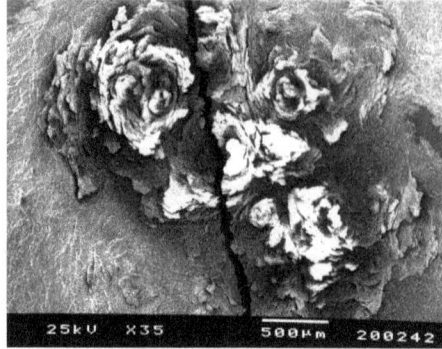

Fig. 9. Enamel. 1000 mj. 10 Hz. Without water cooling. Rose-bud like appearance. Clear evidence of over destruction of enamel with high energy intensity. Enamel lost its integrity in layers around the lased point. (The crack at midline is a result of dehydration during preparation of the specimen for SEM evaluation).

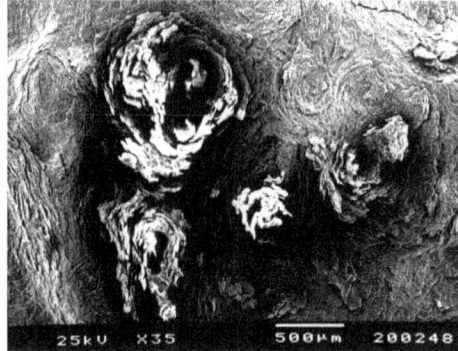

Fig. 10. Enamel. 1000 mj. 10 Hz. Without water cooling. Similar appearance with Fig. 13. Overdestructed and layered surface as a result of excessively heated enamel.

2.2 Morphological analysis of Er:YAG laser treated dentin

Fig. 11. Dentin. 250 mj. 10 Hz. Without water cooling. Swollen dentin orifices.

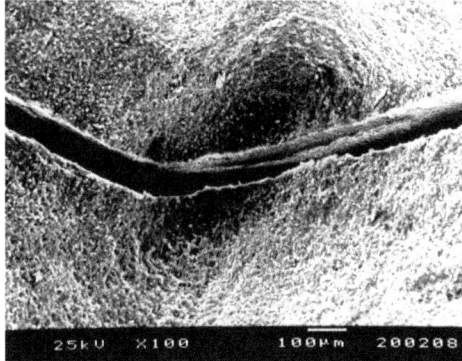

Fig. 12. Dentin. 400 mj. 10 Hz. Without water cooling. Cavitation with charring. (The crack at midline is a result of dehydration during preparation of the specimen for SEM evaluation).

Fig. 13. Dentin. 250 mj. 10 Hz. With water cooling. Partially open dentinal tubules with crater formations.

Fig. 14. Dentin. 500 mj. 10 Hz. Without water cooling. Pop-corn like appearance. One exploded (right) an done over swollen dentin orificies. Evidence of thermal destruction of dentin.

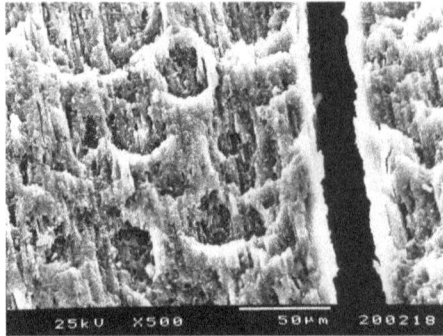

Fig. 15. Dentin. 500 mj. 5 Hz. With water cooling. intertubular Apperent evidence of intertubular dentin being affected dramatically by laser. (The crack on the right is a result of dehydration during preparation of the specimen for SEM evaluation)

Fig. 16. Dentin. 250 mj. 5 Hz. With water cooling. Nearly all dentinal tubules are open. Adequate surface for bonding procedures. Stratified surface due to non-homogenous application of laser. Calcospherite areas which are usually seen following Na(OH) were observed.

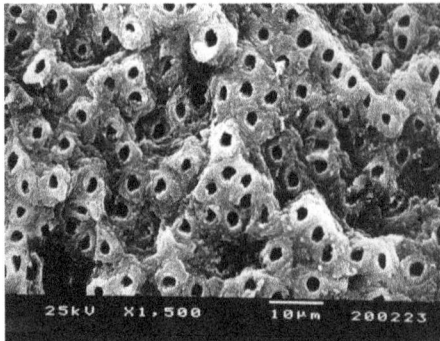

Fig. 17. Dentin. 250 mj. 5 Hz. With water cooling. Higher magnification of the surface in Fig. 16. No signs of thermal damage. No melted and swollen dentin. All dentinal tubules are open. Adequate surface for bonding procedures.

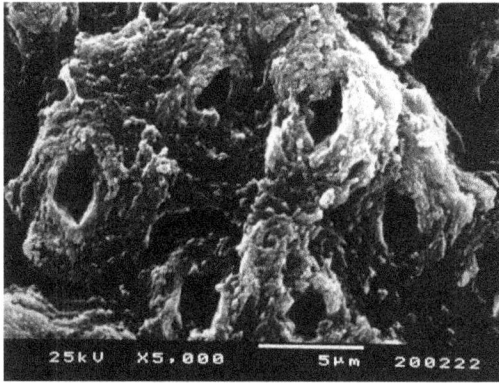

Fig. 18. Dentin. 250 mj. 5 Hz. With water cooling. Higher magnification of the surface in Fig. 17. No signs of thermal damage.

3. Morphological analysis of Er:YAG laser treated all-ceramic materials

A new class of dental framework materials have been introduced to the market for crown and fixed partial denture fabrication such as high-aluminium trioxide (alumina) ceramics, leucite reinforced feldspathic ceramics, castable glass-ceramics, machining and CAD/CAM ceramic systems and yttrium tetrogonal zirconia polycrystal (Y-ZTP; zirconia) (Atsu, et al., 2006; Amaral, et al., 2006; Bottino, et al., 2005; Kim, et al., 2005; Kern & Wegner, 1998). Alumina and zirconia demonstrate high clinical success due to their high cristalline content and are potential substitutes for traditional materials (Cavalcanti, et al., 2009a; Jacobsen, et al., 1997; Haselton, et al., 2000; Toksavul & Toman, 2007; Fradeani & Redemani, 2002).

The tetragonal to monoclinic phase transformation capability of zirconia results in high mechanical properties (Guazzato, et al.,2004). External stresses such as sandblasting, grinding, impact, and thermal aging can trigger this phase transformation mechanism (Karakoca & Yılmaz, 2009).

The clinical succes and survival rates of these restorations depend on several factors such as cementation procedure. To maintain a micromechanical bond, a key factor between restoration and the resin, luting surfaces of the restorations should be conditioned (Awliya, et al., 1998; Özcan, et al., 2001). To achieve reliable adhesion to these new materials, surface pre-treatments usually followed by silanization are required (Atsu, et al., 2006; Amaral, et al., 2006; Bottino, et al., 2005; Kim, et al., 2005; Kern & Wegner, 1998).

To obtain high mechanical bond strength to newer restorations, the inner surfaces are roughened by numerous techniques to increase the luting surface area. Among several methods that have been investigated for surface modification dental restorative materials, grinding, abrasion with diamond rotary instruments, airborne particle abrasion with aluminum oxide particles (sandblasting), chemical etching with different concentrations of hydrofluoric acid (HF), silica coating (Cojet, Rocatec), Silicoater MD, PyrosilPen silanization, selective infiltration-etching technique and combinations of any of these methods are the most common conditioning techniques prior to luting procedures (Amaral, et al., 2006; De Oyague, et al., 2009; Özcan, et al., 2001; Özcan, 2002; Kern & Thompson, 1994; Aboushelib, et al., 2007). Although surface treatments are used to micromechanical retention, they might affect the mechanical properties of zirconia (Sato, et al., 2008).

For chemical etching, different concentrations of HF acid, acidulated phosphate fluoride and ammonium bifluoride are used to condition the restorations (Blatz, et al. 2003; Clauss, 2000; Janda, et al. 2003). Etching dissolves the low fusing glass matrix exposing the cristalline structure and creates a micromechanically retentive surface but also promotes hydroxyl group formation on the etchable ceramic materials (Matinlinna & Vallitu, 2007; Özcan, 2003; Van Noort, 2002a; Özcan, et al., 2001). But some new materials such as zirconia and alumina are non-etchable because of they do not have glassy phase at the cristalline border and it is difficult to form microretentive surfaces to obtain strong and durable bonds with chemical etching techniques (Blatz, et al., 2003; Clauss, 2000; Janda, et al., 2003; Awliya, et al., 1998). Therefore different surface conditioning methods such as sandblasting and silica coating have been suggested for surface pretreatments of alumina and zirconia frameworks to modify the surface properties (Della Bona, et al., 2004; Phark, et al., 2009; Ersu, et al., 2009; Jacobsen, et al., 1997).

Different sizes of alumina particles between 25 and 250 µm are used (Blatz, et al., 2003; Kern & Wegner, 1998; Hummel & Kern, 2004; Curtis, et al., 2006). Sand blasting the surface with aluminum oxide particles cleans the ceramic surface and creates adequate bonding with micromechanical mechanisms to alumina- and zirconia based frameworks (Matinlinna & Vallitu, 2007; Phark, et al., 2009; Blatz, et al., 2003; Kern & Wegner, 1998; Hummel & Kern, 2004; Blatz, et al., 2004). The abrasive process removes loose contaminated layers, increases surface area and improves the wettability (Amaral, et al., 2006; Kümbüloğlu, et al., 2006).

Large abrasive particles result in rougher surface since the abrasion of the surface increases in proportion to the square of the diameter of the particle. Particle size variations and and the high pressure during sandblasting may cause flaws and phase transformation that expedites micro-crack formation and lead to altered mechanical properties of zirconia (Zhang, et al., 2004; Zhang, et al., 2006). Mechanical grinding and sandblasting may create subcritical microcracks and phase transformation within zirconia surface which might negatively affect the mechanical properties (Karakoca & Yılmaz, 2009; Ayad, et al., 2008).

Sandblasting is not recommended to roughen In-Ceram Zirconia frameworks as the aluminum oxide particles used to condition the surface have a hardness similar to that of the aluminum oxide crystals present in the target material (Borges, et al., 2003). Alternatively use of synthetic diamond particles 1-3 µm in size have been advocated to roughen the aluminous ceramics (Sen, et al., 2000).

Another method to inrease the surface energy of ceramic materials is tribochemical silica coating that is based on forming a SiO_2 layer followed by silane application with accelerated silica coated alumina particles on to the ceramic surface, including non-etchable alumina and zirconia (Matinlinna & Vallitu, 2007; Kramer, et al., 1996; Sindel, et al., 1996; Özcan, 2002). Silica coating method also provides micromechanical retention like sandblasting and silica deposition on the luting surface (Kern & Thompson, 1995; Matinlinna & Vallitu, 2007; Özcan, et al., 2001). In a recent study AFM results revealed irregular and heterogeneous surfaces following silica coating and sandblasting of zirconia with the formation of high peaks and shallows while SEM observations showed microretentive grooves in conjuction with Atomic Force Microscope (AFM) results (Subaşı & İnan, 2011).

In addition to currently used conditioning methods, laser-induced modifications of dental materials have also been studied. Lasers have been proposed to modify the surface of materials in relatively safe and easy means (Ersu, et al., 2009; Gökçe, et al., 2007; Akova, et al., 2005; Spohr, et al., 2008; Cavalcanti, et al., 2009b; Jacobsen, et al., 1997). Implant surfaces treated with lasers exhibit high degree of purity with adequate surface roughness (Gaggl, et al., 2000; Cho & Jung, 2003).

Among the several applications of lasers, surface conditioning for bonding have also been reported. Various laser types such as Nd:YAG, Er:YAG, Er,Cr:YSGG and CO_2 have been studied for surface alterations of dental materials (Convissar & Goldstein, 2001). But only limited studies are available on the laser treatment of all ceramic materials (Ersu, et al., 2009; Gökçe, et al., 2007; Akova, et al., 2005; Cavalcanti, et al., 2009b; Jacobsen, et al., 1997; Cavalcanti, et al. (2009a).

Ceramics do not effectively absorb some certain wavelengths such as 1064 nm (Nd:YAG). To increase the energy absorption of this laser the surface of ceramic material can be covered with graphite powder prior to laser irradiation. During laser application the graphite is removed from the surface with microexplosions (Spohr, et al., 2008).

Some lasers are also used for other applications such as forming a glazed surface layer on ceramics, the removal of resin composite filling materials, laser welding of ceramics and metal alloys, including titanium, and increasing the corrosion resistance of metal alloys (Ersu, et al., 2009; Schmage, 2003). Focussed CO_2 laser causes in conchoidal tears (result of surface warming) on ceramic surface that provides mechanical retention between resin composite and ceramics. But sudden temperature changes could create internal tensions that might affect the bond strength (Ersu, et al., 2009). The authors concluded that CO_2 laser surface modification demonstrated higher bond strength than control, sandblasted and chemical etching.

Results of studies that compared the bond strength of resins to CO_2 laser and chemically etched zirconia vary. Obata, et al., (1999) stated that laser etching produced lower bond strength compared to acid ecthing whereas Ural, et al., (2010) proposed higher bond strength. They attribute the high bond strength values to power levels of the laser used in their study. Increased power settings caused micro-cracks and high bond strength (Ural, et al., 2010).

Watanabe, et al., (2009) suggested that Nd:YAG laser irradiation improved the mechanical properties of cast titanium. Nd:YAG laser as an etchant was also used to enhance the bond strength of low-fusing ceramic to titanium (Kim & Cho, 2009).

Nd:YAG laser was also used to roughen In-Ceram Zirconia and feldspathic ceramic (Li, et al., 2000; Spohr, et al., 2008). Li, et al., (2000) reported that SEM images of Nd:YAG laser applied specimens was fovarable to mechanical retention between the feldspathic ceramic and the resin cement and both laser and HF acid etched groups exhibited same shear bond strength. Nd:YAG laser treatment of In-Ceram Zirconia caused surface changes characterized with material removal due to the micro-explosions resulting in formation of voids and fusing and melting of the most superficial ceramic layer followed by solidification to a smooth blister-like surface (Spohr, et al., 2008). Nd:YAG laser irradiation of zirconia causes color change to black with many cracks and reduced oxygen content (Noda, et al., 2010).

Recently roughening capacity of the Er:YAG laser for the inner surfaces of the lithium disilicate material has been introduced (Gökçe, et al., 2007). Ceramic specimens laser etched with low energy levels exhibited similar bond strength that of chemicaly etched specimens. But as the energy level increased bond strength values decreased dramtically. They concluded that their results could be explained by insufficient micro depths of the irregularities formed by high Er:YAG laser power settings, which resulted in limited penetration of silane and low bond strength. Higher power settings resulted in low bond strengths which might be due to over destruction (disassociation) of the crystal and/or matrix phases or heat damaged layer which was poorly attached to the infra layers or increased luting agent thickness due to craters caused by laser pulses (Gökçe, et al., 2007).

Erbium lasers are absorbed mainly by water and their absorption by water-free materials are compromised. To increase the effect of erbium lasers, covering the zirconia surfaces with graphite or hydroxapatite powder was recommended (Cavalcanti, et al., 2009b). Akın, et al., (2011) irradiated zirconia surface with Er:YAG laser and found increased surface roughness and surface irregularities compared to the untreated specimens. The authors used low power settings with water cooling and did not observe microcracks. They concluded that altering the zirconia surface with Er:YAG laser increased the shear bond strength of ceramic to dentin and found to be effective for decreasing microlekage in the adhesive-ceramic interface. Their results were in accordance with the study of Cavalcanti, et al. (2009a). Erdem & Erdem, (2011) studied the effect of Er:YAG laser irradiation with water cooling on zirconia and unlike forementioned resarchers they suggested that laser treatment decreased the bond strength of resin composite to zirconia framework. They observed microcracks throughout the surface in contrast with Akın, et al., (2011). They attributed the low bond strength values to excessively affected surfaces and crack formation which was possibly a result of laser irradiation. Stepped local temperature changes and pressurized water followed by thermocycling could be responsible for low temperature degredation of zirconia resulting in low bond strengths (Erdem & Erdem, 2011). They might have also induced phase transformation (Cavalcanti, et al., 2009a). The microcrack formation and sizes enlarged as the laser intensity increased (Cavalcanti, et al., 2009b). Stübinger, et al., (2008) demonstrated that Er:YAG and CO_2 lasers adversely affected the zirconia implant surfaces. They found crack formations up to 100 µm depth and large grains in blackened areas under SEM evaluation. Excessive power settings shown to be deterious to zirconia and their use for zirconia surface conditioning was questionable (Cavalcanti, et al. 2009b; Navarro, et al., 2010).

Subaşı & İnan, (2011), evaluated the Er:YAG laser treated zirconia with AFM and SEM. AFM and SEM results of lased surfaces revealed similar texture to that of the control group with the exception that sharp peaks formations of the lased surfaces. Cavalcanti, et al. (2009b) also demonstrated that increased laser energy levels increased surface roughness of zirconia. Melting, excessive loss of mass, and the presence of smooth areas surrounded by cracks were observed. Lower energy intensities (200 mj) had milder effect with smaller cracks along with melting, solidification and color changes without loss of structure compared to higher intensities (400 and 600 mj) (Cavalcanti, et al., 2009b). 200 mj irradiation also provided alterations similar to sandblasting. Effect of Nd:YAG laser (100 mj) and Er:YAG laser (200 mj) exhibited similar topographies although the Nd:YAG laser had a totally different target interaction compared with the Er:YAG laser (Da Silveira, et al., 2005; Cavalcanti, et al., 2009b).

There is no consensus about energy levels of Er:YAG laser that could be used to modify the zirconia surface. 400 mj at 10 Hz (Subaşı & İnan, 2011), 150 mj at 10 Hz (Akın, et al., 2011), 200 mj at 10 Hz (Erdem & Erdem, 2011), 200 mj at 10 Hz (Cavalcanti, et al., 2009a); 200 mj, 400 mj, 600 mj at 10 Hz (Cavalcanti, et al., 2009b), 300 mj at 10 Hz (Şen & Ceylan, 2010) were chosen to roughen zirconia surfaces. Besides different methods have been chosen to evaluate the bond strength and surface topography. Therefore it is difficult to compare the results of the studies reviewed.

3.1 Morphological analysis of Er:YAG laser treated and hydroflouric acid etched Li-Disilicate material

SEM evaluations of 9.5% Hydrofluoric acid and Er:YAG laser Li-disilicate material revealed different surface morphologies, depending on the surface conditioning methods.

3.1.1 SEM evaluation before shear bond strength testing

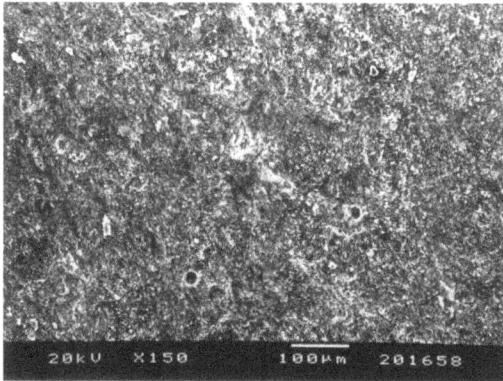

Fig. 19. The untreated surface showing intact glassy phase without any apparent crystals.

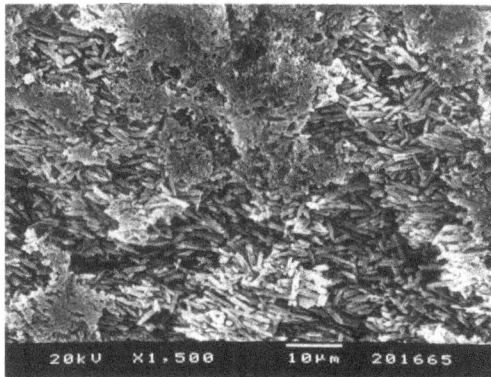

Fig. 20. 9.5% HF, 30 seconds. The surface has both apparent Li-disilicate crystals and glassy matrix. Glass matrix phase could not be completely removed if not applied homogenously and might lead to ill penetration of silane and the adhesive.

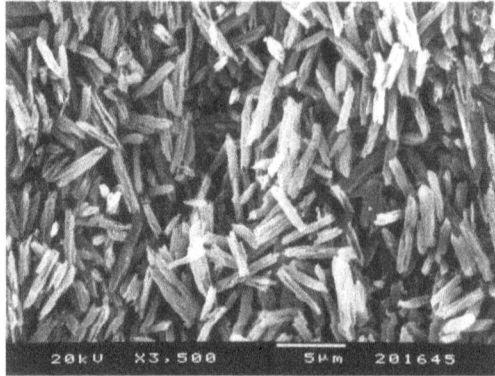

Fig. 21. 9.5% HF, 30 seconds. Visible Li-disilicate crystals. Completely removed glassy matrix. Appropriate etching pattern and surface for adhesive cementation (Gökçe et al., 2007).

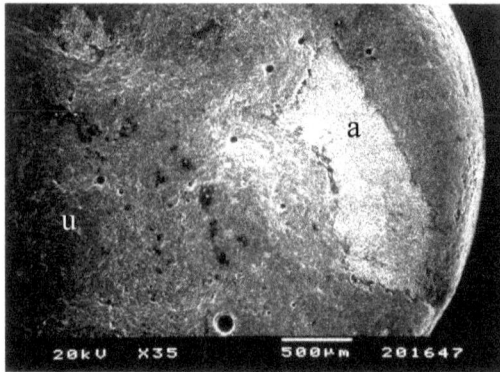

Fig. 22. Er:YAG laser, 300 mj, 10 Hz. Affected (a) and unaffected (u) areas of lased surface.

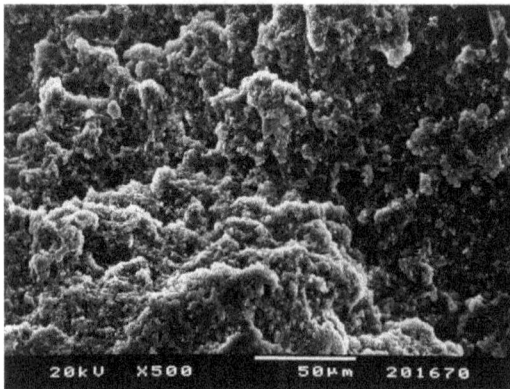

Fig. 23. Er:YAG laser, 300 mj, 10 Hz. Irregular Li-disilicate crystals in smaller sizes (Gökçe et al., 2007).

Fig. 24. Er:YAG laser, 600 mj, 10 Hz. Increased surface irregularities with severely affected and disassociated Li-disilicate crystals (Gökçe et al., 2007).

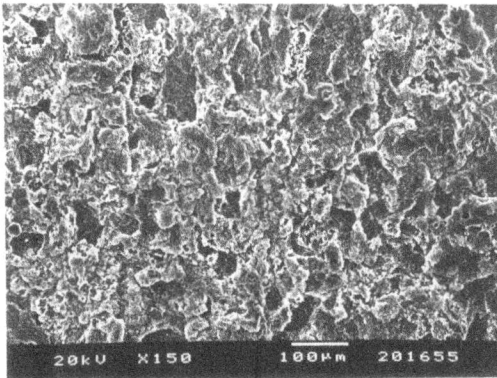

Fig. 25. Er:YAG laser, 900 mj, 10 Hz. Severely affected and disassociated Li-disilicate crystals (Gökçe et al., 2007).

Fig. 26. Er:YAG laser, 1000 mj, 10 Hz. Melted and resolidified surface. This layer is poorly attached to the underlying intact phase.

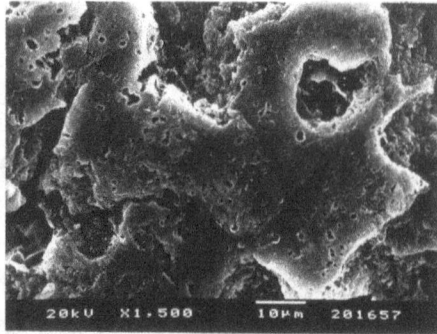

Fig. 27. Er:YAG laser, 1000 mj, 10 Hz. Higher magnification of the surface in Fig. 26.

3.1.2 SEM evaluation after shear bond strength testing

SEM evaluation following shear bond strength of the untreated, HF acid etched and Er:YAG laser conditioned Li-Disilicate material exhibited different failure modes, indicatives of adhesion of the bonding agent and the luting cement (Variolink II).

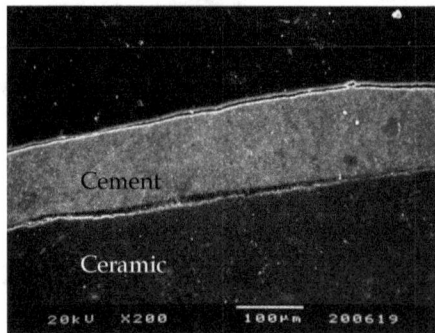

Fig. 28. Untreated ceramic surface. Adhesive failure inbetween the ceramic and the cement. No rough surfaces were noted on the ceramic (Gökçe et al., 2007).

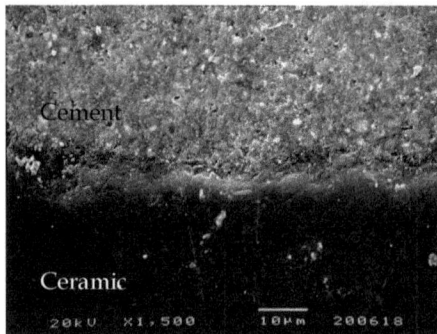

Fig. 29. 9.5% HF, 30 seconds. Good adhesion at the cement-ceramic interface with increased surface roughness. Mainly cohesive failures within the cement (Gökçe et al., 2007).

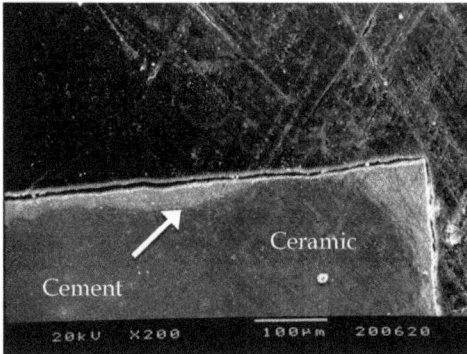

Fig. 30. Er:YAG laser, 300 mj, 10 Hz. No visible cement on the margins, while a cement remnant at the center of the specimen with adhesive+cohesive failures were observed (Gökçe et al., 2007).

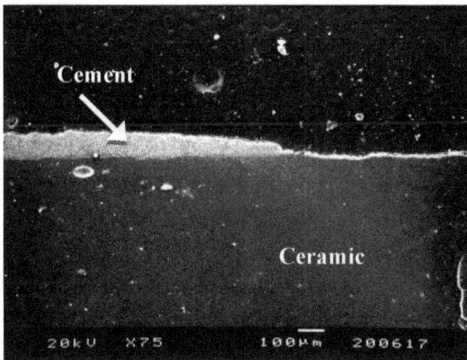

Fig. 31. Er:YAG laser, 600 mj, 10 Hz. Partially delaminated cement surfaces can be observed with adhesive failures (Gökçe et al., 2007).

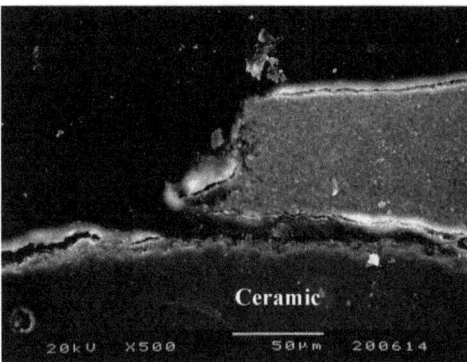

Fig. 32. Er:YAG laser, 900 mj, 10 Hz. Adhesive failure between cement and ceramic. Decreased irregularities and severe effects of laser on the ceramic surface (Gökçe et al., 2007).

3.2 Morphological analysis of sandblasted and Er: YAG laser-roughened alumina material

Fig. 33. Untreated In-Ceram Alumina (Şen, 2010).

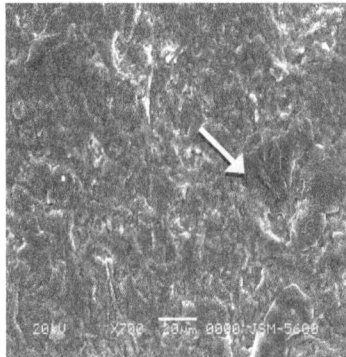

Fig. 34. In-Ceram Alumina. Airborne particle abrasion (110μm Al₂O₃). Affected and rougher surface compared to untreated surface with shallow pits (Şen, 2010).

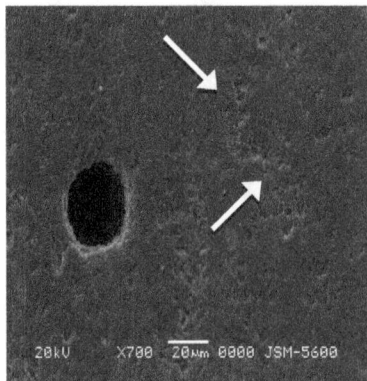

Fig. 35. In-Ceram Alumina. Er:YAG laser, 150 mj at 10 Hz with water cooling. Locally affected points on the surface due ton on homogenous application of the laser (Şen, 2010).

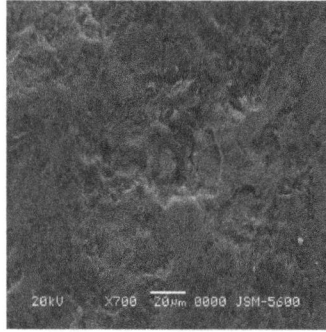

Fig. 36. In-Ceram Alumina. Er:YAG laser, 250 mj at 10 Hz with water cooling. Generalized effect of laser rougher surface compared to untreated and 150 mj laser applied surfaces (Şen, 2010).

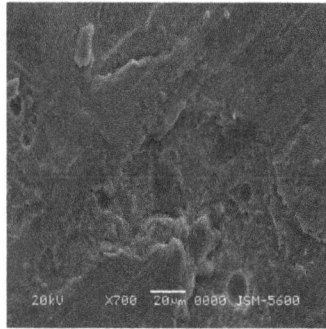

Fig. 37. In-Ceram Alumina. Er:YAG laser, 400 mj at 10 Hz with water cooling. Serrated and smoothened surface by resolidification of melted areas. This resolidified layer might be poorly attached to the underlying material (Şen, 2010).

3.3 Morphological analysis of sandblasted, silica coated and Er: YAG laser-roughened zirconia

3.3.1 SEM evaluation before shear bond strength testing

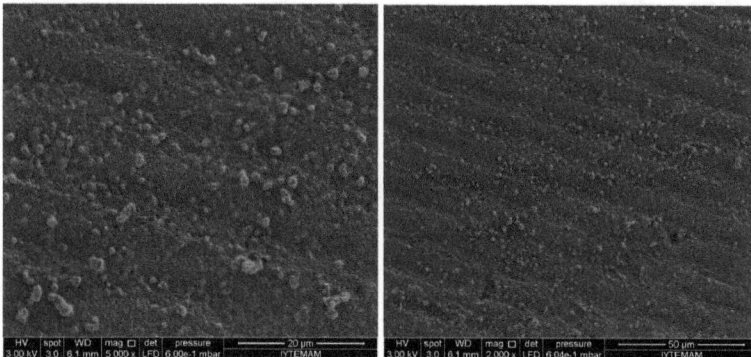

Fig. 38. Untreated zirconia. 2000x (left) and 5000x (right) magnifications (Erdem & Erdem, 2011).

Fig. 39. Sandblasted (particle size 110 μm) zirconia. 2000x (left) and 5000x (right) magnifications. Increased roughness compared to untreated zirconia (Erdem & Erdem, 2011).

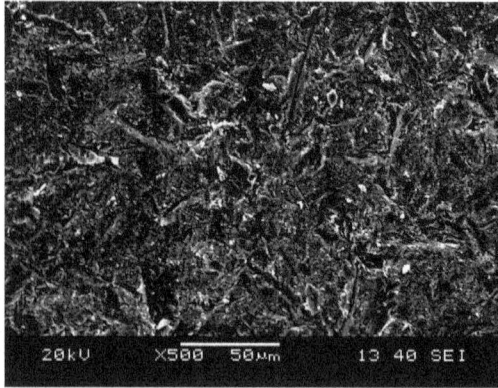

Fig. 40. Sandblasted (particle size 180 μm) zirconia. 500x magnification. Similar texture with the 110 μm air abraded surface.

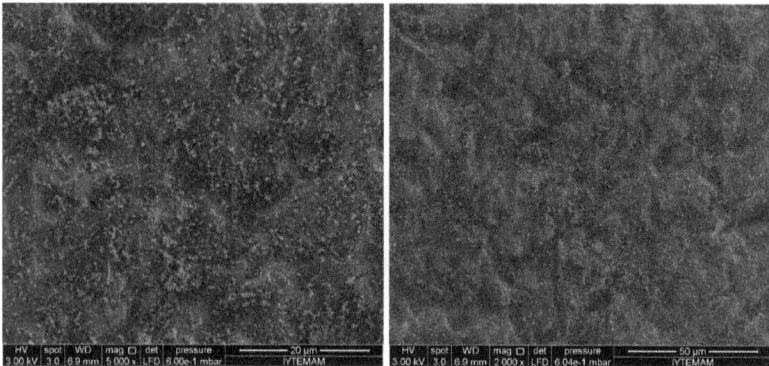

Fig. 41. Silica coated (Rocatec Pre110 μm and Rocatec Soft 30 μm) zirconia. 2000x (left) and 5000x (right) magnifications. Increased roughness similar to sandblasting and silica deposition on the surface can be observed (Erdem & Erdem, 2011).

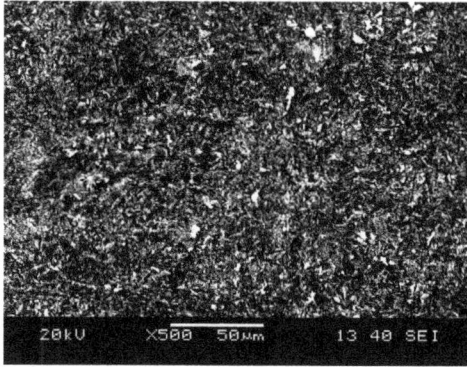

Fig. 42. Silica coated (Rocatec Pre110 μm and Rocatec Soft 30 μm) zirconia. 500x magnification. Increased roughness similar to particle abrasion with aluminum oxide (Şen, 2010).

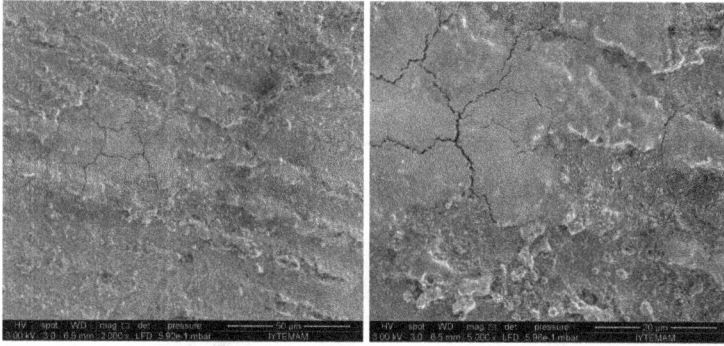

Fig. 43. Graphite coated and lased (200 mj, 10 Hz) zirconia. 2000x (left) and 5000x (right) magnifications. Rough and severely affected appearance with irregular surface (Şen, 2010) with micro cracks (Erdem & Erdem, 2011).

3.3.2 SEM evaluation after shear bond strength testing

| Panavia | Clearfil | RelyX | Multilink | Superbond |

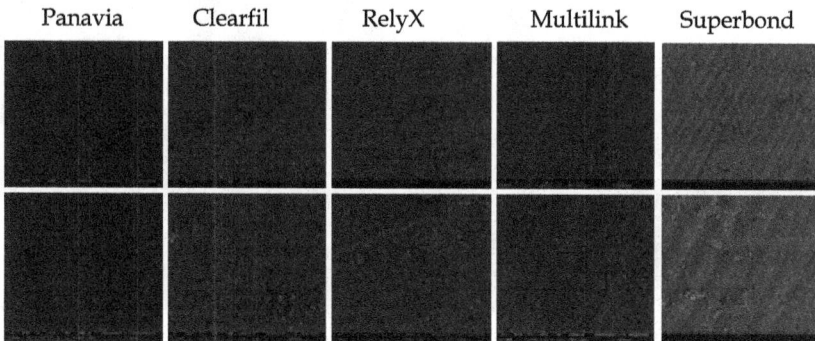

Fig. 44. Untreated zirconia. 2000x (first line) and 5000x (second line) magnifications. No cement retention was observed on untreated zirconia (Erdem & Erdem, 2011).

| Panavia | Clearfil | RelyX | Multilink | Superbond |

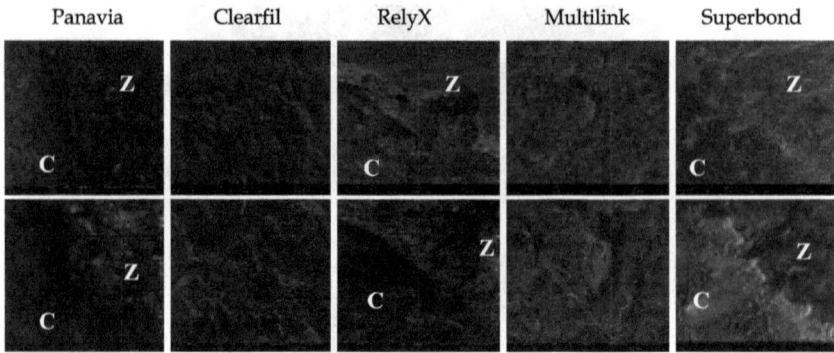

Fig. 45. Sandblasted (particle size 110 µm) zirconia. 2000x (first line) and 5000x (second line) magnifications. Two of the cements tested exhibited both adhesive and cohesive failures. (Z: Zirconia, C: Cement) (Erdem & Erdem, 2011).

| Panavia | Clearfil | RelyX | Multilink | Superbond |

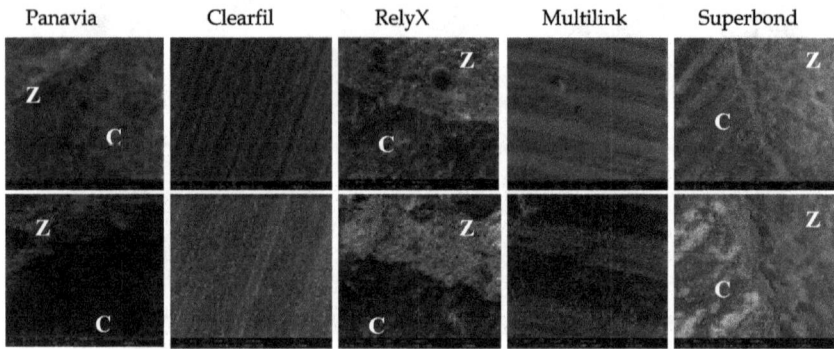

Fig. 46. Silica coated (Rocatec Pre110 µm and Rocatec Soft 30 µm) zirconia. 2000x (first line) and 5000x (second line) magnifications. Similar results with sandblasting was oserved after shear bond strength testing (Z: Zirconia, C: Cement) (Erdem & Erdem, 2011).

| Panavia | Clearfil | RelyX | Multilink | Superbond |

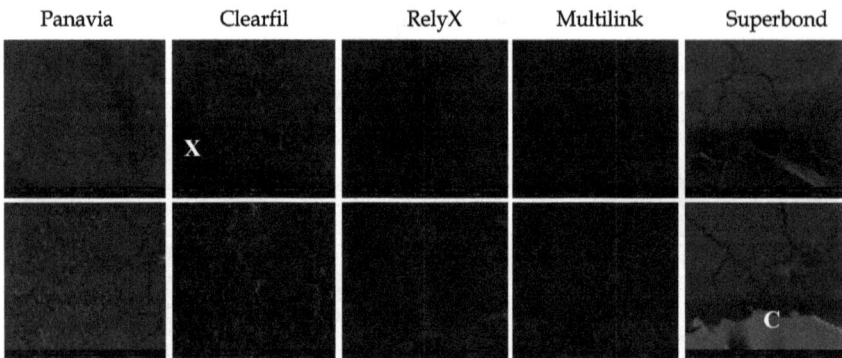

Fig. 47. Graphite coated and lased (200 mj, 10 Hz) zirconia. 2000 (first line) and 5000 (second line) magnifications. Adhesive failures observed in all cement groups. No cement retention on any of the groups. (X: severely affected area, C: Cement) (Erdem & Erdem, 2011).

4. Conclusion

There are many techniques to condition dental hard tissues and luting surfaces of indirect restorations prior to bonding. Operators find it difficult to decide which technique offers better results, and are also uncertain about the factors that might influence their techniques of choice. However micromechanical retention of luting materials to acid etched conditioned dental hard tissues is currently seems to be the most successful and reliable approach for dental bonding. But surface characteristics of Er:YAG lased enamel and dentin are responsible for considering this surface adequate for resin bonding.

It is assumed that the ablation rate of lasers on the dental materials is strongly influenced by the differences in composition and microstructure of the material and the presence of water. In spite of its great potential for ablation, Er:YAG laser effectiveness and safety is also directly related to adequate setting parameters. Power settings, frequency and durations of laser irradiation play an important role to obtain optimum bond strength and roughness values.

Future studies are needed to evaluate the superficial and sub-superficial layers of irradiated dental hard tissues and materials in order to develop new agents that can interact properly with lased substrate. In my opinion, in the near future, 9.6 µm CO2 laser with an adequate delivery system that has the absorption peak in hydroxyapatite will replace many dental hard tissue lasers, which are currently being used. In the presented chapter, the morphological assessment of Er:YAG lased dental hard tissues and materials have been discussed under the light of the current literature.

5. References

Aboushelib, MN.; Kleverlaan, CJ. & Feilzer, AJ. (2007). Selective infiltration-etching technique for a strong and durable bond of resin cements to zirconia-based materials. *Journal of Prosthetic Dentistry*, Vol.98, No.5, pp. 379-388.

Akin, H.; Tugut, F.; Emine, A.G.; Guney, U. & Mutaf, B. (2011). Effect of Er:YAG laser application on the shear bond strength and microleakage between resin cements and Y-TZP ceramics. *Lasers in Medical Science,* Vol 21. [Epub ahead of print]

Akova, T.; Yoldas, O.; Toroglu, MS. & Uysal, H. (2005). Porcelain surface treatment by laser for bracket-porcelain bonding. *American Journal of Orthodontics & Dentofacial Orthopedics*, Vol.128, pp. 630–637.

Amaral, R.; Özcan, M.; Bottino, MA. & Valandro, LF. (2006). Microtensile bond strength of a resin cement to glass infiltrated zirconia-reinforced ceramic: the effect of surface conditioning. *Dental Materials*, Vol.22, pp. 283–290.

Aoki, A.; Ishikawa, I.; Yamada, T.; Otsuki, M.; Watanabe, H.; Tagami, J.; Ando, Y. & Yamamoto, H. (1998). Comparison between Er:YAG laser and conventional technique for root caries treatment in vitro. *Journal of Dental Research*, Vol.77, pp. 1404–1414.

Apel, C.; Meister, J.; Schmitt, N.; Graʾber, H.G. & Gutknecht, N. (2002). Calcium solubility of dental enamel fol- lowing sub-ablative Er:YAG and Er:YSGG laser irradiation in vitro. *Lasers in Surgical Medicine*, Vol.30, pp. 337–341.

Arima, M. & Matsumoto, K. (1993). Effects of ArF excimer laser irradiation on human enamel and dentin. *Lasers in Surgical Medicine,* Vol.13, pp. 97–105.

Ariyaratnam, MT.; Wilson MA. & Blinkhorn, AS. (1999). An analysis of surface roughness, surface morphology and composite/dentin bond strength of human dentin following the application of the Nd:YAG laser. *Dental Materials,* Vol.15, No.4, pp. 223-8.

Armengol, V.; Jean, A.; Rohanizadeh, R. & Hamel, H. (1999). Scanning electron microscopic analysis of diseased and healthy dental hard tissues after Er:YAG laser irradiation: in vitro study. *Journal of Endodontics,* Vol.25, No.8, pp. 543–546

Atsu, SS.; Kilicarslan, MA.; Kucukesmen, HC. & Aka, PS. (2006). Effect of zirconium-oxide ceramic surface treatments on the bond strength to adhesive resin. *Journal of Proshtetic Dentistry,* Vol.95, pp. 430–436.

Awliya, W.; Oden, A.; Yaman, P.; Dennison, JB. & Razzoog, ME. (1998). Shear bond strength of a resin cement to densely sintered high-purity alumina with various surface conditions. *Acta Odontologica Scandinavica,* Vol.56, pp. 9–13.

Ayad, MF.; Fahmy, NZ. & Rosenstiel, SF. (2008). Effect of surface treatment on roughness and bond strength of a heat-pressed ceramic. *Journal of Prosthetic Dentistry,* Vol.99, pp. 123–130

Bader, C. & Krejci, I. (2006). Marginal quality in enamel and dentin after preparation and finishing with an Er:YAG laser. *American Journal of Dentistry,* Vol.19, pp. 337–342.

Barkmeier, WW. & Cooley, RL. (1992). Laboratory evaluation of adhesive systems. *Operative Dentistry,* Supplementary 5, pp. 50-61

Blatz, MB.; Sadan, A. & Kern, M. (2003). Resin-ceramic bonding: a review of the literature. *Journal of Prosthetic Dentistry,* Vol.89, pp. 268–274.

Blatz, MB.; Sadan, A.; Martin, J. & Lang, B. (2004). In vitro evaluation of shear bond strengths of resin to densely-sintered high-purity zirconium-oxide ceramic after long-term storage and thermalcycling. *Journal of Prosthetic Dentistry,* Vol.91, pp. 356–362.

Boehm, R.; Rica, J.; Webster, J. & Janke, S. (1997). Thermal stress effects and surface cracking associated with laser use on human teeth. *Journal of Biomechanical Engineering,* Vol.99, pp. 189-194.

Borges, G.A.; Spohr, A.M.; Goes, M.F.; Correr Sobrinho, L. & Chan, J.D. (2003). Effect of etching and airborne parti- cle abrasion on the microstructure of different dental ce- ramics. *Prosthetic Dentistry,* Vol.89, pp. 479–487.

Bottino, MA.; Valandro, LF.; Scotti, R. & Buso, L. (2005). Effect of surface treatments on the resin bond to zirconium-based ceramic. *International Journal of Prosthodontics,* Vol.18, pp. 60–65.

Buonocore, MG. (1955). A simple method of increasing the adhesion of acrylic filling materials to enamel surfaces. *Journal of Dental Research,* Vol.34, No.6, pp. 849-853.

Burkes, EJ. Jr.; Hoke, J.; Gomes, E. & Wolbarsht, M. (1992). Wet versus dry enamel ablation by Er:YAG laser. *Journal of Prosthetic Dentistry,* Vol.67, No.6, pp. 847–851.

Cardoso, M.V.; Coutinho, E., Ermis, RB.; Poitevin, A.; Van Landuyt, K.; De Munck, J.; Carvalho, RC.; Lambrechts, P. & Van Meerbeek, B. (2008). Influence of Er,Cr:YSGG laser treatment on the microtensile bond strength of adhesives to dentin. *Journal of Adhesive Dentistry*, Vol.10, No.1, pp. 25–33.

Carvalho, AO.; Reis AF de Oliveira, MT.; de Freitas, PM.; Aranha, AC.; Eduardo, CD. & Giannini, M. (2011). Bond Strength of Adhesive Systems to Er,Cr:YSGG Laser-Irradiated Dentin. *Photomedicine and Laser Surgery*, Vol.16, (Epub ahead of print)

Cavalcanti, AN.; Foxton, RM.; Watson, TF.; Oliveira, MT.; Giannini, M. & Marchi, GM. (2009a). Bond strength of resin cements to a zirconia ceramic with different surface treatments. *Operative Dentistry*, Vol.34, No.3, pp. 280–287

Cavalcanti, AN.; Pilecki, P.; Foxton, RM.; Watson, TF.; Oliveira, MT.; Gianinni, M. & Marchi, GM. (2009b). Evaluation of the surface roughness and morphologic features of Y-TZP ceramics after different surface treatments. *Photomedicine and Laser Surgery*, Vol.27, No.3, pp. 473–479.

Ceballo, L.; Toledano, M.; Osorio, R.; Tay, FR. & Marshall, GW. (2002). Bonding to Er–YAG-laser-treated dentin. *Journal of Dental Research*, Vol.81, No.2, pp. 119–122

Cernavin, I. (1995). A comparison of the effects of Nd:YAG and Ho:YAG laser irradiation on dentine and enamel. *Australian Dental Journal*, Vol.40, pp. 79–84.

Cho, SA. & Jung, SK. (2003). A removal torque of the laser-treated titanium implants in rabbit tibia. *Biomaterials*, Vol.24, pp. 4859–4863

Clauss, C. (2000). All-ceramic restoration based on milled zirconia. *Zahn Mund Kieferheilkd Zentralbl*, Vol.18, pp. 436–442.

Convissar, RA. & Goldstein, EE. (2001). A combined carbon dioxide/ erbium laser for soft and hard tissue procedures. *Dentistry Today*, Vol.20, pp. 66–71.

Cooper, LF.; Myers, ML.; Nelson, DG. & Mowery, AS. (1988). Shear strength of composite bonded to laser pre-treated dentin. *J Prosthetic Dentistry*, Vol.60, pp. 45-49.

Corona, S.A.; de Souza, A.E.; Chinelatti, M.A.; Borsatto, M.C.; Pecora, J.D. & Palma-Dibb, R.G. (2007). Effect of energy and pulse repetition rate of Er: YAG laser on dentin ablation ability and morphological analysis of the laser-irradiated substrate. *Photomedicine and Laser Surgery*, Vol.25, pp. 26–33.

Curtis, A.R.; Wright, A.J. & Fleming, G.J. (2006). The influence of surface modification techniques on the performance of a Y-TZP dental ceramic. *Journal of Dentistry*, Vol.34, pp. 195–206.

Da Silveira, B.L.; Paglia, A.; Burnett, L.H.; Shinkai, R.S.; Eduardo, C.P. & Spohr, A.M. (2005). Micro-tensile bond strength between a resin cement and an aluminous ceramic treated with Nd:YAG laser, Rocatec System, or aluminum oxide sandblasting. *Photomedicine and Laser Surgery*, Vol.23, pp. 543–548.

De Munck, J.; Van Meerbeek, B.; Yudhira, R.; Lambrechts, P. & Vanherle, G. (2002). Microtensile bond strength of two Erbium:YAG-lased vs. bur-cut enamel and dentin. *European Journal of Oral* Sciences, Vol.110, pp. 322–329.

De Oliveira, M.T.; de Freitas, P.M.; de Paula Eduardo, C.; Ambrosano, G.M. & Giannini, M. (2007). Influence of di- amond sono-abrasion, air-abrasion and Er:YAG laser irra-diation on bonding of different adhesive systems to dentin. *European Journal of Oral* Sciences, Vol.1, pp. 158–166.

De Oyague, RC.; Monticelli, F.; Toledano, M.; Osorio, E.; Ferrari, M. & Osorio, R. (2009). Influence of surface treatments and resin cement selection on bonding to densely-sintered zirconium-oxide ceramic. *Dental Materials,* Vol.25, pp. 172–179

Della Bona, A.; Shen, C. & Anusavice, KJ. (2004). Work of adhesion of resin on treated lithia disilicate-based ceramic. *Dental Materials,* Vol.20, No.4, pp. 338-344.

Dunn, WJ.; Davis, JT. & Bush, AC. (2005). Shear bond strength and SEM evaluation of composite bonded to Er:YAG laser-prepared dentin and enamel. *Dental Materials,* Vol.21, No.7, pp. 616-624

Erdem A, Erdem A, (2011) Evaluation of bond strength of luting resins to a Y-TZP framework material processed with different surface treatments. PhD Thesis. Ege University School of Dentistry Department of Prosthodontics, Izmir, Turkey

Ersu, B.; Yuzugullu, B.; Ruya, Y.A. & Canay, S. (2009). Surface roughness and bond strengths of glass-infiltrated alumina- ceramics prepared using various surface treatments. *Journal of Dentistry,* Vol.37, pp. 848–856.

Esteves–Oliveira, M.; Zezell, DM.; Apel, C.; Turbino, ML.; Aranha, AC.;Eduardo Cde, P. & Gutknecht, N. (2007). Bond strength of self-etching primer to bur cut, Er,Cr:YSGG, and Er:YAG lased dental surfaces. *Photomedicine and Laser Surgery,* Vol.25, No.5, pp. 373–380.

Fabianelli, A.; Pollington, S.; Papacchini, F.; Goracci, C.; Cantoro, A.; Ferrari, M. & Van Noort, R. (2010). The effect of different surface treatments on bond strength between leucite reinforced feldspathic ceramic and composite resin. *Journal of Dentistry,* Vol.38, No.1, pp. 39–43.

Fradeani, M. & Redemagni, M. (2002). An 11-year clinical evaluation of leucite-reinforced glass-ceramic crowns: a retrospective study. *Quintessence International,* Vol.33, pp. 503–510.

Frentzen, M. & Koort, H.J. (1992). The effect of Er:YAG laser irradiation on enamel and dentin. *Journal of Dental Research,* Vol.71, pp. 571-577.

Frentzen, M.; Koort, H.J. & Thiensiri, I. (1992). Excimer laser in dentistry: future possibilities with advanced technology. *Quintessence International,* Vol.23, pp. 117–133.

Fried, D.; Featherstone, J.D.B.; Visuri, S.R.; Seka, W.D. & Walsh, J.T. (1996). The caries inhibition potential of Er:YAG and Er:YSGG laser radiation. *SPIE Proceedings,* 2672, pp. 73–78.

Fusayama, T.; Nakamura, M.; Kurosaki, N. & Iwaku, M. (1979). Non-pressure adhesion of a new adhesive restorative resin. *Journal of Dental Research,* Vol.58, pp. 1364–1370.

Gaggl, A.; Schultes, G.; Müller, WD. & Kärcher, H. (2000). Scanning electron microscopical analysis of laser-treated titanium implant surfaces–a comparative study. *Biomaterials,* Vol.21, pp. 1067–1073

Gökçe, B.; Özpınar, B.; Dündar, M.; Çömlekoglu, E.; Sen, BH. & Güngör, MA. (2007). "Bond Strengths of All Ceramics: Acid vs Laser Etching". *Operative Dentistry,* Vol.32, pp. 168-173.

Guazzato, M.; Albakry, M.; Ringer, SP. & Swain, MV. (2004). Strength, fracture toughness and microstructure of a selection of all-ceramic materials. Part II. Zirconia-based dental ceramics. *Dental Materials,* Vol.20, pp. 449–456.

Harashima, T.; Kinoshita, J. & Kimura, Y.; Brugnera, A.; Zanin, F.; Pecora JD. & Matsumoto, K. (2005). Morphological comparative study on ablation of dental hard tissues at cavity preparation by Er:YAG and Er,Cr:YSGG lasers. *Photomedicine and Laser Surgery*, Vol.23, No.1, pp. 52–55.

Haselton, DR.; Diaz-Arnold, AM. & Hillis, SL. (2000). Clinical assess- ment of high-strength all-ceramic crowns. *Journal of Prosthetic Dentistry*, Vol.83, pp. 396–401.

Hibst, R. & Keller, U. (1989). Experimental studies of the application of the Er:YAG laser on dental hard substances: I. Measurement of the ablation rate. *Laser in Surgery and Medicine*, Vol.9, No.4 , pp. 338– 344.

Hossain, M.; Nakamura, Y.; Yamada, Y.; Kimura, Y.; Nakamura, G. & Matsumoto, K. (1999). Ablation depths and morphological changes in human enamel and dentin after Er:YAG laser irradiation with or without water mist. *Journal of Clinical Laser in Medical Surgery*, Vol.17, pp. 105– 109.

Hossain, M.; Nakamura, Y.; Kimura, Y.; Yamada, Y.; Ito, M. & Matsumoto, K. (2000). Caries-preventive effect of Er:- YAG laser irradiation with or without water mist. *Journal of Clinical Laser in Medical Surgery*, Vol.18, pp. 61–65.

Hossain, M.; Nakamura, Y.; Yamada, Y.; Suzuki, N.; Mur- akami, Y. & Matsumoto, K. (2001). Analysis of surface roughness of enamel and dentin after Er,Cr:YSGG laser ir- radiation. *Journal of Clinical Laser in Medical Surgery*, Vol.19, pp. 297–303.

Hummel, M. & Kern, M. (2004). Durability of the resin bond strength to the alumina ceramic Procera. *Dental Materials*, Vol.20, No.5, pp. 498–508

Jacobsen, NL.; Mitchell, DL.; Johnson, DL. & Holt, RA. (1997). Lased and sandblasted denture base surface preparations affecting resilient liner bonding. *Journal of Prosthetic Dentistry*, Vol.78, No.2, pp. 153–158

Janda, R.; Roulet, JF.; Wulf, M. & Tiller, H-J. (2003). A new adhesive technology for all-ceramics. *Dental Materials*, Vol.19, No.6, pp. 567–573

Karakoca, S. & Yılmaz, H. (2009). Influence of surface treatments on surface roughness, phase transformation, and biaxial flexural strength of Y-TZP ceramics. *Journal of Biomedical Materials Research Part B: Applied Biomaterials*, Vol.91, No.2, pp. 930–937

Kayano, T.; Ochiai, S.; Kiyono, K.; Yamamato, H.; Nakajima, S. & Mochizuki, T. (1989). Effects of Er:YAG laser irradiation on human extracted teeth. *The Journal of the Stomatological Society, Japan*, Vol.56, No.2, pp. 381–392

Kataumi, M.; Nakajima, M.; Yamada, T. & Tagami, J. (1998). Tensile Bond strength and SEM evaluation of Er:YAG laser irradiated dentin using dentin adhesive. *Dental Materials Journal*, Vol.17, pp. 125–138

Keller, U. & Hibst, R. (1989). Experimental studies of the application of the Er:YAG laser on dental hard substances: II. Light microscopic and SEM investigations. *Lasers in Surgery and Medicine*, Vol.9, No.4, pp. 345–351

Kern, M. & Thompson, V.P. (1994). Sandblasting and silica coating of a glass-infiltrated alumina ceramic: volume loss, morphology, and changes in the surface composition. *Journal of Prosthetic Dentistry*, Vol.71, No.5, pp. 453–461

Kern, M. & Thompson, V.P. (1995). Bonding to glass infiltrated alumina ceramic: adhesive methods and their durability. *Journal of Prosthetic Dentistry*, Vol.73, No.3, pp. 240–249

Kern, M. & Wegner, SM. (1998). Bonding to zirconia ceramic: adhesion methods and their durability. *Dental Materials*, Vol.14, No.1, pp. 64–71

Kim, BK.; Bae, HE.; Shim, JS. & Lee, KW. (2005). The influence of ceramic surface treatments on the tensile bond strength of composite resin to all-ceramic coping materials. *Journal of Prosthetic Dentistry*, Vol.94, No.4, pp. 357–362

Kim, JT. & Cho, SA. (2009). The effects of laser etching on shear bond strength at the titanium ceramic interface. *Journal of Prosthetic Dentistry*, Vol.101, No.2, pp.101–106

Kramer, N.; Popp, S.; Sindel, J. & Frankenberger, R. (1996). Einfluss der Vorbehandlung von ompositinlays auf die Verbundfestigkeit. *Deutsch Zahnarztl Z*, Vol.51, pp. 598–601

Kumbuloglu, O.; Lassila, L.V.; User, A. & Vallittu, P.K. (2006). Bonding of resin composite luting cements to zirconium oxide by two air-particle abrasion methods. *Operative Dentistry*, Vol.31, No.2, pp. 248–255

Kutsch, VK. (1993). Lasers in dentistry: comparing wavelengths. *Journal of the American Dental Association*, Vol.124, No.2, pp.49-54

Lee, B.S.; Lin, P.Y.; Chen, M.H.; Hsieh, TT.; Lin, CP.; Lai, JY. & Lan, WH. (2007). Tensile bond strength of Er,Cr:YSGG laser-irradiated human dentin and analysis of dentin–resin interface. *Dental Materials*, Vol.23, No.5, pp.570–578.

Leinfelder, KF. (2001). Dentin adhesives for the twenty-first century. *The Dental Clinics of North America*, Vol.45, No.1, pp. 1-6

Li, R.; Ren, Y. & Han, J. (2000). Effects of pulsed Nd:YAG laser irradiation on shear bond strength of composite resin bonded to porcelain. *Hua Xi Kou Qiang Yi Xue Za Zhi*, Vol.18, No.6, pp. 377–379.

Li, Z.Z.; Code, J.E. & Van De Merwe, W.P. (1992). Er:YAG laser ablation of enamel and dentin of human teeth: Determination of ablation rates at various fluencies and pulse repetition rates. *Lasers in Surgery and Medicine*, Vol.12, No.6, pp. 625–630

Liberman, R.; Segal, TH.; Nordenberg, D. & Serebro, LI. (1984). Adhesion of composite materials to enamel: Comparison between the use of acids lasing as pretreatment. *Lasers in Surgery and Medicine*, Vol.4, No.4, pp. 232-237

Liu, J.F.; Liu, Y. & Stephen, H.C. (2006). Optimal Er:YAG laser energy for preventing enamel demineralization. *Journal of Dentistry*, Vol.34, No.1, pp. 62–66

Malmström, HS.; McCormack, SM.; Fried, D. & Featherstone, JD. (2001). Effect of CO_2 laser on pulpal temperature and surface morphology: an in vitro study. *Journal of Dentistry*, Vol.29, No.8, pp. 521-529

Marshall Jr., GW.; Marshall, SJ.; Kinney, JH. & Balooch, M. (1997). The dentin substrate: structure and properties related to bonding. *Journal of Dental Research*, Vol.25. No.6, pp. 441-458

Matinlinna, JP. & Vallitu, PK. (2007). Bonding of resin composites to etchable ceramic surfaces – an insight review of the chemical aspects on surface conditioning. *Journal of Oral Rehabilitation*, Vol.34, No.8, pp. 622-630

McCormack, SM.; Fried, D.; Featherstone, JD.; Glena, RE. & Seka, W. (1995). Scanning electron microscope observations of CO2 laser effects on dental enamel. *Journal of Dental Research*, Vol.74, No.10, pp. 1702-1708

Moretto, SG.; Azambuja, N.Jr.; Arana–Chavez, V.E; Reis, AF.; Giannini M.; Eduardo C de P. & De Freitas, PM. (2010). Effects of ultramorphological changes on adhesion to lased dentin-Scanning electron microscopy and transmission electron microscopy analysis. *Microscopy Research Technique*, Vol.74, No.8, pp. 720–726

Nakabayashi, N.; Kojima, K. & Mashura E. (1982). The promotion of adhesion by resin infiltration of monomers into tooth structure. *Journal of Biomedical Materials Research*, Vol.16, pp. 265-273

Navarro, RS.; Gouw-Soares, S.; Cassoni, A.; Haypek, P.; Zezell, DM. & Eduardo, CP. (2010). The influence of erbium:yttrium-aluminum-garnet laser ablation with variable pulse width on morphology and microleakage of composite restorations. *Lasers in Medical Science*, Vol.25, No.6, pp. 881–889

Noda, M.; Okuda, Y.; Tsuruki, J.; Minesaki, Y.; Takenouchi, Y. & Ban, S. (2010). Surface damages of zirconia by Nd:YAG dental laser irradiation. *Dental Materials Journal*, Vol.29, No.5, pp. 536-541

Obata, A.; Tsumura, T.; Niwa, K.; Ashizawa, Y.; Deguchi, T. & Ito, M. (1999). Super pulse CO2 laser for bracket bonding and debonding. *European Journal of Orthodontics*, Vol.21, No.2, pp. 193–198

Oelgiesser, D.; Blasbalg, J. & Ben-Amar, A. (2003). Cavity preparation by Er-YAG laser on pulpal temperature rise. *American Journal of Dentistry*, Vol.16, No.2, pp. 96-98

Oliveira, MT.; Arrais, CA.; Aranha, AC.; Paula Eduardo, C.; Miyake, K.; Rueggeberg, FA. & Giannini, M. (2010). Micromorphology of resin–dentin interfaces using one-bottle etch & rinse and self-etching adhesive systems on laser- treated dentin surfaces: a confocal laser scanning microscope analysis. *Lasers in Surgery and Medicine*, Vol.42, No.7, pp. 662–670

Olsen, ME.; Bishara, SE.; Damon, P. & Jakopsen, JR. (1997a) Comparison of shear bond strength and surface structure between conventional acid etching and air abrasion of human enamel. *American Journal of Orthodontics and Dentofacial Orthopedics*, Vol.112, No.5, pp. 502-506

Olsen, ME.; Bishara, SE.; Damon, P. & Jakopsen, JR. (1997b). Evaluation of Scotchbond multipurpose and maleic acid as alternative methods of bonding orthodontic brackets. *American Journal of Orthodontics and Dentofacial Orthopedics*, Vol.111, No.5, pp. 498-501

Özcan, M.; Alkumru, HN. & Gemalmaz, D. (2001). The effect of surface treatment on the shear bond strength of luting cement to a glass-infiltrated alumina ceramic. *International Journal of Prosthodontics*, Vol.14, No.4, pp. 335–339

Özcan, M. (2002). The use of chairside silica coating for different dental applications: a clinical report. *The Journal of Prosthetic Dentistry*, Vol.87, No.5, pp. 469–472

Özcan M. (2003). Adhesion of resin composites to biomaterials in dentistry: an evaluation of surface conditioning methods. PhD Thesis, University of Groningen, Groningen, The Netherlands

Parker, S. (2004). The use of lasers in fixed prosthodontics. *Dental Clinics of North America*, Vol.48, No.4, pp. 971-998

Pashley, DH. (1992). The effects of acid etching on the pulpodentin complex. *Operative Dentistry*, Vol.17, pp. 229-242

Pashley, DH. & Carvalho, RM. (1997). Dentin permeability and dentin adhesion. *Journal of Dentistry*, Vol.25, pp. 355-372

Phark, JH.; Duarte S, Jr.; Blatz, M. & Sadan, A. (2009). An in vitro evaluation of the long-term resin bond to a new densely sintered high-purity zirconium-oxide ceramic surface. *The Journal of Prosthetic Dentistry*, Vol.101, No.1, pp. 29–38

Ramos, A.C.; Esteves-Oliveira, M., Arana-Chavez, V.E. & de Paula Eduardo, C. (2010). Adhesives bonded to erbium: yttrium–aluminum–garnet laser-irradiated dentin: transmission electron microscopy, scanning electron microscopy and tensile bond strength analyses. *Lasers in Medicine and Science*, Vol.25, No.2, pp. 181–189

Reynold, IR. (1975). A review of direct bonding. *British Journal of Orthodontics*, Vol.2, pp. 171–180

Roberts-Harry, DP. (1992). Laser etching of teeth for orthodontic bracket placement: A preliminary clinical study. *Lasers in Surgery and Medicine*, Vol.12, No.5, pp. 467-470

Sakakibara, Y.; Ishimaru, K. & Takamizu, M. (1998). A study on bond strength to dentin irradiated be Erbium:YAG laser. *The Japanese Journal of Conservative Dentistry*, Vol.41, pp. 207-219

Sasaki, LH.; Lobo, PD.; Moriyama, Y.; Watanabe, IS.; Villaverde, AB.; Tanaka, CS.; Moriyama, EH. & Brugnera A, Jr. (2008). Tensile bond strength and SEM analysis of enamel etched with Er:YAG laser and phosphoric acid: a comparative study in vitro. *Brazilian Dental Journal*, Vol.19, No.1, pp. 57-61

Sato, H.; Yamada, K.; Pezzotti, G.; Nawa, M. & Ban, S. (2008). Mechanical properties of dental zirconia ceramics changed with sandblasting and heat treatment. *Dental Materials Journal*, Vol.27, No.3, pp. 408–414

Schein, MT.; Bocangel, JS.; Nogueira, GE. & Schein, PA. (2003). SEM evaluation of the interaction pattern between dentin and resin after cavity preparation using ER:YAG laser. *Journal of Dentistry*, Vol.31, No.2, pp. 127-135

Schmage, P.; Nergiz, I.; Herrmann, W. & Özcan, M. (2003). Influence of various surface-conditioning methods on the bond strength of metal brackets to ceramic surfaces. *American Journal of Orthodontic Dentofacial Orthopedic*, Vol.123, No.5, pp. 540-546

Sen, D.; Poyrazoglu, E.; Tuncelli, B. & Goller, G. (2000). Shear bond strength of resin luting cement to glass-infiltrated porous aluminum oxide cores. *The Journal of Prosthetic Dentistry*, Vol.83, No.2, pp. 210–215

Şen S & Ceylan G. (2010) The Effects of Different Surface Treatments on the Bond Strength of Zirconium-Oxide Ceramic and Adhesive Resin. PhD Thesis. Ondokuz Mayıs University, School of Dentistry Department of Prosthodontics, Samsun, Turkey.

Sharpe, AN. (1967). Influence of the crystal orientation in human enamel on its reactivity to acid as shown by high resolution microradiography. *Archieves of Oral Biology*, Vol.12, No.5, pp. 583-591

Sindel, J.; Gehrlicher, S. & Petschel, A. (1996). Untersuchungen zur Haftung von Kompositan VMK-Kerakim. *Deutsch Zahnarztl Z.*, Vol.51, pp. 712–716

Spohr, AM.; Borges, GA.; Júnior, LH.; Mota, EG. & Oshima, HM. (2008). Surface modification of In-Ceram Zirconia ceramic by Nd:YAG laser, Rocatec system, or aluminum oxide sandblasting and its bond strength to a resin cement. *Photomedicine and Laser Surgery*, Vol.26, No.3, pp. 203-208

Stern, RH.; Vahl, J. & Sognnaes, RF. (1972). Lased enamel: ultrastructural observations of pulsed carbon dioxide laser effects. *Journal of Dental Research*, Vol.51, No.2, pp. 455-460

Stübinger, S.; Homann, F.; Etter, C.; Miskiewicz, M.; Wieland, M. & Sader, R. (2008). Effect of Er:YAG, CO2 and diode laser irradiation on surface properties of zirconia endosseous dental implants. *Lasers in Surgery and Medicine*, Vol.40, No.3, pp. 223-228

Subaşı, MG. & Inan, O. (2011). Evaluation of the topographical surface changes and roughness of zirconia after different surface treatments. *Lasers in Medical Science*, July 24. [Epub ahead of print]

Tachibana, A.; Marques, MM.; Soler, JM. & Matos, AB. (2008). Erbium, chromium:yttrium scandium gallium garnet laser for caries removal: influence on bonding of a self-etching adhesive system. *Lasers in Medical Science*, Vol.23, pp. 435–441

Toksavul, S. & Toman, M. (2007). A short-term clinical evaluation of IPS Empress 2 crowns. *The International Journal of Prosthodontics*, Vol.20, No.2, pp. 168–172

Trajtenberg, CP.; Pereira, PN. & Powers JM. (2004). Resin bond strength and micromorphology of human teeth prepared with an Erbium:YAG laser. *American Journal of Dentistry*, Vol.17, No.5, pp. 331-336

Ural, Ç.; Külünk, T.; Külünk, Ş. & Kurt, M. (2010). The effect of laser treatment on bonding between zirconia ceramic surface and resin cement. *Acta Odontologica Scandinavica*, Vol.68, No.6, pp. 354-359

Van Meerbeek, B.; De Munck, J. & Yoshida, Y. (2003). Buonocore memorial lecture. Adhesion to enamel and dentin: current status and future challenges. *Operative Dentistry*, Vol.28, pp. 215–235

Van Noort, R. (2002a). Dental Ceramics, In: *An introduction to dental materials*, Van Noort R, (Ed.), 231–246, Elsevier Science, Hong Kong

Van Noort, R. (2002b). Principles of adhesion. In: *An introduction to dental materials*, Van Noort R, (Ed.), 68-78, Elsevier Science, Hong Kong

Visuri, SR.; Gilbert, JL.; Wright, DD.; Wigdor, HA. & Walsh, JT. Jr. (1996a). Shear strength of composite bonded to Er:YAG laser-prepared dentin. *Journal of Dental Research*, Vol.75, pp. 599-605

Visuri, SR.; Walsh, JT.; & Wigdor, HA. (1996b). Erbium laser ablation of dental hard tissue: Effect of water cooling. *Lasers in Surgery and Medicine*, Vol.18, No.3, pp. 294–300

Walsh, LJ. (1994). Clinical evaluation of dental hard tissue applications of carbon dioxide lasers. *Journal of Clinical Laser Medicine & Surgery*, Vol.12, No.1, pp. 11-15

Watanabe, I.; McBride, M.; Newton, P. & Kurtz, KS. (2009). Laser surface treatment to improve mechanical properties of cast titanium. *Dental Materials*, Vol.25, No.5, pp. 629-633

White, JM.; Goodis, HE.; Setcos, JC.; Eakle S.; Hulscher BE. & Rose CL. (1993). Effects of pulsed Nd:YAG laser energy on human teeth: a three-year follow- up study. *Journal of American Dental Association*, Vol.124, No.7, pp. 45-51

Wigdor, H.; Abt, E.; Ashrafi, S. & Walsh JT. Jr. (1993). The effect of lasers on dental hard tissues. *Journal of American Dental Association*, Vol.124, No.2, pp. 65-70

Wilder-Smith, P.; Lin, S.; Nguyen, A.; Liaw LH.; Arrastia AM.; Lee JP. & Berns MW. (1997). Morphological effects of ArF excimer laser irradiation on enamel and dentin. *Lasers in Surgery and Medicine*, Vol.20, No.2, pp. 142–148

Zachrisson, BU. & Buyukyılmaz, T. (1993). Recent advances in bonding to gold, amalgam and porcelain. *Journal of Clinical Orthodontics*, Vol.27, pp. 661-675

Zhang, Y.; Lawn, BR.; Rekow, ED. & Thompson, VP. (2004). Effect of sandblasting on the long-term performance of dental ceramics. *Journal of Biomedical Materials Research. Part B Applied Biomaterials*, Vol.71, No.2, pp. 381–386

Zhang, Y.; Lawn, BR.; Malament, KA.; Van Thompson, P. & Rekow, ED. (2006). Damage accumulation and fatigue life of particle-abraded ceramics. *The International Journal of Prosthodontics*, Vol.19, No.5, pp. 442–448

11

Diversity of Lips and Associated Structures in Fishes by SEM

Pinky Tripathi and Ajay Kumar Mittal
Banaras Hindu University,
India

1. Introduction

Lips are specialized structures that cover the jawbones, and border the anterior orifice of alimentary canal, the mouth. In general, lips and structures associated with them in different fish species may be considered as mainly concerned with the selection, capture, deglutition and pre digestive preparation of food. The effectiveness of these structures is dependent on modifications in relation to food and feeding habits of the fishes and environmental niches inhabited by them.

Morphological data are also key to understanding fish nutrition in ecology and aquaculture, and during development as well as mechanisms for physiological adaptations to a changing environment. A number of the multifunctional roles of the fish lips and associated structures that are discussed incorporate distinctive morphological features that will be highlighted in this chapter. The lips and associated structures represent a significant vertebrate innovation and are highly diversified.

Therefore, present work was undertaken to investigate diversity of the epithelia of lips and associated structures in different fresh-water fish species with the aim to elucidate the surface architecture using Scanning Electron Microscope. The functional aspects of the lips and associated structures in family Gobiidae, Cobitidae, Belontiidae and few species of Cyprinidae show considerable variation and exhibit unique morphological modifications associated with their lips and other structures around the mouth regarding information on the level of surface architecture as seen under SEM in relation to various food and feeding habits and ecological niches.

When we started to survey, collect and organize the current knowledge on lips and associated structures for the invited chapter on SEM, we soon realized that such a study would lead to a greater understanding only if the lips were discussed as incorporate distinctive morphological features.

The successful maintenance of fish populations in challenging environments requires responsive adjustments in their behaviour, morphology and physiology and these have been reflected by modifications at the level of their organ systems, organs and tissues. The lips are no exception to this. The importance of food in daily life of a fish is obvious, and is reflected in the form of the mouth, lips, jaws and so on. These structures present more diverse modifications than any other organ of the body.

Lips and the structures intimately associated with them in different fish species are in direct contact with a complex ever-changing aquatic environment and ecological conditions in which fish inhabits. They are highly sensitive, serve a variety of functions and are characteristically modified in different groups of fishes. These modifications may be associated in some way either with the diet or the method of feeding.

In general, the upper and the lower jaws bear relatively simple and thin lips. These may be thick, fleshy and fimbriated or even unculiferous. In some cases one or both of the lips may regress or fail entirely to develop. The rostral cap is probably present in most fish species, although it may be so reduced as to be overlooked easily. In many forms it lies well above the upper lip and plays no direct role in feeding, while in others it is greatly enlarged, partially or completely overlies the upper lip and plays a major role in gathering food from the substrate. In the older literature the rostral cap and horny jaw sheaths frequently are confused with the lips.

The gross and fine structure of the lips, the rostral caps and the horny jaw sheaths is extremely varied. This involves, among other things, formation of unculiferous fimbriae, tubercles, unculi, papillae, or ridges and grooves of variable height and distribution on the lips and the rostral cap, and sharp cutting edge, cone shaped structure or unculi on the horny jaw sheaths.

Literatures pertaining to the morpho-anatomical structures of the lip in freshwater teleosts are fragmentary and many authors while studying the alimentary canal, briefly described the morphology and structural organisation of the lips of different fish species (Vanajakshi, 1938; Kapoor, 1958; Khanna, 1961, 1962; Pasha, 1964 a, b, c; Saxena & Bakhshi, 1964; Lal et al., 1964; Chitray, 1965; Sehgal, 1966; Moitra & Bhowmik, 1967; Lal, 1968; Sehgal & Salaria, 1970; Moitra & Sinha, 1971; Sinha, 1975; Sinha & Moitra, 1975, 1976, 1978; Kapoor et al.,1975). Suzuki (1956) described the histological organisation of lips of bottom feeding scythe fish *Pseudogobio esocinus*. At the surface of lips, he reported the presence of characteristic processes, with numerous taste buds, playing an important role as food finders. Miller & Evans (1965) studied relationship between the external morphology of brain and lips with emphasis on the distribution of taste buds. Branson & Hake (1972) described morphological organisation of the lips of *Piaractus nigripinnis* and reported that the lips in this fish are adapted for accessory respiratory function. Kiyohara et al. (1980) gave an account of the distribution of taste buds on the lips of a minnow *Pseudorasbora parva*. Ono (1980) reported epidermal projections associated with taste buds on the lips in some loricariid catfishes.

Agrawal & Mittal (1991, 1992 a, b, c) reviewed the literature and described the structural organisation of the epithelia of lips and associated structures of three Indian major carps – a surface plankton and detritus feeder, *Catla catla*; a herbivorous column feeder, *Labeo rohita* and an omnivorous bottom feeder, *Cirrhina mrigala;* and a sluggish, bottom dwelling, carnivorous catfish, *Rita rita*. Mittal & Agrawal (1994) reported the structural organisation of the epithelia of lips and associated structures of an active predatory fish *Channa striata*.

Scanning Electron microscope (SEM) reveals the details of surface architecture of tissues to an extent not possible by other procedures. In spite of this fact, the review of literature reveals that not much attention has hitherto been paid to study the surface architecture of fish lips and structures associated with them using SEM. In view of this, Roberts (1982) who examined a variety of fish species using scanning electron microscope, reported that differences in morphology of the lips and associated structures include:

1. Degree of development and specialisation of the lips,
2. Degree of development and specialisation of the rostral cap,
3. Presence or absence of horny jaw sheaths on the jaws in addition to, or in place of the normal lips, and
4. Form and distribution of unculi on the rostral cap, lips and horny jaw sheaths.

Recently, Ojha & Singh (1992), using SEM described functional morphology of the anchorage system and food scrapers of *G. lamta*. Again, Pinky et al. (2002) made a detailed report on lips and associated structures of the same fish *G. lamta*. Yashpal et al. (2009) briefly reported the presence of unculi on the upper jaw epithelium of *Cirrhinus mrigala*. More recently, Tripathi & Mittal (2010) made a detailed report on lips and associated structures of the fish *Puntius sophore*.

Our current knowledge on aspects of modifications in lips and associated structures in fish arises from too little data to arrive at general trends without running the risk of confusing variability, both real and experimental with adaptive phenomenon. What will be required to remedy this situation is a more extensive examination of a larger variety of species, particularly species adapted to different living conditions. This is an area of research for which the extreme biodiversity of fishes will be a powerful tool. Recently, the lips and associated structures of the Labeonini cyprinids have traditionally been identified as important characters in their classification (Yang & Mayden 2010).

The present work has, therefore, been undertaken with the aim to make a comparative study of the organisational pattern of lips and structures associated with them, at scanning electron microscopic levels, in certain fresh-water fish species having different feeding habits and inhabiting varied ecological niches, to address following specific questions:

> Do the lips and the structures associated with them in fish species with different feeding habits and inhabiting varied ecological niches show modifications in their organisational pattern?
> Does the surface architecture of epidermis of lips and associated structures of the fish species show adaptive modifications in relation to their ecophysiological status and varied feeding habits?

This chapter treats the morphology of lips and associated structures of the family Gobiidae, Cobitidae, Belontiidae and few species of Cyprinidae. The fresh-water fish species inhabiting different ecological niches and having different feeding habits selected for this study are *Glossogobius giuris*, *Noemacheilus botia*, *Colisa fasciata*, *Garra lamta*, *Puntius sophore* and *Cyprinus carpio*.

Glossogobius giuris (Hamilton, 1822), belongs to the Family Gobiidae, suborder Gobioidei and Order Perciformes. It is predominantly a fresh water fish found throughout the plains of India and sometimes also found in brackish waters. The members of this genus, in general, are bottom fishes and food chiefly consists of small bottom living animals (Günther, 1989). *G. giuris* is a carnivorous surface feeder and feeds on small crustaceans, insects, molluscs, smaller fishes and tadpoles (Hora & Mukerji, 1953; Khanna, 1993).

Noemacheilus botia (Hamilton, 1822) belongs to the family Cobitidae, sub-order Cyprinoidei, and order Cypriniformes. It is a bottom dweller fish, which usually live under stones, and in currents swim from stone to stone (Nikolsky, 1963). It is an omnivorous very undemanding species, which accepts any kind of food and feeds on algal films as well.

Colisa fasciata (Bloch & Schneider, 1801) belongs to the family Belontiidae, suborder Anabantoidei and order Perciformes (Graham, 1997). It is an air breathing fish and the natural habitats of the members of this genus include very weedy rivers, streams and ponds, irrigation ditches, flooded rice fields and also very dirty accumulations of water such as drains (Günther, 1989). It is an omnivorous fish and feeds on nearly equal amount of plant and animal material (Khanna, 1993). It has been regarded a good larvicidal form and is a very efficacious fish (Hora & Mukerji, 1953).

Garra lamta (Hamilton, 1822) belongs to the family Cyprinidae, sub-order Cyprinoidei and order Cypriniformes. It is a hill stream fish and is predominantly adapted to life in swift-flowing waters. Behind the ventral mouth there is a sucking disc which enables the fish to hold fast in strong currents, mountain streams and rapids. Riverbeds comprising mainly rocks, boulders, stones and gravel form a useful hiding and anchoring substratum for the fish. These fishes have "stone clinging" and "stone licking" habit. Food mainly consists of algal felts, mats and periphyton that they scrape off stones.

Puntius sophore (Hamilton, 1822) belongs to the family Cyprinidae, sub-order Cyprinoidei and order Cypriniformes. It inhabits in shallow ponds and streams and is often found in large numbers in the polluted waters where large drains enter the main river (Hora & Mukerji, 1953). It is an omnivorous column feeder (Khanna, 1993) and likes to grub about the bottom (Günther, 1989). Food of the fish consists of much larger amount of plant material than that of animal.

Cyprinus carpio var. communis (Linnaeus, 1758) belongs to the family Cyprinidae, sub-order Cyprinoidei and order Cypriniformes. It is a cold water fish, but being very hardy, easily adapts to warm water. It is an omnivorous feeder. It browses on the shallow bottom and margins, takes in vegetable debris, insects, worms, crustaceans and also planktonic algae (Alikunhi, 1957). The species has been implicated in degradation of the aquatic environment mainly through its habit of rooting in the bottom that causes turbidity and deoxygenated conditions (Welcomme, 1988). In searching for worms and insect larvae, it burrows hole into the embankments (Hora & Pillay, 1962).

2. Materials and methods

Live specimens of *Glossogobius giuris* (approximately 60 ± 10 mm in length), *Noemacheilus botia* (approximately 35 ± 5 mm in length), *Colisa fasciata* (approximately 55 ± 5 mm in length), *Puntius sophore* (approximately 50 ± 10 mm in length), *Cyprinus carpio var communis* (approximately 40 ± 5 mm in length) were collected from river Ganges and ponds at Varanasi, Uttar Pradesh, India. *Garra lamta* (approximately 60 ± 10 mm in length) were collected from hill streams, Jonha Falls at Ranchi, Bihar, India. The fishes were maintained in the laboratory conditions at controlled room temperature (25 ± 2⁰C). The fishes were cold anaesthetised following (Mittal & Whitear, 1978). Pieces of lips and associated structures were excised and rinsed in physiological saline, dipped briefly in 0.1% solution of S-carboxymethyl- L-cysteine to remove mucus (Whitear and Moate, 1994) and fixed in 3%

glutaraldehyde in 0.1M sodium cacodylate buffer, at pH 7.4 for 4h at 4°C. Following fixation, the tissues were washed in 0.1M sodium cacodylate buffer (pH 7.4) and dehydrated at 4°C with graded ethyl alcohol in ascending concentrations. The tissues were then treated with ethyl alcohol and acetone in the ratios 3:1, 1:1 and 1:3, anhydrous acetone and critical point dried using a critical point dryer (BIO-RAD, England) with liquid carbon dioxide as the transitional fluid. Tissues were glued to stubs, using conductive silver preparation (Eltecks Corporation, India), coated with gold using a sputter coater (AGAR, B7340, England) and examined with a Scanning Electron Microscope (Leo, 435 VP, England).

3. Results

In all the fish species investigated the upper jaws, in general, show a variable degree of protrusion. The upper lip (UL), borne on upper jaw, is associated with the rostral cap (RC) (rostralkappe = Minzenmay, 1933), through a fold of skin (FSUR), that in turn continues with the dorsal head skin. The FSUR, in general, is thin and membranous, and shows remarkable capacity of extendibility. The FSUR, when mouth is closed, lies in a deep groove between the UL and the RC.

The lower jaws in the fish species, in contrast, are only slightly protrusive. The lower lip (LL) is borne on lower jaw. Generally, it continues with the ventral head skin directly at the narrow middle region and through a fold of skin (FSLS) at the lateral sides of the lower jaw. In *G. lamta*, however, a specialised structure - the adhesive pad (AP) is located between the LL and the ventral head skin. The AP is separated from the LL by a narrow groove lined by a thin fold of skin (FSLA) and from the ventral head skin by a deep cleft lined by a thin and extensive fold of skin (FSAV).

In *N. botia*, *C. fasciata*, *G. lamta*, *P. sophore* and *C. carpio* the UL and the LL, on the side facing the mouth opening, are associated with the horny upper jaw sheath (HUJS) and the horny lower jaw sheath (HLJS) respectively. The horny jaw sheaths are absent in *G. giuris*. Several papilliform teeth are also observed in the jaws of *G. giuris* and *C. fasciata*.

In the fish species investigated, in general, the epithelia of the UL, the RC, the FSUR, the LL, the FSLS, the AP, the FSLA and the FSAV are mucogenic. The epithelia of the HUJS and the HLJS, in contrast, are keratinized. In the epithelia of the RC and the AP in *G. lamta* and those of the HUJS and the HLJS in *C. fasciata*, however, both mucogenic and keratinized regions are observed.

In the present account unculi or keratinization are recorded in a total of 5 fishes, and SEM observations are recorded for all of the above mentioned 6 fishes. These observations may be resulted as follows:

4. *Glossogobius giuris*

In *G. giuris* the mouth is terminal (Fig 1 a). The lower jaw remains slightly projected beyond the upper jaw. The upper jaw, as compared to the lower jaw, is highly protrusive (Fig. 1 b, c). The UL and the LL covering the upper jaw and the lower jaw respectively are thick and prominent.

The epithelia of the UL, the LL, the RC, the FSUR and the FSLS are mucogenic and are covered by a mosaic of irregularly polygonal epithelial cells of varied dimensions (Fig. 2 a, b, c). The surface architecture of the epithelial cells is characterised by the presence of a series of micro-ridges. The boundaries between adjacent epithelial cells are demarcated by smooth well-defined uninterrupted double row of closely approximated micro-ridges.

The oral cavity of G. *giuris* is prominent in the strong dentition of their jaws, armored by several papilliform teeth (Fig.1c).

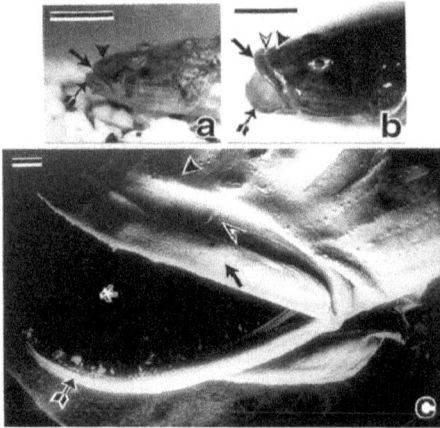

Fig. 1.

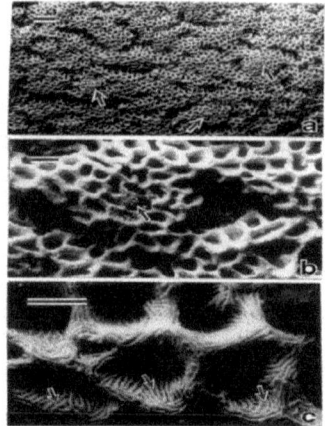

Fig. 2.

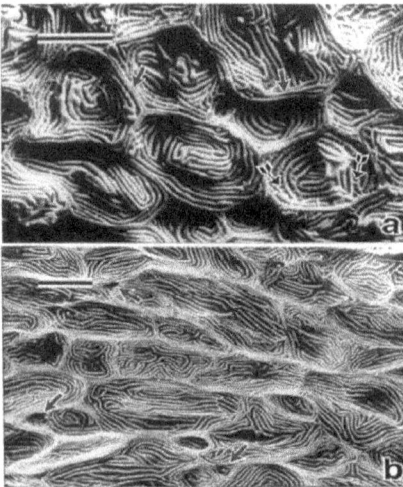

Fig. 3.

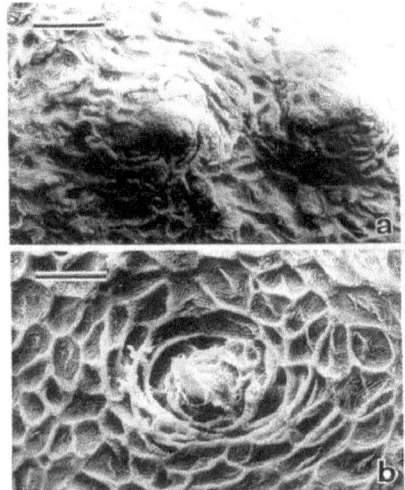

Fig. 4.

4.1 Epithelial cells

The surface of the epithelia of the UL and the LL characteristically appears like that of honeycomb. The epithelia appear folded and differentiated in to wide ridges separated by shallow and narrow furrows. The epithelial cells show a variable degree of invagination and thus their surfaces characteristically appear as concave depressions of varied depth (Fig. 2 a, b, c).

The micro-ridges on the surface of the epithelial cells in the epithelia of the UL and the LL appear smooth, extensive, uninterrupted, and are separated by wide furrows. In general, these appear systematically arranged parallel to each other often traversing towards the deeper regions of concave depressions in each cell (Fig. 2 b, c). The micro-bridges interconnecting the adjacent micro-ridges are prominent and are often located close to each other (Fig. 2 c).

The surfaces of the epithelial cells in the epithelium of the RC, in contrast, show only a slight concavity and appear as shallow depressions. The micro-ridges appear smooth, extensive, uninterrupted, at times branched and are separated by wide furrows. In general these are arranged systematically in a concentric manner, traversing almost parallel to the boundary of the cell forming intricate patterns. The micro-bridges interconnecting the adjacent micro-ridges are prominent similar to those in the epithelia of the UL and the LL (Fig. 3 a).

The epithelial cells in the epithelia of the FSUR and the FSLS, unlike those in the epithelia of the UL, the LL and the RC, appear flattened and do not show depressions at their surfaces. The micro-ridges on the surface of the epithelial cells in these regions though are extensive and often traverse parallel to each other are not interconnected by micro-bridges (Fig. 3 b).

4.2 Mucous cells

Interspersed between the epithelial cells in the epithelia of the FSUR and the FSLS mucous cell apertures of varied dimensions are observed similar to those in *P. sophore, C. carpio, N. botia* and *C. fasciata* (Fig. 3 b). In the epithelia of the UL, the LL and the RC, however, the mucous cell openings could not be clearly distinguished.

4.3 Taste buds

There are two types of TBs in *G. giuris,* type I-minute taste buds could be located on careful search in the epithelia of the UL and the LL. The epithelial cells around each taste bud are concentrically arranged (Fig. 2 a, b).

On the other hand type II-big taste buds on the surface of the epithelium of the RC are observed. At intervals, rounded mounds of epithelial cells bearing this type of comparatively big taste buds (Fig. 4 a). In the central region, the microvilli are arranged in the form of characteristic rosettes (Fig. 4 a). The microvilli of each rosette probably represent sensory hairs originating from sensory cells of the taste buds (Fig. 4 a).

4.4 Superficial neuromasts

Superficial neuromasts sunk slightly in the epithelium are observed. The central region of each such superficial neuromast is characterised by the presence of a characteristic structure consisting of tall-elongated closely approximated projections. This could represent the cupula of the superficial neuromast. The epithelial cells surrounding the superficial neuromasts are concentrically arranged to form a characteristic ring like pattern. (Fig. 4 b).

5. *Noemacheilus botia*

In *Noemacheilus botia* the mouth is small, inferior or sub-terminal, transverse and semicircular (Fig. 5 a). The upper jaw is highly protrusive (Fig. 5 b, c) and the lower jaw is only slightly protrusive. The HUJS and HLJS in this fish are prominent and are characteristically modified. Further, the UL and the LL in this fish are thick and plicate (Fig. 5 a, b, c).

5.1 Mucogenic epithelia

The epithelia of the UL, the LL, the RC, the FSUR and the FSLS are mucogenic. The epithelia of the UL and the LL are thrown in to distinctive protuberances of variable dimensions delineated by narrow furrows (Fig. 6 a, b). Further, the epithelia of the FSUR and the FSLS are characteristically pleated (Fig. 6 a, b, c).

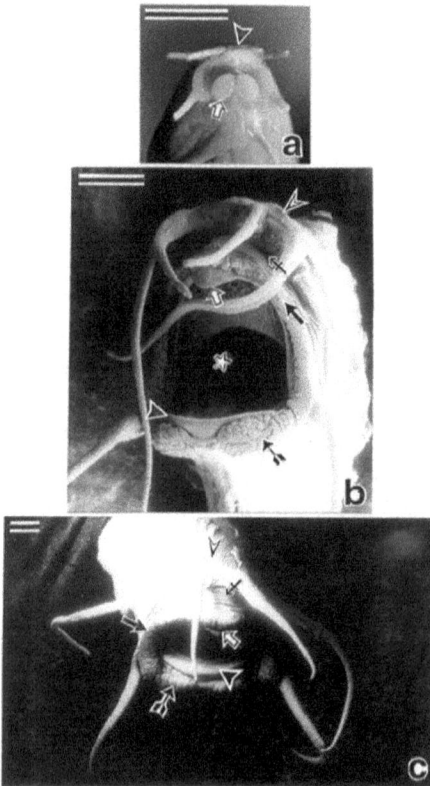

Fig. 5.

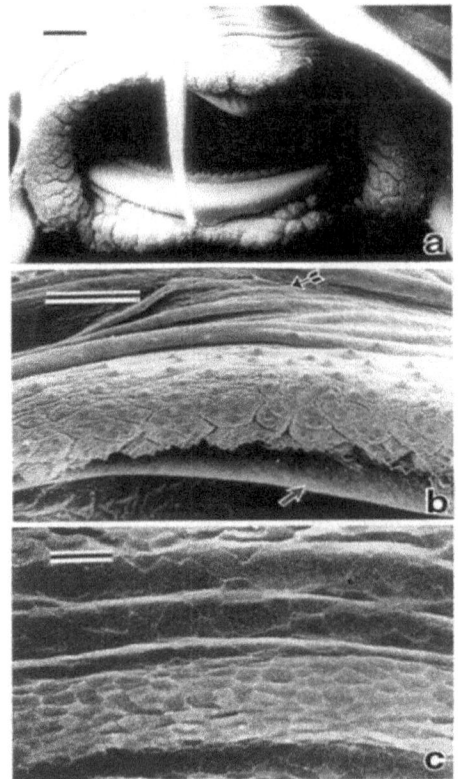

Fig. 6.

5.1.1 Epithelial cells

The epithelia of the UL, the LL, the RC, the FSUR and the FSLS in *N. botia* like those of *G. lamta*, *P. sophore* and *C. carpio* are covered by a mosaic of irregularly polygonal epithelial cells of varied dimensions (Fig. 6 c). The surface architecture of the epithelial cells is characterised by the presence of a series of micro-ridges separated by prominent irregular spaces. The micro-ridges in general appear sinuous having roughed surface, short with abrupt ends and irregularly interwoven to form web like patterns (Fig. 7 a, b). The boundaries between adjacent epithelial cells are demarcated by a well-defined double row of micro-ridges, which are often connected by transverse strands to give a braided appearance (Fig. 7 a, b).

5.1.2 Mucous cells

Mucous cell openings, seen as wide, rounded apertures or crypts, often containing blobs of mucus, are interspersed between the epithelial cells. Generally, such apertures occur where the boundaries of 3 or more epithelial cells meet (Fig. 7 a, b).

5.1.3 Taste buds

The most conspicuous surface feature of the epithelia of the UL and the LL in *N. botia* is the presence of distinct protuberances each studded with a large number of small, protrusions or elevations (Fig. 6 a, b; Fig. 8 a, b, c). These elevation appear conical and extend beyond the general epithelial surface (Fig. 8 a). The slopes of each elevation or papilla are covered by a mosaic pavement of concentrically arranged epithelial cells. The epithelial cells covering the surface of these papillae and at their vicinity, in general, are arranged concentrically (Fig. 8 b, c). The apex of each elevation is characterised by the presence of numerous closely packed microvilli (Fig. 8 b, c). These microvilli represent the taste hairs originating from taste cells of the buds and are projected through the rounded taste pore at the summit of these elevations.

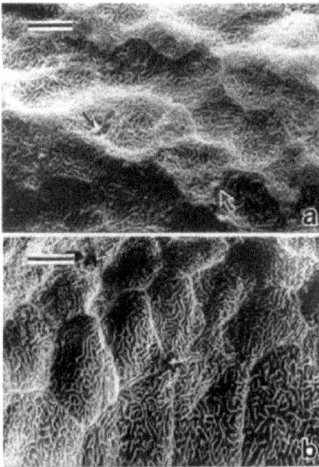 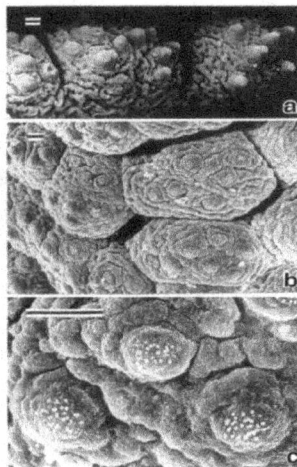 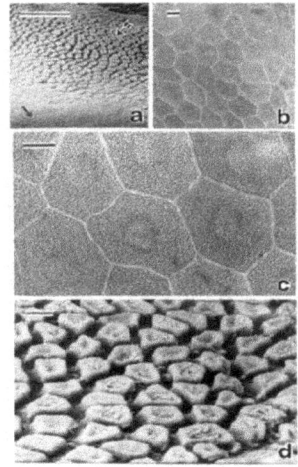

Fig. 7.   Fig. 8.   Fig. 9.

5.2 Keratinized epithelia

The epithelia of the HUJS and the HLJS are keratinized and resemble with each other in their surface architecture. The surface epithelial cells of the HUJS and the HLJS, in general, towards the proximal regions gradually get transformed in to truncated specialised structures - the unculi. Thus the surface architecture of the horny jaw sheaths at their distal regions i.e. the regions boarding the mouth is strikingly distinct from those at the proximal regions i.e. towards the buccal cavity (Fig. 9 a).

The surface of the epithelia at the distal regions of the HUJS and the HLJS are covered by a mosaic pavement of irregularly polygonal epithelial cells of varied dimensions (Fig. 9 b). The free surface of these cells is characterised by the presence of compactly arranged micro-ridges separated by narrow irregular spaces. The micro-ridges, in general, are short, sinuous, beaded, branched with abrupt ends and are irregularly interwoven to form maze like pattern (Fig. 9 c). The boundaries between adjacent epithelial cells are very prominent and appear slightly raised from the general surface of the epithelia (Fig. 9 c). These are demarcated by well-defined double rows of micro-ridges, which appear either lying very close to each other or fused. The central regions of these epithelial cells, in general, show rounded bulge at the surface. Each bulge is further demarcated by a narrow depression around them in the form of a ring (Fig. 9 b, c). These bulges could represent the nuclei of these epithelial cells, which appear greatly flattened in cross sections.

The surface of the epithelia at the proximal regions of the HUJS and the HLJS are studded with characteristic truncated, polygonal unculi (Fig. 9 d). The unculi, in general, appear uniform in dimensions and remain projected at the free surface. Each unculus represents modified surface relief of fine projections of a superficial layer epithelial cell. The unculi appear irregularly arranged and the central part of each unculus is wide and often irregularly distributed projections. The boundaries of the adjacent epithelial cells bearing the unculi are distinctly demarcated.

6. *Colisa fasciata*

In *C. fasciata* the mouth is oblique or slightly upturned (Fig. 10 a). The upper jaw is highly protrusive and extends forward to a great extent (Fig. 10 b, c). The lower jaw, however, is only slightly protrusive. Further, the UL and the LL, in general, are thin. The UL is, however, relatively less conspicuous than the LL. The lips are relatively thick and appear lobular at the lateral sides of the mouth. The lobes are separated by shallow grooves. The HUJS and the HLJS are very conspicuous and their distal regions are modified into prominent sharp cutting edges. At intervals papilliform teeth like structures are observed protruding on the surface of the jaw sheaths (Fig. 10 d).

The epithelia of the UL, the LL, the RC, the FSUR and the FSLS are mucogenic. In contrast, the epithelia of the HUJS and HLJS are keratinized and the dead keratinized epithelial cells at the surface are commonly visualised to be lifted up from the underlying tissues. They are probably in the process of being exfoliated. Further, the jaw sheaths are characterised by the presence of papilliform teeth like structures, which protrude at intervals from their surfaces facing the mouth opening (Fig. 11 a, b, c). The surface of the UL epithelium shows slight infoldings, which are visible even in stretched conditions. The folds on the surface of the LL, in contrast, are more distinct and are distinguished in to prominent ridges separated by shallow gutter-like depressions. Generally, these ridges run parallel to each other along the surface bordering the mouth. (Fig. 11 a, c).

6.1 Mucogenic epithelia

6.1.1 Epithelial cells

The epithelia of the UL, the LL, the RC, the FSUR and the FSLS in *C. fasciata* like those of *G. lamta, P. sophore, C. carpio* and *N. botia* are covered by a mosaic of irregularly polygonal epithelial cells of varied dimensions (Fig. 12 a). The surface architecture of the epithelial cells is characterised by the presence of a series of micro-ridges. The boundaries between adjacent epithelial cells are demarcated by smooth well-defined uninterrupted double row of closely approximated micro-ridges (Fig. 12 a, b, c).

The micro-ridges on the surface of the epithelial cells in the epithelia of the UL and the LL are generally short, straight or sinuous and smooth often arranged in the form of small groups (Fig. 12 c; Fig. 13 b). Several such groups of micro-ridges may be observed on the surface of each cell. Adjacent groups of micro-ridges are delineated from each other by extensive micro-ridges, which are often branched and encircle each group. The micro-ridges within a group are generally arranged parallel to each other either linearly or concentrically. The adjacent micro-ridges are interconnected with each other by fine transverse connections, the micro-bridges (Fig. 12 b).

The micro-ridges on the surface of the epithelial cells in the epithelium of the RC, in contrast, appear smooth, extensive, uninterrupted, at times branched and are separated by wide furrows. In general these are arranged systematically in a concentric manner, traversing almost parallel to the boundary of the cell forming intricate patterns. In the narrow central region of these cells, the micro-ridges are often either indistinct or fragmented (Fig. 12 a, b). The micro-ridges on the surface of the epithelial cells in the epithelia of the FSUR and the FSLS are relatively few, extensive and are located parallel to each other at long intervals (Fig. 13 a). Further micro-bridges could not be located.

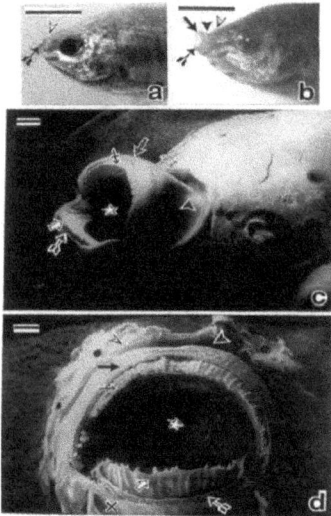 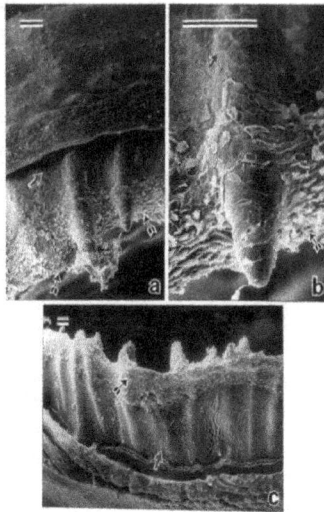 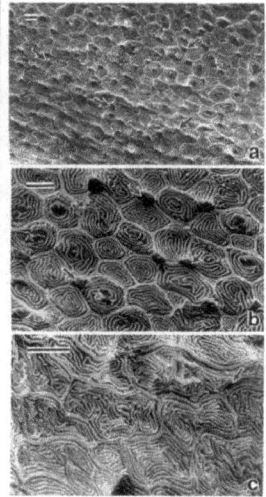

Fig 10.   Fig. 11.   Fig. 12.

6.1.2 Mucous cells

Interspersed between the epithelial cells in the epithelia of the UL, the LL, the RC, the FSUR and the FSLS mucous cell apertures of varied dimensions are observed similar to those in *G. lamta, P. sophore, C. carpio* and *N. botia* (Fig. 12 a, b, c).

6.1.3 Taste buds

Taste buds are located on the ridges at the surface of the UL and the LL. Further, the epithelial surface in the regions where taste buds are located is thrown into papillae like projections protruding beyond the general surface of the epithelia. At the summit of each such papilla several microvilli representing the taste hairs of the taste buds are located. The arrangement of the epithelial cells at and around each papilla bearing a taste bud is concentric and the appearance of the taste buds at the summit of these papillae are similar to those in *G. lamta, P. sophore, C. carpio* and *N. botia* (Fig. 13 a, b).

6.2 Keratinized epithelia

The HUJS and HLJS are covered with a pavement of epithelia of close packed polygonal cells of irregular shape and size. The epithelial cells, however, show regional variations in their surface architecture.

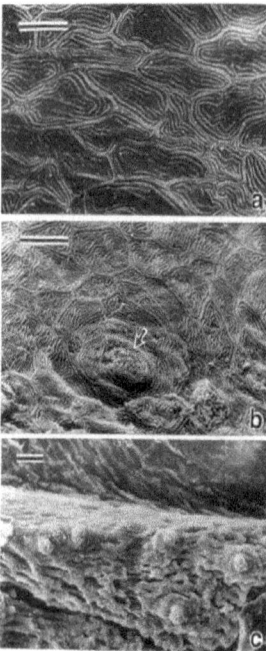
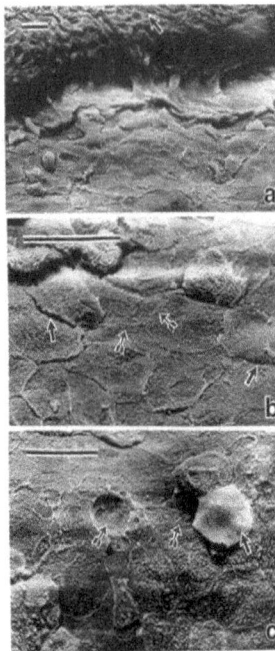

Fig. 13.   Fig. 14.   Fig.15.

At the proximal regions of the jaw sheaths, the surface of the epithelial cells appears scrawly. The micro-ridges are small, low, irregular and ill defined. Further, the pattern formed by these at the cell surfaces seems indistinctive. The boundary between the adjacent epithelial cells is delineated either by shallow separating clefts or by a double row of micro-ridges separated by distinct spaces (Fig. 14 a, b).

At the distal regions of the jaw sheaths the micro-ridges at the surface of the epithelial cells are frequently punctated (Fig. 14 c) and are separated by wide spaces. In addition short, sinuous, branched micro-ridges interwoven to form characteristic patterns are also observed (Fig. 15 a). The boundaries between adjacent epithelial cells are demarcated by double row of closely approximated micro-ridges.

At and near the apical margins of the horny jaw sheaths the epithelial cells generally, exhibited a surface relief of fine closely approximated micro-ridges such being often prominent in the central parts of the cells. The micro-ridges at the narrow peripheral portions of the cells were relatively short being more widely spaced and irregularly located. The boundaries of the epithelial cells were demarcated by prominent continuous marginal elevations of adjacent cells, sometimes with an inconspicuous gap between them.

The epithelial cells at the distal regions including the apical margins are frequently observed to be lifted up from the underlying tissues. They were probably in the process of being sloughed (Fig. 14 c; Fig. 15 a, b).

7. *Garra lamta*

In *Garra lamta* mouth is sub-terminal and is situated on the ventral side of the head. In this fish the upper jaw and the lower jaw are only slightly protrusive. The UL and the LL are rudimentary and are represented by slight thickening of the epithelia, at narrow regions, covering the upper jaw and the lower jaw respectively. In the regions bordering the lateral margins of the mouth, however, the lips are distinguished as small, stumpy, papillae like structures (Pinky et al., 2002, 2008).

7.1 Mucogenic epithelia

The RC is very prominent and greatly enlarged. Its epithelium may be distinguished in to a keratinized belt towards the mouth opening and a major mucogenic region towards dorsal head skin. At the apical margins of the RC mucogenic islands are observed between the keratinized regions. These non-keratinized and keratinized regions show characteristic alternate arrangements.

The epithelium of the AP like that of the RC is distinguished in to mucogenic and keratinized regions. The epithelium of the major central region of the AP is mucogenic. The narrow peripheral regions of the AP are, however, keratinized.

7.1.1 Epithelial cells

The surface of the mucogenic epithelium of the RC is covered by a mosaic pavement of irregularly polygonal epithelial cells of varied dimensions. The free surface of each epithelial cell is characterised by the presence of a series of compactly arranged micro-ridges separated by narrow irregular spaces. The micro- ridges, in general, appear sinuous, having smoothed surface, short with abrupt ends and irregularly interwoven to form maze like patterns. The boundaries between adjacent epithelial cells are demarcated by well-defined double row of micro- ridges, which are often interwoven to give a braided appearance.

7.1.2 Mucous cells

Interspersed between the epithelial cells rounded or irregular shaped crypts could be observed. These crypts often contain blobs of mucus and represent mucous cell openings. Generally, these apertures are located at the points where the boundaries of three or more epithelial cells meet.

7.1.3 Taste buds

A large number of epithelial protrusion or elevations that extends beyond the epithelial surface are located at irregular intervals. Each epithelial elevation is characteristically associated with a taste bud. The epithelial cells covering the surface of the elevations and at their vicinity, in general, are arranged concentrically. At the apical surface of each epithelial elevation, closely packed microvilli are observed. These microvilli appear to represent the taste hairs originating from the sensory cells of the taste buds.

The mucogenic islands in between the keratinized regions at the apical margin of the RC are characterised by the presence of several stumpy epithelial protrusions lying close to each other. Each epithelial protrusion is associated with a taste bud.

The surface relief of the epithelial cells in the mucogenic region of the AP, in general, is similar to that of the epithelial cells in the mucogenic epithelium of the RC. Further, the mucogenic epithelium of the AP resembles with that of the RC in the distribution of a large number of taste buds and in the presence of mucous cell openings.

7.2 Keratinized epithelia

7.2.1 Rostral cap

The surface of the epithelium of the RC in the keratinized regions, in contrast, to that in mucogenic regions appears shaggy. In general, it is matted with rounded projections or excrescencies in an organised array that are separated by shallow grooves.

The surface of each excrescence is represented by a cluster of several (15-25 or even more) prominent somewhat curved spine like unculi each having a broad base and a narrow apical end. Each unculus is projected at the free surface and represents modified surface relief of fine projections of a superficial layer epithelial cell.

These projections show a gradual increase in their height from the peripheral margin to the centre of the cell and in general appear compactly arranged or fused. The apical end of an unculus is either blunt or conical and the surface is rough with vertically oriented micro-villous projections. Between the unculi, the boundaries of the adjacent epithelial cells demarcated by distinct rows of micro-ridges may be observed.

The epithelial cells in shallow grooves between the excrescencies also show modified surface relief of fine projections. These are, however, less prominent and are not differentiated in to unculi like structures.

7.2.2 Adhesive pad

The keratinized epithelium at the posterior and lateral margins of the AP is characterized by the presence of rounded projections or excrescencies similar to those in the keratinized epithelium of the RC. In the keratinized epithelium at the anterior margin of the AP, in contrast, these excrescencies appear relatively prominent and tall. In general, these appear inverted cone shaped or basket like, each with a narrow proximal base, which gradually becomes relatively wide at the distal region. Like in the keratinized epithelium of the RC, the distal surfaces of these projections are characterised with the presence of a cluster of unculi, which represent modified surface relief of fine projections of superficial layer epithelial cells. The boundaries of the adjacent epithelial cells are often clearly demarcated by well-defined uninterrupted rows of micro-ridges.

7.2.3 Horny upper jaw sheath & Horny lower jaw sheath

The surface sculpture of the epithelia of the HUJS and the HLJS is similar to each other and are characteristically studded with tall, truncated, polygonal unculi. These unculi, in general, appear uniform in dimensions and shape and remain projected at the free surface. The unculi are arranged diagonally in parallel rows in an organised manner to form a characteristic pattern on the surface of the horny jaw sheaths. Each unculus, like that of the RC and the AP, represents modified surface relief of fine projections of a superficial layer epithelial cell. In contrast, these projections appear more developed smooth and prominent at the margins of the cells and show a gradual decline in their height towards the central part of the cell. This results in the formation of a characteristic sharp edge at the margin and a deep depression at the central region of each unculus. Each unculus thus appear very much like a tooth.

The rudimentary UL and the LL, and the delicate FSUR, the FSLA, the FSACAP and the FSAV remain concealed and thus the surface architecture could not be visualized because these regions are deeper in position.

8. *Puntius sophore*

In *Puntius sophore* the mouth is terminal (Fig. 16 a, b, c). The upper jaw is highly protrusive. In contrast, the lower jaw is only slightly protrusive. Further, the UL and the LL are prominent. The UL is however, thin and the LL, in contrast, is very thick (special permission for figures 16-19 have been taken from Tissue and Cell).

8.1 Mucogenic epithelia

8.1.1 Epithelial cells

Surface architecture of the mucogenic epithelia of the UL, the LL, the RC, the FSUR and the FSLS, in general, resembles with each other. The surface of the epithelia is covered by a mosaic pavement of irregularly polygonal epithelial cells of varied dimensions. The free surface of each epithelial cell is characteristically thrown in to a series of micro-ridges having smooth surface. The micro-ridges, in contrast to those of *G. lamta*, are separated by wide furrows and are extensive, at times branched and traverse almost parallel to the boundary of the cell forming intricate patterns (Fig. 17 a, b). Further, the adjacent micro-ridges are interconnected with each other by fine transverse connections, the micro-bridges.

The boundaries between adjacent epithelial cells are demarcated by smooth well-defined double row of closely approximated micro-ridges (Fig. 17 a, b). In addition, epithelial cells with compactly arranged sinuous, short micro-ridges with abrupt ends to form maze like pattern or with micro-ridges giving a punctated appearance to the surface are frequent in the epithelia on the apical side of the UL and the LL (Fig. 17 c).

8.1.2 Mucous cells

Crypts representing the mucous cell openings, often containing blobs of mucus, are frequently observed at the borders of 3 or 4 epithelial cells (Fig. 17 a). The crypts are relatively conspicuous, large, rounded and frequent in the epithelia of the UL, the LL, and the RC, than those of the FSUR and the FSLS.

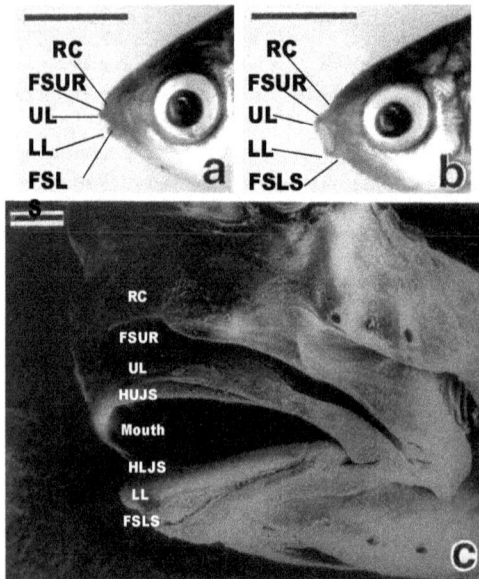

Fig. 16.

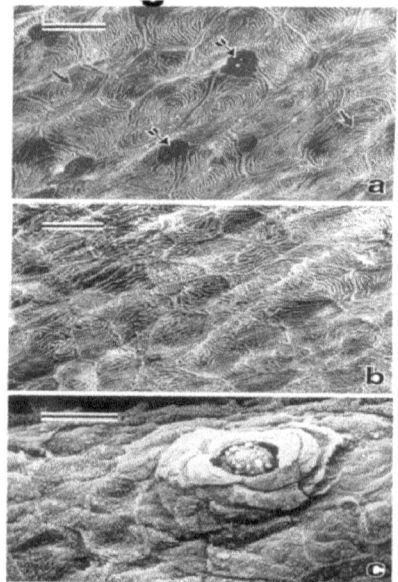

Fig. 17.

8.1.3 Taste buds

In the epithelia of the UL and the LL a large number of taste buds are observed (Fig. 18 a, b). Each taste bud is situated on a small epithelial papilla projecting at the surface. The epithelial cells covering the surface of these papillae and at their vicinity, in general, are arranged concentrically (Fig. 17 c; Fig. 18 a, b). At the summit of each papilla closely packed microvilli are observed (Fig. 17 c). These microvilli appear to represent the taste hairs originating from the sensory cells of the taste buds.

8.2 Keratinized epithelia

The epithelial surface of the HUJS and the HLJS at their distal regions are studded with characteristic truncated, polygonal unculi (Fig. 18 b). The unculi, in general, appear uniform in dimensions and remain projected at the free surface (Fig. 18 c; Fig. 19 a). Each unculus represents modified surface relief of fine projections of a superficial layer epithelial cell and resemble in their shape and organisation to those in the epithelia at the proximal regions of the HUJS and the HLJS of *N. botia*. The unculi in this fish appear relatively less orderly arranged. Further, the central part of each unculus is much wide and often shows the presence of characteristic irregularly distributed projections (Fig. 19 a, b).

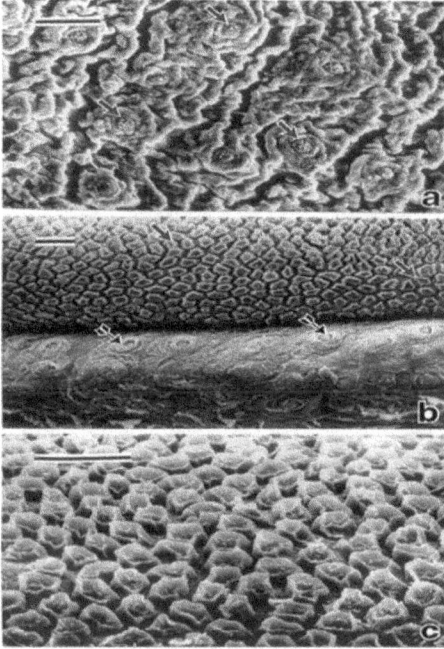 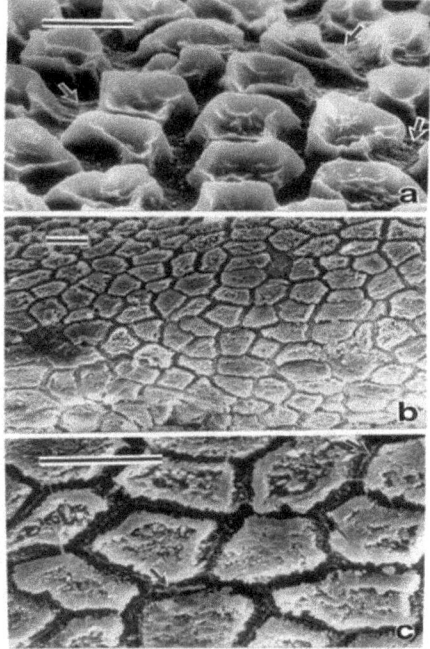

Fig. 18.   Fig. 19.

The unculi show a gradual decrease in their height towards the proximal regions of the HUJS and the HLJS. At these regions the surface relief of each unculus in contrast to those at the distal regions appears scraggy (Fig. 19 b, c). The micro-villous projections at the peripheral region on the surface of each cell appear fused to form the outer boundary of each unculus with scrawly surface (Fig. 19 c). The major central part of each unculus is occupied with micro-villous projections, which often appear fused awkwardly to give a scrambled or frothy appearance to the surface (Fig. 19 c).

Between the unculi both at the distal and proximal regions of the HUJS and the HLJS the boundaries of the adjacent epithelial cells demarcated by distinct rows of micro-ridges are observed (Fig. 19 a, c).

9. *Cyprinus carpio*

In *C. carpio* the mouth is terminal (Fig. 20 a, b). The upper jaw is highly protrusive (Fig. 20 c, d) and the lower jaw is slightly protrusive. The UL and the LL covering the upper jaw and the lower jaw respectively, in contrast, are thick and prominent and are associated with relatively inconspicuous HUJS and the HLJS.

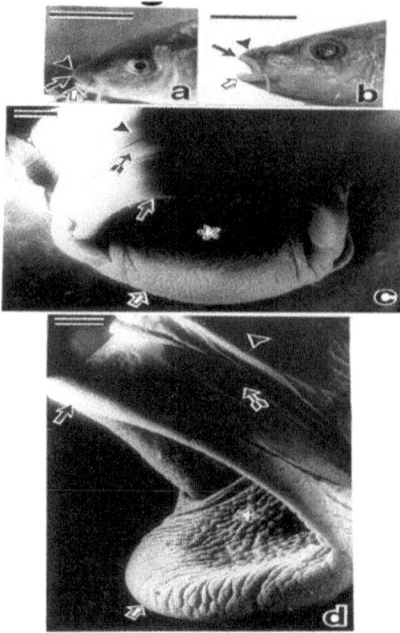

Fig. 20.

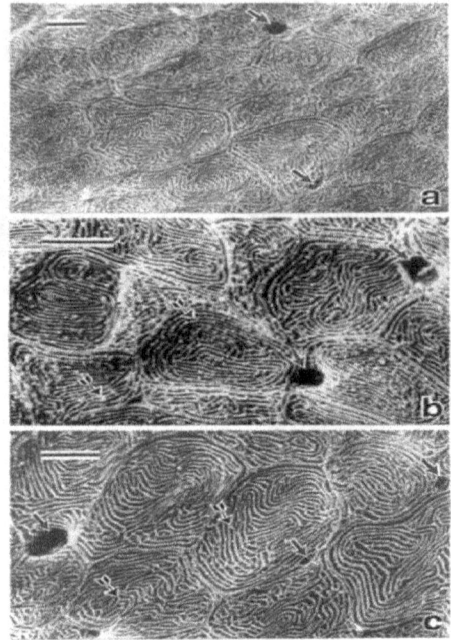

Fig. 21.

9.1 Mucogenic epithelia

9.1.1 Epithelial cells

The surface of the mucogenic epithelia of the UL, the LL, the RC, the FSUR and the FSLS, like those of *P. sophore,* is covered by a mosaic pavement of epithelial cells. The micro-ridges on the surface of the epithelial cells of these epithelia are separated by wide furrows and are extensive. In contrast, the micro-ridges in this fish often appear beaded, branched and interlocked to form intricate patterns (Fig. 21 a, b, c). Further, the micro-bridges interconnecting the adjacent micro-ridges are more prominent and frequent. The boundaries between adjacent epithelial cells are demarcated by well-defined double row of closely approximated micro-ridges often interconnected by transverse connections (Fig. 21 b, c).

9.1.2 Mucous cells

Interspersed between the epithelial cells in the epithelia of the UL, the LL, the RC, the FSUR and the FSLS are observed wide, rounded crypts or pores of varied dimensions, like in *P. sophore.* Generally these crypts or pores occur where the boundaries of three or four epithelial cells meet. These could represent the openings of the mucous cells. This is further confirmed by the presence of blobs of mucus in most of these crypts (Fig. 21 a, b, c).

9.1.3 Taste buds

In the RC epithelium, small taste buds are located individually at long intervals (Fig. 22 a). In the epithelia of the UL and the LL, however, a large number of taste buds are observed, each located on a small epithelial papilla projecting at the surface (Fig. 22 a, b). The taste buds at the major distal potion of the lips are characteristically located in-groups arranged in parallel rows (Fig. 22 b, c). The arrangement of the epithelial cells at and around each papilla bearing a taste bud and the appearance of the taste buds at the summit of these papillae are similar to those in *P. sophore* and *G. lamta*.

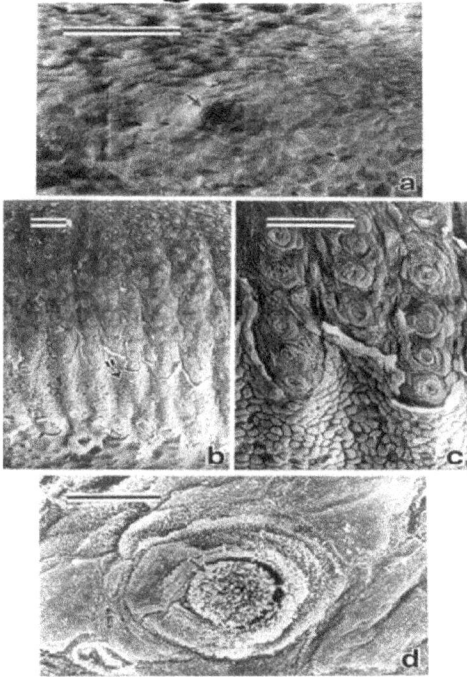
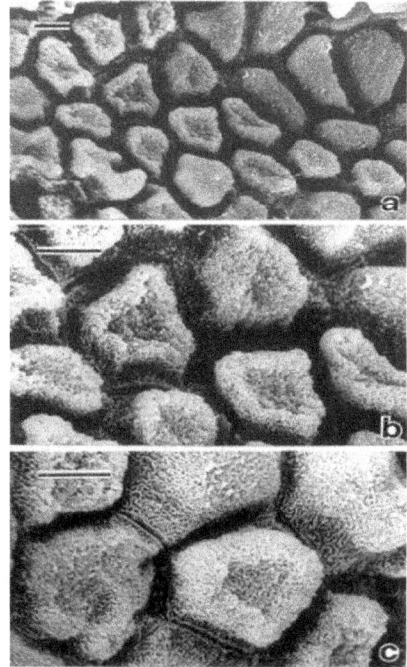

Fig. 22.   Fig. 23.

9.2 Keratinized epithelia

The epithelial cells at the surface of the HUJS and the HLJS are studded with characteristic polygonal unculi (Fig. 23 a, b, c). The surface relief of each unculus appears truncated. Each unculus is raised significantly from the general surface as a narrow bend at the periphery and has a shallow depression at its major central region (Fig. 23 a, b, c). Further, the surface relief appears scraggy similar to those at the proximal regions of the HUJS and the HLJS of *P. sophore*. In contrast, the micro-ridges at the surface of each unculus are prominent, separated by wide spaces and do not fuse with each other. These micro-ridges are interwoven forming web-like patterns (Fig. 23 b, c). The boundaries of the adjacent epithelial cells, modified as unculi, are demarcated by distinct double rows of micro-ridges (Fig. 23 c). The space between these micro-ridges is relatively prominent and wide.

10. Explanation of figures

Figure 1 (a - b) Photographs of a part of the dorso-lateral side of the head of *G. giuris,* with closed mouth (a) and with open mouth (b) showing the RC (arrowhead), the UL (arrow), the FSUR (white arrowhead) and the LL (winged arrow) (Scale bar = 5 mm) **(c)** Scanning electron photomicrograph of a part of the dorso-lateral side of the head of *G. giuris,* showing the RC (arrowhead), the FSUR (white arrowhead), the UL (arrow), the LL (winged arrow) and the mouth (white star). Several papilliform teeth are also observed inside the mouth. (Scale bar = 400 µm)

Figure 2 (a – c) Scanning electron photomicrographs showing surface architecture of the epithelia of the UL of *G. giuris.* (Scale bar = (a) 30 µm, (b, c) 10 µm). **(a)** Showing the epithelial cells having a variable degree of invagination and small type I taste buds (arrows). **(b)** Same as (a) in higher magnification. **(c)** Same as (b) in higher magnification, showing the surface of the epithelial cells characterised with smooth, extensive, uninterrupted micro-ridges traversing towards the deeper regions of concave depressions and are separated by wide furrows. Note fine micro-bridges interconnecting the adjacent micro ridges (arrows).

Figure 3 (a - b) Scanning electron photomicrographs showing surface architecture of the epithelia of the RC (a), the FSUR (b) of *G. giuris.* (Scale bar = 10 µm). **(a)** Showing the surface of the epithelial cells is characterised with smooth, extensive, uninterrupted sometimes branched micro-ridges separated by wide furrows. Note boundary is demarcated by double row of micro-ridges (arrows) and fine micro-bridges interconnecting the adjacent micro-ridges (winged arrows). Note each epithelial cell shows slight concavity, compare with plate 51 (b, c). **(b)** Showing micro-ridges forming characteristic pattern on the epithelial cell surface and small mucous cell apertures (arrows).

Figure 4 (a - b) Scanning electron photomicrographs showing surface architecture of the epithelia of the RC of *G. giuris.* (Scale bar = 20 µm) **(a)** Showing big type II taste buds located on the mounds of epithelial cells. Note the epithelial cells at each mound are characteristically arranged concentrically. **(b)** Showing superficial neuromast sunk slightly in the epithelium with characteristic cupula like structure.

Figure 5 (a) Photograph of the head region of *N. botia* showing the UL (arrow), the LL (white arrow) and the RC (arrowhead). (Scale bar = 5 mm). **(b - c)** Scanning electron photomicrograph of a part of the head of *N. botia* showing the RC (white arrowhead), the FSUR (barred arrow), the UL (arrow), the HUJS (white arrow), the HLJS (arrowhead), the LL (winged arrow) and mouth (white star) (Scale bar = 400µm).

Figure 6 (a) Scanning electron photomicrograph of a part of the head of *N. botia* showing the RC, the FSUR, the UL, the HUJS, the HLJS and the LL. On the UL and the LL protuberances of various dimensions separated by furrows are discernible. (Scale bar = 200 µm). **(b)** Scanning electron photomicrograph of the epithelium of the UL showing protuberances of variable dimensions delineated by narrow furrows. A part of the HUJS (arrow) and the FSUR (winged arrow) is also discernible. (Scale bar = 200 µm). **(c)** Scanning electron photomicrograph of the epithelium of the FSUR showing characteristic pleats and a mosaic pavement of surface epithelial cells. (Scale bar = 20 µm).

Figure 7 (a, b) Scanning electron photomicrographs of the epithelium of the RC (a) and the FSUR (b) of *N. botia*. (Scale bar = 5 μm). Showing intricate patterns of micro-ridges at the surface of the epithelial cells. Note the presence of mucous cell apertures (arrows).

Figure 8 (a - c) Scanning electron photomicrographs of LL of *N. botia*. (Scale bar = 10 μm). **(a)** Showing distinct protuberances separated by furrows. Each protuberance is studded with a large number of epithelial elevations. **(b)** Same as (a) in higher magnification. Showing taste buds at the apex of each epithelial elevation. **(c)** Same as (b) in still higher magnification. Showing the apex of each epithelial elevation characterised by the presence of numerous closely packed microvilli, which represent the taste hairs originating from taste cells of the taste buds. Note the epithelial cells at the epithelial elevations are concentrically arranged.

Figure 9 (a – d). Scanning electron photomicrographs of HLJS of *N. botia*. (Scale bar = (a) 50 μm, (b, c, d) 5μm). **(a)** Showing the superficial layer epithelial cells at the distal region (arrow) that gradually get transformed into truncated unculi towards the prox imal region on the buccal cavity side of the HLJS (winged arrow). **(b)** Showing a mosaic pavement of irregularly polygonal epithelial cells at the distal region of the HLJS. Note distinct boundaries between the epithelial cells and characteristic rounded bulge in the central region of the epithelial cells. **(c)** Same as (b) in higher magnification, showing double row of micro-ridges demarcating the boundaries of the adjacent cells. Note the micro-ridges on the surface of the epithelial cells are short, sinuous and compactly arranged. **(d)** Proximal region of the HLJS, towards buccal cavity showing the epithelial cells modified into truncated unculi.

Figure 10 (a – b) Photographs of the head region of *C. fasciata* with closed mouth (a) and with open mouth (b) showing the RC (white arrowhead), the FSUR (arrowhead), the UL (arrow) and the LL (winged arrow). (Scale bar = 5 mm). **(c - d)** Scanning electron photomicrograph of a part of head region of *C. fasciata* in dorso-lateral view (c) and in front view (d), showing the RC (white arrowhead), the FSUR (arrowhead), the UL (arrow), the HUJS (barred arrow), the HLJS (white arrow), the LL (winged arrow), the FSLS (cross) and mouth (white star). (Scale bar = 400 μm).

Figure 11 (a – c) Scanning electron photomicrograph showing the surface architecture of the HUJS (a, b) and the HLJS (c) of *C. fasciata*. (Scale bar = 50 μm) **(a)** Note the difference in the surface architecture of the epithelia at the proximal region (arrow) and the distal region (winged arrows) of the HUJS. The epithelial cells in the distal region exfoliate at the surface. Note the presence of papilliform teeth like structures protruding at the surface. **(b)** Same as (a) in higher magnification. **(c)** Note the difference in the surface architecture of the epithelium at the proximal region (arrow) and the distal region (winged arrow) of the HLJS. The epithelial cells in the distal region exfoliate at the surface. Note the presence of papilliform teeth like structures protruding at the surface.

Figure 12 (a – c) Scanning electron photomicrograph showing the surface architecture of the RC (a, b), the UL (c) of *C. fasciata*. (Scale bar = 5 μm). **(a)** Showing mosaic pavement of irregularly polygonal epithelial cells interspersed with mucous cell apertures (arrows). Epithelial cells with a series of micro-ridges and the boundaries demarcated by smooth well-defined un-interrupted double row of micro-ridges. **(b)** Same as (a) in higher magnification. **(c)** The micro-ridges are short, straight or sinuous and smooth often arranged in the form of small groups.

Figure 13 (a - c) Scanning electron photomicrograph showing surface architecture of the FSUR (a), the LL (b, c) of *C. fasciata*. (Scale bar = 10 µm). **(a)** Showing mosaic pavement of irregularly polygonal epithelial cells. **(b)** The micro-ridges are short, straight or sinuous and smooth often arranged in the form of small groups. Note a taste bud at the summit of a papilla like projection (arrow). The epithelial cells are arranged concentrically around the taste bud. **(c)** Showing taste buds at the summit of papilla like projections on the surface of the LL thrown in to papillae like projections protruding beyond the general surface of the epithelia. At the summit of each such papilla several microvilli representing the taste hairs of the taste buds are located.

Figure 14 (a - c) Scanning electron photomicrographs showing surface architecture of the epithelia of the HLJS of *C. fasciata*. (Scale bar = 20 µm) **(a)** Showing a mosaic pavement of irregularly polygonal epithelial cells. A part of the lower lip is also visible (arrow). **(b)** Showing ill-defined micro-ridges on the surface of the epithelial cells at the proximal region of the HLJS. Note the boundaries between the adjacent epithelial cells are delineated either by shallow clefts (arrows) or by double row of micro-ridges (winged arrows). **(c)** Showing the epithelium at the transitional zone between the proximal region and the distal region of the HLJS. Note an epithelial cell separated from the underlying epithelial cell before its exfoliation (arrow). Impressions left on the surface (winged arrows) of the epithelium is due to the exfoliated epithelial cells. The surface of the epithelial cells at the distal region is characterised with punctated micro-ridges.

Figure 15 (a – b) Scanning electron photomicrographs showing surface architecture of the epithelia of the distal region of the HUJS of *C. fasciata*. (Scale bar = 5 µm) **(a)** Micro-ridges at the surface are either punctated or are short and sinuous. Note an exfoliated epithelial cell (arrow). **(b)** Showing prominent and raised micro-ridges in the central region of the epithelial cells (arrows).

Figure 16 (a – b) Photographs of the head region of *P. sophore* with closed mouth (a) and with open mouth (b), showing the upper lip (UL), the lower lip (LL), the rostral cap (RC) ,the fold of skin between UL and RC (FSUR) and the fold of skin between LL and ventral head skin (FSLS). (Scale bar = 5 mm). **(c)** Scanning electron photomicrograph of a part of the lateral side of mouth of *P. sophore* (Scale bar = 400 µm).

Figure 17 Scanning electron photomicrographs showing surface architecture of the epithelia of the UL **(a)**, the FSUR **(b)** and the LL **(c)** of *P. sophore*. (Scale bar = 10 µm). **(a)** Showing micro-ridges arranged characteristically (arrows) at the surface of the epithelial cells. Note wide mucous cell apertures (winged arrows). **(b)** Showing characteristic pattern of micro-ridges at the surface epithelial cells. **(c)** Showing a taste bud at the apex of a mound of epithelial cells. Note the epithelial cells surrounding the taste bud are arranged concentrically.

Figure 18 Scanning electron photomicrographs showing surface architecture of the epithelia of the UL **(a)**, distal region of HLJS and the LL **(b)** and distal region of the HLJS **(c)** of *P. sophore*. (Scale bar = 10 µm). **(a)** Showing a large number of taste buds (arrows). **(b)** Showing truncated unculi at the HLJS (arrows) and taste buds at the LL (winged arrows). **(c)** Showing truncated unculi with central concavity and raised margins.

Figure 19 Scanning electron photomicrographs showing surface architecture of the epithelia at the distal region of the HLJS **(a)** and at the proximal region of the HLJS **(b, c)** of *P. sophore*. (Scale bar = 5 µm). **(a)** Same as 21 c, in higher magnification, showing the superficial layer epithelial cells modified in to truncated unculi. Note the boundaries of the adjacent epithelial cells are demarcated by distinct micro-ridges (arrows). **(b)** Showing scraggy surface relief of the unculi. **(c)** Same as (b) in higher magnification, showing the scrawly surface of the unculi. The microvillar projections occupying the central part of each unculus give a frothy appearance to the surface. Note, between the unculi the boundaries of the adjacent epithelial cells are demarcated by distinct micro-ridges (arrows).

Figure 20 (a – b) Photographs of the head region of C. carpio showing the UL (arrow), the LL (white arrow), the RC (arrowhead). (Scale bar = 5 mm). **(c – d)** Scanning electron photomicrograph of a part of the head with open mouth showing the UL (arrow), the LL (white arrow), the RC (arrowhead) and the FSUR (winged arrow) and mouth (white star) of *C. carpio*. (Scale bar = 400 µm). **(c)** front view **(d)** lateral view.

Figure 21 (a – c). Scanning electron photomicrographs showing surface architecture of the epithelia of the UL (a, c) and the RC (b) of *C. carpio*. (Scale bar = 5 µm). **(a)** Showing mosaic pavement of irregularly polygonal epithelial cells with micro-ridges forming intricate patterns. Note mucous cell apertures (arrows) at the boundary of the epithelial cells.   **(b)** Showing micro-ridges forming intricate patterns at the surface of the epithelial cells. Note fine transverse connections in between the micro-ridges (winged arrows) and mucous cell apertures (arrows). **(c)** Showing micro-ridges forming intricate patterns at the surface of the epithelial cells. Note fine transverse connections in between the micro-ridges (winged arrows) and mucous cell apertures (arrows).

Figure 22 (a - d) Scanning electron photomicrographs showing surface architecture of the epithelia of the RC (a) and the LL (b, c, d) of *C. carpio*. (Scale bar = (a, b, c) 50 µm and (d) 10 µm). **(a)** Showing a mosaic pavement of the epithelial cells and a taste bud at the apex of a mound of epithelial cells (arrow). **(b)** Showing major mucogenic region characterised with the presence of a large number of taste buds arranged in parallel rows. Note a narrow keratinized region, the surface of which appears rough and studded with pebble like structures (winged arrows). **(c)** Same as (b) in higher magnification. **(d)** Showing a taste bud (arrow) at the apex of a mound of epithelial cells.

Figure 23 (a – c) Scanning electron photomicrographs showing surface architecture of the epithelium of HLJS of *C. carpio*. (Scale bar = 5 µm). **(a)** Showing truncated polygonal unculi with concavity in the central region of each unculus. **(b)** Same as (a) in higher magnification. Note prominent micro-ridges at the surface of the unculi, separated by wide spaces. **(c)** Same as (b) in higher magnification. Note the boundaries of the adjacent unculi are demarcated by double row of micro-ridges.

11. Discussion

Research on lips and associated structures began about 200 years ago, as described by Anson, 1929 in his manuscript "The comparative anatomy of the lips and labial villi of vertebrates". He made an attempt to define lips and on Danforth's interpretation of homology, homologous lips are found at certain stages of development in some representatives of all classes of vertebrates. The primary lips characteristic of selachians, after the maxillary and premaxillary bones have developed within the territory of the upper lip (toadfish, cod), may disappear (trout, Spelerpes), accompanied by a forward migration of the lower jaw. The secondary lips of higher forms are first indicated in certain teleosts and amphibians. Lips vary in structure to accord with their physiological functions, whether sensory, prehensile, or adhesive (Anson, 1929). By precise comparative morphology and gene expression analyses, a possibility was inferred that ammocoete lips may not be identical to gnathostome jaws (Kuratani, 2003).

The surface architecture of the superficial layer of epithelial cells in the lips and associated structures is characterised by specialised structures, the micro-ridges forming different patterns in different fish species investigated in this study. These structures in other fishes, have been described as cytoplasmic folds (Merrilees, 1974), microvilli (Harris & Hunt, 1975), microfolds (Hunter & Nayudu, 1978) or ridges (Iger et al., 1988). Insofar as these structures appeared as micro-ridges under SEM and microvilli under TEM. The term "microridges" is used in this study following Whitear & Mittal (1986), Whitear (1990), Suzuki (1992) and Whitear & Moate (1998) and seems appropriate.

The micro-ridges on the surfaces of the mucogenic epithelia form characteristic maze like patterns in different fish species. Fishelson, (1984) correlated the variations in micro-ridge pattern with the locomotory activity of the fish. He suggested that in faster swimming fishes, micro-ridges are most developed and serve to trap mucus on the epithelial surface. The present study, however, is not in support of this since micro-ridges are well developed and conspicuous on the free surface of the lips and associated structures in all the six fish species investigated showing significant difference in their habits and habitats.

The retention of secretion has been the most popular hypothesis of micro-ridge function (Hughes & Wright, 1970; Hughes, 1979; Tillman et al., 1977; Meyer-Rochow, 1981; Fishelson, 1984). Modifications in the pattern of micro-ridges can also be caused by various intrinsic, e.g. hormonal (Schwerdtfeger, 1979 a, b), or extrinsic factors e.g. temperature (Ferri, 1982), salinity (Ferri, 1983), mercury salts (Pereira, 1988), organic pollutants, (Iger et al., 1988), handling and ectoparasites (Whitear, 1990). Some speculations about the function of micro-ridges have centred on mechanical considerations (Lanzing & Higginbotham, 1974; Hawkes, 1974; Sibbling & Uribe, 1985). The provision of reserve apical membrane to allow for distortion was postulated by Zeiske et al., (1976) but Sperry & Wassersug (1976) found no change of pattern after stretching fish oesophageal epithelium and suggested that spread of mucus from goblet (mucous) cells might be guided by the direction of ridges.

Presence of conspicuous micro-ridges on the surfaces of the mucogenic epithelia in the fish species investigated could be considered to reflect high secretory activity of the surface epithelial cells in these regions. Secretion of glycoproteins (GPs), shown histochemically (Pinky et al., 2008, Tripathi & Mittal, 2010), in the surface epithelial cells in the mucogenic regions is in support of this. Further, Whitear (1990) proposed that the form of micro-ridges correlate with the type and rate of secretion at the cell apex. The development of micro-ridges would then be a consequence of arrival of new membranes as vesicles carrying the secretion fuse with the apical plasmalemma and high ridges would indicate a rapid sequence of arrival of secretory vesicles at the surface.

In the epithelia of lips and associated structures of all fish species studied, the adjacent micro-ridges are often interconnected with each other by fine cross connections i.e. micro-bridges. It should be pointed out that such specific structures connecting micro-ridges have also been reported previously in fish epidermis (Whitear, 1990) and have been variously named as interconnections (Reutter *et al.*, 1974), micro-bridges (Karlsson, 1983), cross-bridges (Avella & Ehrenfeld, 1997). Previous workers have not commented on the functional significance of these structural peculiarities. It is possible that these structures may provide mechanical strength to the micro-ridges. However, it is open to other interpretations.

Secretions elaborated by the epithelial cells and the mucous cells in the mucogenic epithelia could be regarded as an adaptation to lubricate and protect the epithelia from abrasion (Pinky et al., 2002). The role of mucus was likewise postulated previously to inhibit the invasion and proliferation of pathogenic micro-organisms and to prevent their colonisation in fish epidermis (Nigrelli, 1937; Nigrelli et al., 1955; Hildemann, 1962; Liguori et al., 1963; Lewis, 1970).

In the oral cavity the lining mucous membrane becomes keratinized to varying degrees in different animals and also in different areas of the mouth (Adams, 1976). Some of the most dramatic advances made over the past 2-3 decades in epidermal research have come about through the utilization of newly developed biochemical investigative techniques, examples of which include the use of gene cloning to study the organization of the keratin gene family, and the use of immuno-fluorescence with monoclonal antibodies to discern when various keratin proteins appear during differentiation. In SEM studies, the Keratinized surfaces of the fishes studied are shaggy and are matted with an organised array of horny projections separated by shallow grooves. The boundaries of the adjacent epithelial cells (the surfaces of which are modified into unculi) are demarcated by well-defined and distinct rows of microridges.

Horny projections from single cells of lips and associated structures have been reported in a wide variety of fishes by various workers (Leydig, 1895; Rauther, 1911; 1928; Hora, 1922; Minzenmay, 1933; Saxena, 1959; Thys, 1961; Kaiser, 1962; Lal *et al.*, 1966; Saxena & Chandy, 1966; Roberts, 1982). Girgis (1952), in a herbivorous bottom feeder *Labeo horie*, observed horny protuberances on lips and two sharp horny cutting edges lying in the upper and lower borders of mouth immediately inside the lips. Mester (1971) made a morpho-histological analysis of the buccopharyngeal cavity in *Noemacheilus barbatulus,* and reported horny plates on the inner surface of the lips that are used by the fish for trituration of nutrients. Verighina (1971) while describing the structure of the digestive tract, reported the presence of horny cutting edge on the lower lip of periphyton eater fish *Chondrostoma nasus variabile.* Agrawal & Mittal (1992 a) reported the presence of keratinized unculi on the ventral side of the upper lip and keratinized cone like structure with sharp cutting edge on the horny lower jaw sheath associated with the lower lip of an omnivorous bottom feeder, *Cirrhina mrigala.* Agrawal & Mittal (1992 b) observed keratinized unculi on the ventral side of the upper lip and on the dorsal side of the lower lip facing the mouth opening. In addition, they observed keratinized cone like structure on the horny upper jaw sheath and on the horny lower jaw sheath associated with the lips of a herbivorous column feeder, *Labeo rohita.*

The present study shows that in *N. botia, C. fasciata, G. lamta, P. sophore* and *C. carpio* the UL and the LL, on the side facing the mouth opening, are associated with the horny upper jaw sheath (HUJS) and the horny lower jaw sheath (HLJS) respectively. The horny jaw sheaths are absent in *G. giuris* having few villiform teeth.

Keratinization occurs in the structures associated with lips of fish *Garra lamta* by histochemical investigations (Pinky et al., 2004). The unculi also observed in the upper jaw of an herbivorous fish, *Cirrhinus mrigala* by SEM studies. More recently, in *Puntius sophore*, the HUJS and the HLJS are keratinized. The surface epithelial cells in these regions are modified to form single cell modification—the unculi, give positive results for keratin with histochemical reactions. The unculi appear functionally significant on the lips and associated structure in this fish species lacking jaw teeth (Tripathi & Mittal, 2010).

In the epithelia of the RC and the AP in *G. lamta* and those of the HUJS and the HLJS in *C. fasciata*, however, both mucogenic and keratinized regions are observed. In *C. fasciata* the apical regions of the horny jaw sheaths are modified into cone like structures each consisting of several superimposed keratinized epithelial cells. The spine like or conical unculi in the keratinized regions of the RC and AP of *G. lamta* may be considered to facilitate clinging or adherence of the fish to the substratum by engaging irregularities on the surface of rocks or stones. Truncated unculi on the surface of HUJS and HLJS in 4 fish species investigated except *G. giuris* and apical cone like structure of the horny jaw sheaths in *C. fasciata* may have a function to assist the fish in scooping mud from the bottom in search of food or to act as sharp cutting edge assisting the fish to browse upon or scrape the food attached with the substratum for feeding.

The keratinized epithelia are mainly composed of the epithelial cells only. The mucous cells and the taste buds are not observed. The absence of the gland cells in the keratinized epithelia suggests an inverse relationship between the degree of keratinization and slime secretion.

In most vertebrates the sense of taste is used as a close range receptor for food item discrimination. Fish are unique among vertebrates in having taste buds widely distributed over various regions. The present study shows that, the taste buds in the epithelia of the lips and structures associated with them are conspicuous in *C. fasciata* and are small and inconspicuous in *G. giuris*. In both these fish species, however, the taste buds are few and could be located at long intervals. This indicates that the cutaneous gustatory function in these fish species is probably of less importance. Low density of taste buds, in *G. giuris* having active predatory habit could be associated with the presence of superficial neuromasts and canal neuromasts in the localisation of its prey. This is supported by reports that the taste buds are absent on the lips of the ox-eye herring *Megalops cyprinoides* (Pasha, 1964 c), are poorly developed on lips of *Channa striata* (Agarwal & Mittal, 1994) and are relatively few on the outer surface of the body than those on the lips and palatal organs and gills in *Pseudorasbora parva* (Kiyohara *et al.*, 1980) all these fishes are active predators.

In the lips and associated structures of *G. lamta*, *P. sophore*, *C. carpio* and *N. botia* taste buds, as compared to those in *C. fasciata* and *G. giuris*, are very prominent and are distributed in large numbers. This indicates the development of acute gustatory function, an adaptation to the peculiar mode of life of these fishes. A similar correlation between the distribution of taste buds and gustatory feeding has been shown in two races of the Mexican characin, *Astyanax mexicanus*; one lives in caves and is sightless, and the other lives in river and is visually normal. Schemmel (1967) has shown that the cave-dwelling race has a larger number and a more extensive distribution of external buds than the river-dwelling race, corresponding to acute gustatory feeding behaviour.

Presence of large number of taste buds in *C. carpio* and *N. botia* could be due to its habit to live at the muddy bottom of water bodies. *C. carpio* browses on the shallow bottom and margins, takes in vegetable debris, insects, worms, crustaceans and also planktonic algae. *N. botia* is a bottom dwelling fish, which accepts any kind of food and feeds on algal films as well. At muddy bottoms the visibility, in general, is poor owing to (i) depth and (ii) increase in turbidity caused by the disturbance of the bottom mud due to the movements of the fish, which is in habit of suddenly burrowing into the mud or sand on the bottom to protect itself from predators. The presence of a large number of taste buds increases the probability of detecting and locating accurately prey concealed by darkness or turbidity and may also permit the accurate location of small food particles, which would be missed otherwise.

The taste buds in the lips and associated structures of all the six fish species investigated remain encircled by characteristic concentric whorls of epidermal cells. Harvey & Batty (1998), suggested that it was sometimes possible to locate and count taste buds by the presence of the characteristic ring of epidermal cells surrounding the sensory apex, even when the apex itself was damaged or missing.

Earlier SEM studies reported that fish taste buds fall into three categories based on their external surface morphology (Reutter *et al.*, 1974; Ezeasor, 1982). In addition to the three types of taste buds previously described from various teleost fish, a fourth type comprising very small buds, was found in some cardinal fish (Fishelson, 2004). The taste buds in the mouth cavity of *Rita rita* are of three types which are elevated from the epithelium at different levels, which may be useful for ensuring full utilization of the gustatory ability of the fish, detection and analysing of taste substances, as well as for assessing the quality and palatability of food, during its retention in the mouth cavity (Yashpal et al., 2006). In *C. mrigala*, there is only one type of taste buds observed in mouth cavity (Yashpal et al., 2009).In the present study only one type of taste buds are observed in the 5 fish species whereas in *G. giuris* there are two types of taste buds are observed in different regions of lips and associated epithelia.

In *G. giuris*, in addition to the taste buds, specialised sensory structures the superficial neuromasts are also observed in the epithelium of the RC and in the epithelia of the FSLS close to the ventral head skin. Neuromasts can be found on the entire body surface including the tail with the cupulae extending into the water (Schellart & Wubbels, 1998; Eastman & Lannoo, 2003; Tarby and Webb, 2003; Sane et. al., 2009).

The fishes studied are characterised by the peculiar trophic niche they occupy: many scrape epilithic or epiphytic algae and other food items from submerged substrates. This specialized feeding type is possible thanks to the remarkably formed, ventrally placed suckermouth of *G. lamta* that allows itself to attach to a surface while scraping and eating the food attached to it. In spite of this highly specialized feeding apparatus, diversity in both thickness of the different regions of lips and in shape of lips exists, and these fishes actually feed on a broad range of food. As such, Cyprinidae are the most specialized and successful fish family within the order Cypriniformes. More basal families within the teleosts like Gobiidae, Cobitidae and Belontiidae mostly display a more general (non-specialized) feeding mode, with a typical feeding apparatus suitable for finding and processing insects and other food items that abound in the water column or in the bottom. Some knowledge exists on the trophic, evolutionary trend in the group, but detailed studies dealing with the morphology of lips and associated structures or the feeding apparatus are few and often fragmentary.

12. References

Adams, D. (1976) Keratinization of the oral epithelium. *Annals of the Royal College of Surgeons of England* 58, 351-358.

Agrawal, N., Mittal, A. K. (1991) Epithelium of lips and associated structures of the Indian major carp, *Catla catla. Japan. J. Ichthyol.* 37, 363-373.

Agrawal, N., Mittal, A. K. (1992 a) Structural modifications and histochemistry of the epithelia of lips and associated structures of a carp – *Labeo rohita. Eur. Arch. Biol.* 103, 169-180.

Agrawal, N., Mittal, A. K. (1992 b) Structural organisation and histochemistry of the epithelia of lips and associated structures of a carp – *Cirrhina mrigala. Can. J. Zool.* 70, 71-78.

Agrawal, N., Mittal, A. K. (1992 c) Structure and histochemistry of the epithelia of lips and associated structures of a catfish *Rita rita. Japan. J. Ichthyol.* 39, 93-102.

Alikunhi, K. H. (1957). *Fish culture in India, Farm Bulletin No.* 20. New Delhi: Indian Council of Agricultural Research.

Anson, B. J. (1929) The comparative anatomy of the lips and labial villi of vertebrates. *J Morph. Physiol.* 47, 2, 335-413.

Avella, M. & Ehrenfeld, J. (1997). Fish gill respiratory cells in culture: A new model for Cl⁻ - secreting epithelia J. *Membrane Biol.* 156, 8-97.

Bloch, M. E. & Schneider, J. G. (1801). *Syst. Ichthyologie.* i-ix, 1-584.

Branson, B. A. & Hake, P. (1972). Observation on an accessory breathing mechanism in *Piaractus nigripinnis* (Cope) (Pisces: Teleostomi: Characidae). *Zool. Anz.* Leipzig. 189, 292-297.

Chitray, B. B. (1965). The anatomy and histology of the alimentary canal of *Puntius sarana* (Ham.) with a note on feeding habits. *Ichthyologica* 4, 53-62.

Eastman, J. T. and Lannoo, M. J. (2003) Diversification of Brain and Sense Organ Morphology in Antarctic Dragonfishes (Perciformes: Notothenioidei: Bathydraconidae). J Morphol 258:130–150.

Ezeasor, D. N. (1982). Distribution and ultrastructure of taste buds in the oropharyngeal cavity of the rainbow trout, *Salmo gairdneri* Richardson. *J. Fish Biol.* 20, 53-68.

Ferri, S. (1982). Temperature induced transformation of teleost (*Pimelodus maculatus*) epidermal cells. *Gegenbaurs Morph. Jahrb.* Leipzig. 128, 712-731.

Ferri, S. (1983). Modification of microridge pattern in teleost (*Pimelodus maculatus*) epidermal cells induced by NaCl. *Gegenbaurs Morph. Jahrb.* Leipzig. 129, 325-329.

Fishelson, L. (1984). A comparative study of ridge-mazes on surface epithelial cell/membranes of fish scales (Pisces, Teleostei). *Zoomorphologie.* 104, 231-238.

Fishelson, L. Delarea, Y. and Zverdling, A. (2004) Taste bud form and distribution on lips and in the oropharyngeal cavity of cardinal fish species (Apogonidae, Teleostei), with remarks on their dentition. J. Morph. 259:316–327.

Girgis, S. (1952). On the anatomy and histology of the alimentary tract of an herbivorous bottom-feeding Cyprinoid fish, *Labeo horie* (Cuvier). *J. Morph.* 90, 317-362.

Graham, J. B. (1997). *Air-breathing Fishes.* Academic Press. California, USA.

Günther, S. (1989). *Fresh water fishes of the world Vol. I & II* (Translated and revised by Tucker, D. W.), New Delhi: Falcon books, Cosmo Publications.

Hamilton, F. B. (1822). *An account of fishes found in the river Ganges and its branches.* Edinburg and London, pp. VIII+405 39 pls.

Harris, J. E. & Hunt, S. (1975). The fine structure of the epidermis of two species of salmonid fish, the Atlantic salmon (*Salmo salar* L.) and the brown trout (*Salmo trutta* L.). I. General organisation and filament containing cells. *Cell Tissue Res.* 157, -553-565.

Harvey, R. & Batty, R. S. (1998). Cutaneous taste buds in cod. *J. Fish Biol.* 53, 138-149.

Hawkes, J. W. (1974). The structure of fish skin. I. General organisation. *Cell Tissue Res.* 149, 147-158.

Hildemann, W. H. (1962). Immunogenetic studies of poikilothermic animals. *Am. Nat.* 96, 195-204.

Hora S. L. & Mukerji, D. D. (1953). Table for the identification of Indian fresh water fishes,with description of certain families and observation on the relative utility of the probable larvivorous fishes of India. (Revised by T.J. Job) *Health bulletin* No.12. *Malaria bureau*, No. 4. Simla, Govt of India Press.

Hora, S. L. & Pillay, T. V. R. (1962). *Handbook of fish culture in the Indo-Pacific region.* FAO Fish. Biol. Tech. Pap. No. 14. Fisheries Division, Biology Branch, Food and Agriculture Organisation of the United Nations, Rome.

Hora, S. L. (1922). Structural modifications in the fish of mountain torrents. *Rec. Indian Mus.* 24, 31-61.

Hughes, G. M. (1979). Scanning electron microscopy of the respiratory surfaces of trout gills. *J. Zool. Lond.* 188, 443-453.

Hughes, G. M. & Wright, D. E. (1970). A comparative study of the ultrastructure of the water/blood pathway in the secondary lamellae of teleost and elasmobranch fishes - benthic forms. *Z. Zellforsch. mikrosk. Anat.* 104, 478-493.

Hunter, C. R. & Nayudu, P. L. (1978). Surface folds in superficial epidermal cells of three species of teleost fish. *J. Fish Biol.* 12, 163-166.

Iger, Y., Abraham, M., Dotan, A., Fattal, B. & Rahamim, E. (1988). Cellular responses in the skin of carp maintained in organically fertilised water. *J. Fish Biol.* 33, 711-720.

Kaiser, P. (1962). Hornzahnchen als Lippenbewaffnug bei Jungfischen von Cypriniden. *Zool Anz.*, Leipzig 169, 158-161.

Kapoor, B. G. (1958). The anatomy and histology of the alimentary tract of a plankton-feeder, *Gadusia chapra* (Ham.). *Ann. Mus. Stor. nat.* Geneva. 70, 8-32.

Kapoor, B. G., Smit, H & Verighina, I. A. (1975). The alimentary canal and digestion in teleosts. *Adv. Mar. Biol.* 13, 109-239.

Karlsson, L. (1983). Gill morphology in the zebra fish, *Brachydanio rerio* (Hamilton-Buchanan). *J. Fish Biol.* 23, 511-524.

Khanna, S. S. (1961). Alimentary canal in some teleostean fishes. *J. Zool. Soc. India.* 13, 206-219.

Khanna, S. S. (1962). A study of bucco-pharyngeal region in some fishes. *Ind. J. Zool.* 3, 21-48.

Khanna, S. S. (1993). *An introduction to fishes,* Allahabad: Central Book Depot.

Kiyohara, S., Yamashita, S. & Kitoh, J. (1980). Distribution of taste buds on the lips and inside the mouth in the minnow, *Pseudorasbora parva. Physiol. Behav.* 24, 1143-1147.

Kuratani, S. (2003) Evolution of the vertebrate jaw: homology and developmental constraints. *Paleontological research* 7, 1, 89-102.

Lal, M. B. (1968). Studies on the anatomy and histology of the alimentary canal of a carp, *Tor putitora* (Ham). *Proc. Nat. Acad. Sci.* India. 38B, 127-136.

Lal, M. B., Bhatnagar, A. N. & Kailc, R. K. (1964). Studies on the morphology and histology of the digestive tract and associated structures of *Chagunius chagujnio* (Ham). *Proc. Nat. Acad. Sci. India.* 34B, 160-172.

Lal, M. B., Bhatnagar, A. N. & Uniyal, J. P. (1966). Adhesive modifications of a hill stream fish *Glyptothorax pectinopterus* (McClelland). *Proc. natl. Acad. Sci.* India (B) 36, 109-116.

Lanzing, W. J. R. & Higginbotham, D. R. (1974). Scanning microscopy of surface structures of *Tilapia mossambica* (Peters) scales. *J. Fish Biol.* 6, 307-310.

Lewis, R. W. (1970). Fish cutaneous mucus: a new source of skin surface lipids. *Lipids* 5, 947-949.

Leydig, F. (1895). Integument und Hautsinnesorgane der Knochenfishc. *Zool. Jb. Anat.* 8, 1-152.

Liguori, V. R., Ruggieri, G. D., Baslow, S. J. M. H., Stempien, M. F. & Nigrelli, R. F. (1963). Antibiotic and toxic activity of the mucus of the pacific golden striped bass *Grammistes sexlineatus*. *Am. Zool.* 3, 546.

Linnaeus, C. (1758). *Systema Naturae*. 10th edn. Vol. 1. Regnum Animale. Stockholm: Salvius. (Facsimile reprint (1956). London: British Museum (Natural History).)

Merrilees, M. J. (1974). Epidermal fine structure of the teleost *Esox americanus* (Esocidae, Salmoniformes). *J. Ultrastruct. Res.* 47, 272-283.

Mester, L. (1971). Studiul cavitatii buco-faringiene, La *Noemacheilus barbatulus* L. (Pisces, Cobitidae). *St. si. cerc. Biol. seria Zoologie Bucuresti.* 23, 439-444.

Meyer-Rochow, V. B. (1981). Fish tongues - surface fine structures and ecological considerations. *Zoo. J. Linn. Soc.* 71, 413-426.

Miller, R. J. & Evans, H. E. (1965). External morphology of the brain and lips in Catostmid fishes. *Copeia.* 4, 467-487.

Minzenmay, A. (1933). Die Mundregion der Cypriniden. *Zool. Jb. Anat.* 57, 191-286.

Mittal, A. K. & Agrawal, N. (1994). Modifications in the epithelia of lips and associated structures of the predatory murrel (*Channa striata*). *J. Appl. Ichthyol.* 10, 114-122.

Mittal, A. K. & Whitear, M. (1978). A note on cold anaesthesia of poikilotherms. *J. Fish Biol.* 13, 519-520.

Moitra, S. K. & Bhowmik, M. L. (1967). Functional histology of the alimentary canal of the young *Catla catla* (Ham.), an omnivorous surface-feeding fish of Indian fresh-waters. *Vestnik Cs. spol. Zool.* 31, 41-50.

Moitra, S. K. & Sinha, G. M. (1971). Studies on the morphohistology of the alimentary canal of a carp, *Chagunius chagunio* (Ham.) with reference to the nature of taste buds and mucous cells. *Inland Fish.Soc India.* 3, 44-56.

Nigrelli, R. F. (1937). Further studies on the susceptibility and acquired immunity of marine fishes to *Epibdella melleni*, a monogenetic trematode. *Zoologica, N.Y.* 22, 185-192.

Nigrelli, R. F., Jakowska, S. & Padnos, M. (1955). Pathogenicity of epibionts in fishes. *J. Protozool.* 2 (suppl.) 7.

Nikolsky, G.V. (1963). *The Ecology of Fishes*. (Translated by L. Birkett). Academic Press, London and New York.

Ojha, J. & Singh, S. K. (1992). Functional morphology of the anchorage system and food scrapers of a hill stream fish, *Garra lamta* (Ham.) (Cyprinidae, Cypriniformes). *J. Fish Biol.* 41, 159-161.

Ono, D. R. (1980). Fine structure and distribution of epidermal projections associated with taste buds on the oral papillae in some Loricariid catfishes (Siluroidei: Loricariidae). *J. Morph.* 164, 139-159.

Pasha, S. M. K. (1964 a). The anatomy and histology of the alimentary canal of an omnivorous fish, *Mystus gulio. Proc. Ind. Acad. sci.* 59B, 211-221.

Pasha, S. M. K. (1964 b). Anatomy and histology of the alimentary canal of a herbivorous fish, Tilapia *mosambica. Proc. Ind. Acad. Sci.* 59B, 340-349.

Pasha, S. M. K. (1964 c). The anatomy and histology of the alimentary canal of a carnivorous fish, *Megalops cyprinoides. Proc. Ind. Acad. Sci.* 60B, 107-115.

Pereira, J. J. (1988). Morphological effects of mercury exposure on windowpane flounder gills as observed by scanning electron microscopy. *J. Fish Biol.* 33, 571-580.

Pinky, Mittal, S., Ojha, J. Mittal, A. K., 2002. Scanning electron microscopic study of the structures associated with lips of an Indian hill stream fish *Garra lamta* (Cyrinidae, Cyriniformes) *European Journal of Morphology* 40, 161-169.

Pinky, Mittal S, Yashpal M, Ojha J, Mittal AK. 2004. Occurrence of keratinization in the structures associated with lips of a hill stream fish *Garra lamta* (Hamilton) (Cyprinidae. Cypriniformes). *J Fish Biol.* 65, 1165–1172.

Pinky, Mittal S, Mittal AK. 2008. Glycoproteins in the epithelium of lips and associated structures of a hill stream fish *Garra lamta* (Cyprinidae. Cypriniformes): A histochemical investigation. *Anat Histol Embryol* 37,101–113.

Rauther, M. (1911). Beiträge zur Kenntnis der Panzerwelse. *Zool. Jb. Anat.* 31, 497-528.

Rauther, M. (1928). Der Saugmund von *Discognathus*. *Zool. Jb. Anat.* 45, 45-76.

Reutter, K., Breipohl, W. & Bijvank, G. J. (1974). Taste bud types in fishes. II Scanning electron microscopical investigations on *Xiphophorus helleri* Heckel (Poeciliidae, Cyprinodontiformes, Teleostei). *Cell Tissue Res.* 153, 151-165.

Roberts, T. R. (1982). Unculi (Horny projections arising from single cells), an adaptive feature of the epidermis of Ostariophysan fishes. *Zoologica Scripta.* 11, 55-76.

Sane, S. P. and McHenry, M. J. (2009) The biomechanics of sensory organs. *Integrative and Comparative Biology* 1–16.

Saxena, D. B. & Bakshi, P. L. (1964). Functional anatomy of the alimentary canal of a torrential stream fish *Botia birdi* (Choudhari). *Kashmir sci.* 1, 76-86.

Saxena, S. C. & Chandy, M. (1966). Adhesive apparatus in certain Indian hill stream fishes. *J. Zool.* 148, 315-340.

Saxena, S. C. (1959). Adhesive apparatus of a hill stream cyprinid fish *Garra mullya* (Sykes). *Proc. Natn. Inst. Sci. India.* 25, 205-214.

Schellart, N. A. M. & Wubbels, R. J. (1998). The auditory and mechanosensory lateral line system. In *The physiology of fishes.* (Evans, D. H. ed.) pp. 283-312. New York: CRC Press.

Schemmel, C. (1967). Vergleichende Untersuchungen an den Hautsinnesorganen ober-und unterirdisch lebeder Astyanax-Formen. *Z. Morph. Okol. Tiere* 61, 253-316.

Schwerdtfeger, W. K. (1979 a). Morphometrical studies of the ultrastructure of the epidermis of the guppy, *Poecilia reticulata* Peters, following adaptation to sea-water and treatment with prolactin. *Gen.Comp. Endocrinol.* 38, 476-483.

Schwerdtfeger, W. K. (1979 b). Qualitative and quantitative data on the fine structure of the guppy (*Poecilia reticulata* Peters) epidermis following treatment with thyroxine and testosterone. *Gen. comp. Endocr.* 38, 484-490.

Sehgal, P. (1966). Anatomy and histology of the alimentary canal of *Labeo calbasu* (Ham). *Res. Bull. Punjab Univ. Sci.* 17, 257-266.

Sehgal, P. & Salaria J. (1970). Functional anatomy of histology of the digestive organs a *Cirrhina mrigala* (Cuvie and Val.) *Proc. nat. Acad. Sci. India.* 40B, 212-222.

Sibbing, F. A. & Uribe, R. (1985). Regional specialisations in the oropharyngeal wall and food processing in the carp (*Cyprinus carpio* L.). *Neth. J. Zool.* 35, 377-422.

Sinha, G. M. (1975). On the origin development and probable function of taste buds in the lip and bucco-pharyngeal epithelia of an Indian freshwater major carp, *Cirrhinus mrigala* (Hamilton) in relation to food and feeding habits. *Z. mikrosk -anat Forsch,* Leipzig. 82, 294-304.

Sinha, G. M. & Moitra, S. K. (1975). Functional morpho-histology of the alimentary canal of an Indian fresh water major carp *Labeo rohita* (Hamilton) during its different life history stages. *Anat. Anz.* 138, 222-239.

Sinha, G. M. & Moitra, S. K. (1976). Studies on the morpho-histology of the alimentary canal of fresh water fishes of India. Part I. The alimentary canal of young *Cirrhinus reba* (Ham.) with a comparison with that of the adult in relation to food. *Vest. Cs. Spol. Zool.* 40, 221-231.

Sinha, G. M. & Moitra, S. K. (1978). Studies on the comparative histology of the taste buds in the alimentary tract of a herbivorous fish, *Labeo calbasu* (Ham.) and a carnivorous fish, *Clarius batrachus* (Linn.) in relation to food and feeding habits. *Zool. Beitr.* 24, 43-57.

Sperry, D. G. & Wassersug, R. J. (1976). A proposed function for microridges on epithelial cells. *Anat. Rec.* 185, 253-258.

Suzuki, N. (1992). Fine structure of the epidermis of the mudskipper, *Periophthalmus modestus* (Gobiidae). *Japan. J. Ichthyol.* 38, 379-396.

Suzuki, Y. (1956). A histological study of the granular processes on the lips of scythe fish *Pseudogobio esocinus* (T. et S.) *Jap. J. Ichthyol.* 5, 12-14.

Tarby ML, Webb JF. 2003. Development of the supraorbital and mandibular lateral line canals in the cichlid, *Archocentrus nigrofasciatus*. J Morphol 255, 44–57.

Thys (van den Audenaerde), D. F. E. (1961). L' anatomie de phractolaemus ansorgei Blgr et la position Systematique des phractolaemidae. *Anuls. Mus. r. Afr. Cent. Ser. 8vo (Zool).* 103, 99-167.

Tillmann, B., Pietzsch-Rohrschneider, I. & Huenges H. L. (1977). The human vocal cord surface. *Cell Tissue Res.* 185, 279-283.

Tripathi, P. and Mittal A. K. (2010) Essence of Keratin in Lips and Associated Structures of a Freshwater Fish *Puntius sophore* in Relation to its Feeding Ecology: Histochemistry and Scanning Electron Microscope Investigation. *Tissue and Cell.* 42, 223-233.

Vanajakshi, T. P. (1938) Histology of the digestive tract of *Sacchobranchus fossilis* and *Macrones vittatus*. *Proc. Indian Acad. Sci.* 7 (B), 61-79.

Verighina, I. A. (1971). The structure of the digestive tract of the Volga under mouth *Chondrostoma nasus* variable Jak. *Voprosy Ikhtiologi.* 11, 311-318.

Welcomme, R.L. (1988). *International introductions of inland aquatic species.* FAO fisheries technical paper 294, pp 1-318. Food and Agriculture Organisation of the United Nations, Rome.

Whitear, M. (1990). Causative aspects of microridges on the surface of fish epithelia. *J. Submicrosc. Cytol. Pathol.* 22, 211-220.

Whitear, M. (1986). Epidermis. *In Biology of the Integument.* Vol. 2, *Vertebrates* (Bereiter-Hahn, J. Matoltsy, A. G. & Richards, K. S., eds.), pp. 8-38. Berlin: Springer-Verlag.

Whitear, M. & Moate, R. (1998). Cellular diversity in the epidermis of *Raja clavata* (Chondrichthyes). *J. Zool., Lond.* 246, 275-285.

Whitear, M. & Moate, R. M. (1994). Microanatomy of taste buds in the dogfish, *Scyliorbinus canicula*. *J. Submicrosc. Cytol. Pathol.* 26, 357-367.

Yang, L. and Mayden, R. L. (2010) Phylogenetic relationships, subdivision, and biogeography of the cyprinid tribe Labeonini (sensu Rainboth, 1991) (Teleostei: Cypriniformes), with comments on the implications of lips and associated structures in the labeonin classification. *Molecular Phylogenetics and Evolution* 54, 254–265.

Yashpal, M., Kumari, U., Mittal, S., Mittal, A.K. (2006) Surface architecture of the mouth cavity of a carnivorous fish *Rita rita* (Hamilton, 1822) (Siluriformes, Bagridae). *Belg. J. Zool.* 136 (2), 155-161.

Yashpal, M., Kumari, U., Mittal, S., Mittal, A.K. (2009) Morphological specialization of the buccal cavity in relation to the food and feeding habit of a carp *Cirrhinus mrigala*: A scanning electron microscopic investigation. *J. Morphol.* 270, 714 – 728.

Zeiske, E., Melinkat, R., Breucker, H. & Kux J. (1976). Ultrastructural studies on the epithelia of the olfactory organ of Cyprinodonts (Teleostei, Cyprinodontoide). *Cell Tissue Res.* 172, 245-267.

12

Scanning Electron Microscopy Imaging of Bacteria Based on Nucleic Acid Sequences

Takehiko Kenzaka and Katsuji Tani
Osaka Ohtani University, Faculty of Pharmacy,
Japan

1. Introduction

Scanning electron microscopy (SEM) has been widely used in environmental microbiology to characterize the surface structure of biomaterials and to measure cell attachment and changes in morphology of bacteria. Moreover, SEM is useful for defining the number and distribution of microorganisms that adhere to surfaces. Traditionally, inability to provide phylogenetic or genetic information about microorganisms has been one limitation of SEM in environmental microbiology.

Modern molecular studies based on DNA and RNA sequence analysis have led to an understanding of the microbial diversity and composition of bacterial communities in various environments. Introduction of the concept of in situ hybridization (ISH) with RNA- or DNA-targeted fluorescent probes has led to important research regarding the identification and quantification of individual cells and has demonstrated great potential in the analysis of the composition of bacterial communities in the environment. Combining morphological study with SEM and ISH techniques (SEM-ISH) has provided new insights into the understanding of the spatial distribution of target cells on various materials.

2. Application of ISH techniques to SEM

2.1 Concepts of ISH

ISH is a method of detecting and localizing specific nucleic acid sequences in morphologically preserved tissues or cell preparations by hybridizing the complementary strand of a nucleotide probe to the sequence of interest. To detect target cells, the permeability of the cell and the visibility of the nucleotide sequence to the probe must be increased without destroying the structural integrity of the cell. The type of probe to be used and how to label it to yield better resolution with the highest accuracy should be taken into consideration.

In environmental microbiology, fluorescent in situ hybridization (FISH) with rRNA-targeted oligonucleotide probes has provided information about the absolute abundance, morphology, and cell size of bacteria with defined phylogenetic affiliations and had been applied to the investigation of community composition in lakes, river, oceans, activated sludge, drinking water, etc. (Amann et al., 1995). FISH analysis of bacterial communities with mRNA- or DNA-targeted probes presents a unique challenge because of its low sensitivity and resolution (Moraru et al., 2010; Pernthaler & Amann, 2004). The concepts of ISH can be incorporated into scanning electron microscopy imaging of microorganisms (Kenzaka et al., 2005a, 2009).

2.2 Metal labeling of target cells for SEM-ISH

To identify cells of interest by SEM, cells carrying specific DNA or RNA sequences must be labeled with a metal instead of fluorescent molecules. Target cells that were hybridized with oligonucleotide or polynucleotide probes can be labeled with gold or platinum (Fig. 1). In case of gold labeling of the target sequence, hybridization was performed with biotin-labeled oligonucleotide probes, and then target microbes were identified by nanogold-labeled streptavidin (Hacker, 1998; Fig. 1a). In case of the platinum probe complex, hybridization was performed with platinum-labeled oligonucleotide probes (Fig. 1b). Platinum-labeled probes can be prepared by allowing the platinum complex to bind to the probe at guanine-N7 atoms (Brabec & Leng 1993; Dalbiès et al., 1994). To amplify the signal in the both cases, gold enhancement was performed to enlarge the gold particles. In this reaction, gold ions in solution are catalytically deposited onto nanogold particles as metallic gold (Au^0). These particles grow in size as development time elapses. Consequently, hybridized cells contain gold up to tens of nm in size inside the hybridized cells. These hybridized cells release a strong backscatter electron signal (BSE) because of the accumulation of gold atoms inside cells (Kenzaka et al., 2005a, 2009).

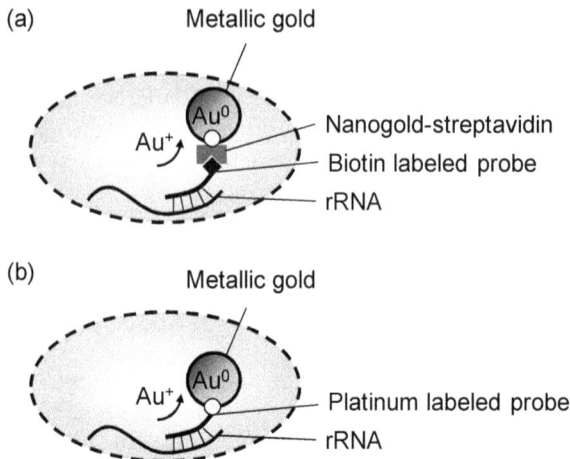

Fig. 1. Systematic representation of metal labeling for SEM-ISH. (a) Biotin labeled probes are hybridized with rRNA, followed by streptavidin-gold labeling, gold enhancement. (b) Platinum labeled probes are hybridized with rRNA, followed by gold enhancement.

High-vacuum SEM images of a mixture of bacterial strains after ISH with an rRNA-targeted probe and labeling with gold are shown in Fig. 2. Secondary electron (SE) image demonstrate the surface topography of all cells (Fig. 2a). In BSE image, *Escherichia coli* cells hybridized with the ES445 probe which was targeted for *Escherichia-Shigella* (Kenzaka et al., 2001) were identified as bright spots because of the enhanced signal as a result of gold labeling (Fig. 2b). In hybridized cells, the amount of gold in the cells was greater than that on the cell surface; thus, this labeling resulted in a higher BSE signal. In the same microscopic field, both images could be viewed side by side.

(a)   (b)

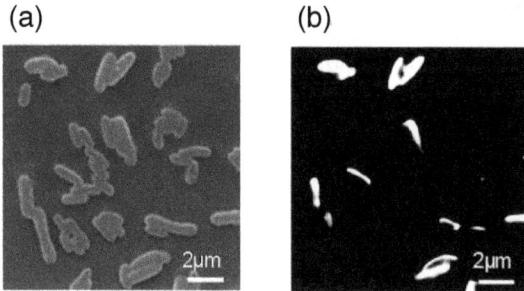

Fig. 2. High-vacuum SEM images of bacterial cells after ISH: Mixture of *Escherichia coli* and *Aeromonas sobria* cells hybridized with ES445 probe (targeted for *Escherichia-Shigella*). The same microscopic fields are shown with SE (a) and BSE images (b).

3. Experimental protocols for SEM-ISH

Experimental protocols for SEM-ISH are similar to FISH except metal labeling, and both techniques have the same challenges and limitations. For success in ISH, several issues need to be considered before proceeding with experiments: permeabilization and pretreatment, hybridization condition (composition of the hybridization solution, temperature, sodium concentration, and presence of organic solvents), washes, controls, etc. Here we discuss the general protocols for ISH in comparison to FISH.

3.1 Preparation of materials for ISH

Before processing for ISH, the specimens must be fixed and permeabilized to allow the penetration of the probes and reagents into the cells and protect the target RNA or DNA from degradation by nucleases. Fixation conditions may vary according to the target bacteria and type of sample. Optimal fixation results in good material penetration as well as the maintenance of cell integrity and morphological detail. For FISH, 3%–4% (v/v) formaldehyde or paraformaldehyde is generally suitable for gram-negative bacteria. In case of gram-positive bacteria, cells were fixed with 50% ethanol (Amann et al., 1995).

Before or after fixation, cells are usually prepared on glass slides or trapped on a polycarbonate filter. For aggregated samples or biofilms, species are fixed in formalin and then embedded in paraffin before being sectioned (Sekiguchi et al., 1999). For SEM-ISH, similar protocols with paraformaldehyde can be employed.

For better attachment of the specimens to the slide or polycarbonate filter, the surfaces were treated with a coating agent such as gelatin (Amann et al., 1990b), poly-L-lysine (Lee et al, 1999), or agarose (Pernthaler et al., 2002a). If peroxidase-labeled molecules were used for signal enhancement, an additional enzymatic treatment with lysozyme was required (Pernthaler et al., 2002b; Schönhuber et al., 1997). Some cases require further enzymatic treatment to open the peptidoglycan layer. Lysozyme and pancreatic lipase for enterococci (Waar et al., 2005), lysozyme and lysostaphin for staphylococci (Kempf et al., 2000), and lysozyme and achromopeptidase for actinobacteria (Sekar et al., 2003) have been previously employed. When peroxidase-labeled probes and antibodies were used, diethyl pyrocarbonate treatment or other additional treatments were required to inactivate intracellular peroxidase (Pernthaler et al., 2002a). Microbial communities in the natural environment are complex, and the permeability of their cell walls is not uniform. The application of mixed enzymes or other chemical treatments may be required (Pernthaler et al., 2002b), but it remains difficult to sufficiently permeabilize the cell walls of all complex bacterial communities.

3.2 Probes types

Probes are sequences of nucleotide bases complimentary to the specific DNA or RNA sequence of interest. These probes can be as small as 15–30 nucleotides or up to 1000 nucleotides. The strength of the binding between the probe and the target molecules is crucial in hybridization. This strength is affected by the various hybridization conditions described below.

Several types of probes can be used in performing ISH: oligonucleotide DNA probes, polynucleotide DNA probes, polyribonucleotide probes, peptide nucleic acid (PNA) probes, locked nucleic acid (LNA) probes, etc. Stability, availability, speed, expense, ease of use, specificity, cell wall penetration ability, and reproducibility should be considered for selecting a probe type.

3.2.1 Oligonucleotide DNA probes

An oligonucleotide DNA probe is a short sequence of nucleotides that are synthesized to match a target. They are synthetically produced, commercially available, and economical. The target nucleotide sequence must be known. The probes used to detect bacteria are small, generally approximately 15–30 bp. A small-sized probe allows for easy penetration into the bacterial cells of interest. For effective hybridization, the thermodynamics of nucleic acid hybridization based on Gibbs free energy change should be considered. The affinity of the probe to the target site is defined as the overall Gibbs free energy change for intramolecular DNA and RNA interactions that take place during ISH (Yilmaz & Noguera, 2004).

3.2.2 Polynucleotide DNA probes

Polynucleotide DNA probes have similar advantages to oligonucleotide DNA probes, but they are much larger, typically 50–1000 base long. These are synthetically produced by reverse transcription of RNA, or by PCR, or by fragmentation of the PCR product (Niki & Hiraga, 1998). Fragmented chromosomal DNA can be used as a probe (Lanoil & Giovannoni, 1997). However, synthesizing this type of probe requires time and expensive reagents. Polynucleotide DNA probes can be labeled at multiple sites with fluorescent dyes, digoxigenin, or biotin and thus are used to amplify probe-derived signals.

3.2.3 Polyribonucleotide probes

RNA probes have the advantage that RNA–RNA hybrids are considerably thermostable and resistant to digestion by RNases. In vitro transcription of plasmid DNA with RNA polymerase can be used to produce RNA probes (Delong et al., 1999; Pernthaler et al., 2004; Zwirglmaier et al., 2004). These probes, however, can be very difficult to work with as they are highly sensitive to ubiquitous RNases.

3.2.4 Peptide nucleic acid (PNA) probes

PNAs are the synthetic analogs of DNA. DNA and RNA have deoxyribose and ribose sugar backbones, respectively, whereas the backbone of a PNA backbone comprises repeating N-(2-aminoethyl)-glycine units that are linked by peptides (Nielsen et al., 1991). The backbone of PNAs contains no charged phosphate groups. Less electrostatic repulsion occurs when the PNA probe hybridizes to DNA or RNA sequences. The PNA–DNA or PNA–RNA complex is more stable than the natural nucleic acid complexes. PNA is not easily identified by either nucleases or proteases, making them resistant to enzyme degradation. Because of their higher binding strength, it is not necessary to design long PNA oligomers. Such oligomers are chemically synthesized and commercially available. PNA-FISH also has broad applications in clinical microbiology (Stender, 2003).

3.2.5 Locked nucleic acid (LNA) probes

LNAs are a class of analogs that contain an extra bridge connecting the 2' oxygen and 4' carbon (Obika et al., 1997). LNA nucleotides can be mixed with DNA or RNA residues in the oligonucleotide whenever desired. The LNA–DNA heteroduplex is thermostable. DNA probes with LNA have the advantage of higher affinity and specificity than normal DNA probes, and greater design flexibility and lower costs than PNA probes (Silahtaroglu et al., 2003). Such oligomers are chemically synthesized and commercially available. LNA nucleotides are used to increase the sensitivity and specificity of expression in DNA microarrays, FISH probes, real-time PCR probes, and other molecular biology techniques.

3.3 Hybridization

Hybridization must be performed under stringent conditions to allow the binding of the probe to the target sequence. The representative components in a hybridization buffer are shown in Table 1. For hybridization of oligonucleotide DNA probes to rRNA, sodium chloride (NaCl), tris(hydroxymethyl)aminomethane-HCl (Tris-HCl), sodium dodecyl sulfate (SDS), and formamide were the major components. For hybridization of polyribonucleotide probes to rRNA, the components were slightly modified. For polynucleotide DNA probes, NaCl, sodium citrate and formamide were major components. A buffer solution containing NaCl, Tris-HCl, and SDS or SSC buffer (containing NaCl and sodium citrate) was used as the wash solution.

The preheated hybridization buffer was applied to the sample containing probes complementary to the target sequence. In case of rRNA-targeted ISH, stringency can be adjusted by varying either the formamide concentration or the hybridization temperature. Formamide decreases the melting temperature by weakening the hydrogen bonds, thus enabling lower temperatures to be used with high stringency. The salt immobilizes hybrid molecules and is used instead of formamide to reduce toxic waste. The general hybridization conditions used for ISH with rRNA-targeted probes are shown in Table 2. The concentration of NaCl ranged from 0 to 900 mM and that of formamide ranged from 0% to 50%, and the hybridization temperature ranged from 37° to 55°C. Hybridization time ranged from 30 min to 16 h. After hybridization, the slides or filters are rinsed with the appropriate buffer to remove the unbound probe.

Components in hybridization buffer	Probe type	Target molecule	References
NaCl, Tris-HCl, SDS	Oligonucleotide DNA	rRNA	Amann et al., 1990a, 1995
NaCl, Tris-HCl, SDS, formamide	Oligonucleotide DNA	rRNA	Amann et al.,1995; Manz et al., 1996
NaCl, Tris HCl, dextran sulfate, SDS, formamide, blocking reagent	Oligonucleotide DNA	rRNA	Pernthaler et al., 2002b
NaCl, Tris-HCl, EDTA, poly(A), formamide, dextran sulfate	Polyribonucleotide	rRNA	Delong et al., 1999
NaCl, Tris-HCl, dextran sulfate, SDS, formamide, *E.coli* tRNA, salmon sperm DNA, blocking reagent	Polyribonucleotide	rRNA	Pernthaler et al., 2002a
NaCl, sodium citrate, formamide, dextran sulfate, blocking reagent, Denhardt's solution, yeast RNA, salmon sperm DNA	Polyribonucleotide	mRNA	Pernthaler & Amann, 2004.
NaCl, Tris-HCl, SDS, formamide	Polyribonucleotide	DNA	Zwirglmaier et al., 2004
NaCl, sodium citrate, formamide, salmon sperm DNA	Polynucleotide DNA (fragmented PCR product)	DNA	Niki & Hiraga, 1998
NaCl, sodium citrate, formamide, dextran sulfate	Polynucleotide DNA (fragmented genomic DNA)	DNA	Lanoil & Giovannoni, 1997

Table 1. Representative components of hybridization buffer.

	Formamide (%)	NaCl (mM)	Temperature	Time (h)
Condition	0-50	0-900	37-55	0.5 -16

Table 2. General hybridization conditions for rRNA-targeted ISH.

Bouvier & del Giorgio (2003) investigated factors that influenced the detection of bacterial cells in rRNA-targeted FISH. They collected experimental conditions for FISH and environmental factors based on published reports and found that both NaCl and formamide in the hybridization buffer and wash solution significantly influence the performance of rRNA-targeted FISH. These two chemicals are used to adjust the stringency conditions of hybridization and wash steps.

Appropriate controls play an important role in optimizing hybridization and wash conditions. In general, target bacteria include perfect match sequence as positive control. As a negative control, non-target bacteria include a known mismatched sequence.

4. Problems and pitfalls in ISH

4.1 False positive results

The accuracy and reliability of ISH is highly dependent on the specificity of the probe. The sequence design and evaluation of the new probe are essential. Appropriate positive and negative control strains should be included in every ISH experiment. Probe sequences should be carefully examined using the latest version of sequencing data. The stringency conditions of hybridization and wash steps affect both false positive and negative results The mild conditions result in nonspecific probe binding to mismatched sequences or cell structures. Newly designed probes should be evaluated in laboratories by molecular microbiological methods such as ISH, dot blot hybridization, melting curve analysis etc.

4.2 False negative results

Low signal intensity may be a result of insufficient penetration of the probe into the target cells. It depends on the structure of the cell wall or membrane of the bacterial cells. Permeabilization conditions need to be optimized so that all reagents can penetrate the cell. The permeability of the bacterial cell wall structures is not uniform, and different permeabilization procedures have been employed for different cells. In general, gram-negative bacteria tend to be permeable under well-known permeability conditions. For gram-positive bacteria, special fixation and pretreatment is required as described above.

In case rRNA is the target molecule, loop and hairpin formation of target RNA hampers hybridization, leading to differential accessibility of these probes. Self-annealing and self-hairpin formation of the probe itself can also lead to low signal intensity (Fuchs et al., 1998). In addition, the rRNA content of bacterial cells may vary considerably within species as well as strains depending on the physiological state in the given environment. Low rRNA content may result in low signal intensity or false negative results. Various strategies have been used to overcome this difficulty, including FISH combined with direct viable counting, use of multiply-labeled polynucleotide DNA or RNA probes, enzymatic signal amplification, and in situ DNA amplification (see 5. Enhancement of signal intensity).

To test false negative results because of methodological problems, universal probes such as EUB338 (Amann et al., 1990a) have been commonly used as positive control probes. If the control with the universal probe yields good results in ISH, then fixation, probe permeabilization, and rRNA content of the target cells are not the problem. A commonly used negative control probe is NON338, which is complimentary to EUB338 (Wallner et al.,

1993). Using this probe, the non-specific binding of the probe to cell structures other than target nucleic acids can be evaluated.

5. Enhancement of signal intensity

The signal from hybridized cells is highly dependent on the content of target molecules. Despite its potential, the application of ISH in targeting rRNA of resident bacteria in oligotrophic environments is hampered by the low copy number of target molecules. To increase signal intensity, two major environmental microbiological approaches were employed, signal amplification and target nucleic acids amplification. The representative methods are shown in Table 3.

Category	Methods	References
Signal amplification	Tyramid signal amplification	Schönhuber et al., 1997
	Two-pass Tyramid signal amplification	Kubota et al., 2006
	HNPP/Fast Red	Yamaguchi et al., 1996
	Multiply-labeled polyribonucleotide	Delong et al., 1999
Target DNA amplification	In situ PCR	Hodson et al., 1995
	LAMP	Maruyama et al., 2003
	CPRINS	Kenzaka et al., 2005b
	In situ RCA	Maruyama et al., 2005
Target RNA amplification	Direct viable counting	Kenzaka et al., 2001

Table 3. Signal amplification and target nucleic acids amplification methods.

5.1 Signal amplification

Enzymatic signal amplification using a tyramide signal amplification system (TSA) or HNPP/Fast Red was combined with oligonucleotide/polynucleotide probes to increase the sensitivity of ISH (Kenzaka et al., 1998; Kubota et al., 2006; Schönhuber et al., 1997). In the alkaline phosphatase-HNPP/Fast Red system, a digoxigenin-labeled oligonucleotide probe was detected by an alkaline phosphatase-conjugated anti-digoxigenin antibody. Fluorescent molecules accumulate in the target cells because of the activity of alkaline phosphatase (Yamaguchi et al., 1996). In horseradish peroxidase (HRP)-TSA system, oligonucleotide probes were labeled with HRP that generates fluorescent molecules in cells when fluorescent tyramide was used as a substrate (Schönhuber et al., 1997). These enzymatic amplification systems resulted in an 8–20-fold amplification of signal intensity. In case of a two-pass TSA, tyramide tagged with dinitrophenyl was generated around the probe/target site by the activity of HRP, and then HRP-labeled anti-dinitrophenyl antibody further accumulated in cells. Tyramide-Cy3 was deposited by the activity of accumulated HRP (Kawakami et al., 2010).

The signal in SEM–ISH can be amplified using a similar system to detect low copy number target DNA sequences in individual cells (Kenzaka et al., 2009). In the study, digoxigenin-labeled polynucleotide probes were hybridized with plasmid DNA. Peroxidase-labeled anti-digoxigenin antibody was bound to digoxigenin. By using tyramide signal amplification, biotin molecules were accumulated inside target cells. Target cells were identified by streptavidin bound to a gold immunoprobe. Gold particle enhancement was performed to amplify probe signals from hybridized cells. Low vacuum SEM images of a mixture of *E. coli* JM109 harboring plasmid pT7GFP (tagged with *gfp* and *bla* genes) cells and *E. coli* Okayama O27 cells after ISH with *gfp* and *bla* probes are shown in Fig. 3a, 3b, respectively. *E. coli* JM109 harboring pT7GFP were approximately 5-μm long and rod shaped. The long rod-shaped cell with the target gene showed a strong signal because of the high density of gold in one portion of the cell (shown as arrows in Fig. 3a and 3b).

(a)   (b)

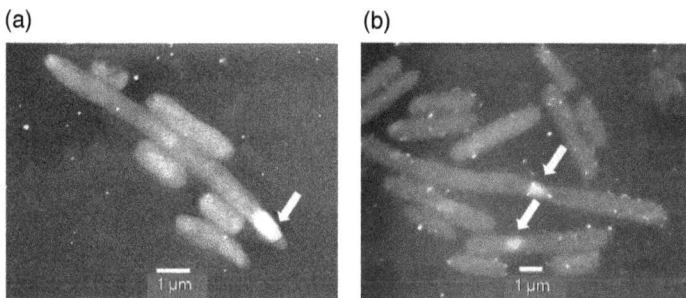

Fig. 3. Low-vacuum SEM images of a mixture containing *E. coli* JM109 harboring pT7GFP cells and *E. coli* Okayama O27 cells after ISH with *gfp* probe (a) and *bla* probe (b). Arrow indicates target *E. coli* JM109 cells harboring pT7GFP.

5.2 Target nucleic acids amplification

In situ DNA amplification techniques based on fluorescent labeling have been successfully applied to identify individual genes in a single bacterial cell. The basic approach is in situ PCR in which target DNA sequences are amplified inside cells (Hodson et al., 1995; Tani et al., 1998). The application of in situ PCR with functional probes provides a powerful tool for detection of genes or gene products in individual cells. This method, however, cannot be applied to diverse species in the natural environment mainly because of permeability, the leakage of amplified products, and less effective concentration of target cells. Longer amplified products after in situ DNA amplification would give better results because these products are less likely to leak out from the cell.

In situ loop-mediated isothermal amplification (LAMP) generates long tandem repeats of the target sequence, preventing amplicons from leaking outside the cell (Maruyama et al., 2003). The mild permeabilization conditions and low isothermal temperature used in the in situ LAMP method causes lesser cell damage than in situ PCR.

Cycling primed in situ amplification (CPRINS) uses a single primer and results in linear amplification of the target DNA. The amplicons are long, single-stranded DNA and are thus retained within the permeabilized microbial cells. ISH with a multiply labeled probe set enabls significant reduction in a nonspecific background while maintaining high signals of target bacteria (Kenzaka et al., 2005b).

In situ rolling circle amplification (RCA) require one short target sequence (less than 40 mer) and generate large, single-stranded, and tandem repeats of target DNA as amplicons. The circularizable probes are approximately 90 mers, comprising short complementary sequences of the target DNA at the 3' and 5' ends, respectively, with an arbitrary sequence in the middle. These probes are labeled with a phosphate group at the 5' end and circularized by ligation when hybridized to the target sequences. The RCA primers amplify the complementary sequence of the circularized probe by hybridization to a specific region of the probe. The amplicons are a single-stranded tandem repeats of the circularized probe sequence. It can be detected with labeled oligonucleotide probes. (Maruyama et al., 2005). CPRINS and in situ RCA can be performed on polycarbonate filters, which allow the effective concentration of target cells from aquatic samples and enhance the quantitative analysis.

Target rRNA molecules can be increased by the direct viable counting (DVC) method (Kenzaka et al., 2001). This method is based on the incubation of samples with antimicrobial agents and nutrients. The antibiotic cocktail acts as a specific inhibitor of DNA synthesis and prevents cell division without affecting other metabolic activities. The resulting cells can continue to metabolize nutrients and elongate and/or become fattened after incubation. By employing these techniques, SEM-ISH will lead to further improvements in sensitivity.

6. Applicability of SEM-ISH to complex microbial communities

Fluorescence microscopy and confocal laser scanning microscopy are important tools in effectively examining complex microbial communities attached to various materials in the natural environment. Potential problems with these fluorescent techniques include autofluorescence, which results from natural substances within plant tissue, organic debris, soil particles, etc. This may hamper the observation of target microbes in complex microbial communities.

SEM allows the visualization of cells attached to materials (e.g., sediment particles) without the interference of autofluorescence and without requiring the ultrathin sectioning of materials. In our experiments, E. coli JM109 cells expressing green fluorescent protein (GFP) were introduced into natural river water samples and subjected to FISH with an ES445 probe. Figs. 4a and 4b show the fluorescence micrographs of bacterial cells hybridized with a Cy3-labeled probe. Under blue excitation, the inoculated cells expressing GFP were identified (Fig. 4a).

Although these cells were expected to show bright Cy3 fluorescence under green excitation, the E. coli cells attached to organic debris in river water samples were masked by the nonspecific binding of the probe to organic debris in river water samples (Fig. 4b). Consequently, this hampered the identification and accurate enumeration of target cells by FISH.

To enhance the reliability of enumerating target cells, SEM-ISH was employed for the same river water samples containing inoculated E. coli JM109 cells. Fig. 4c shows the SE image of organic debris with E. coli cells on the surface. The advantage of SEM-ISH is to enable clear observation of the cell surface structure, detritus, etc., under higher magnification. Fig. 4d shows the portion of Fig. 4c that was magnified. Inoculated E. coli cells were approximately 5-μm long and rod shaped. The magnification of the SE image allowed clear differentiation between the target cells (shown as arrow in Fig. 4d) and the other cells. By comparing the

BSE image from the same microscopic field, the probe signal was detected from the hybridized cells (Fig. 4e). Even when SEM-ISH was applied to the river water samples, the problem of nonspecific binding of probes to organic debris as shown in Fig. 4b was not completely resolved. However, high magnification allowed target cells to be detected and distinguished from others, even with a high background. As a result, SEM-ISH proved better than fluorescence-based methods in discriminating between target cells and others in the water samples (Fig. 4d and 4e).

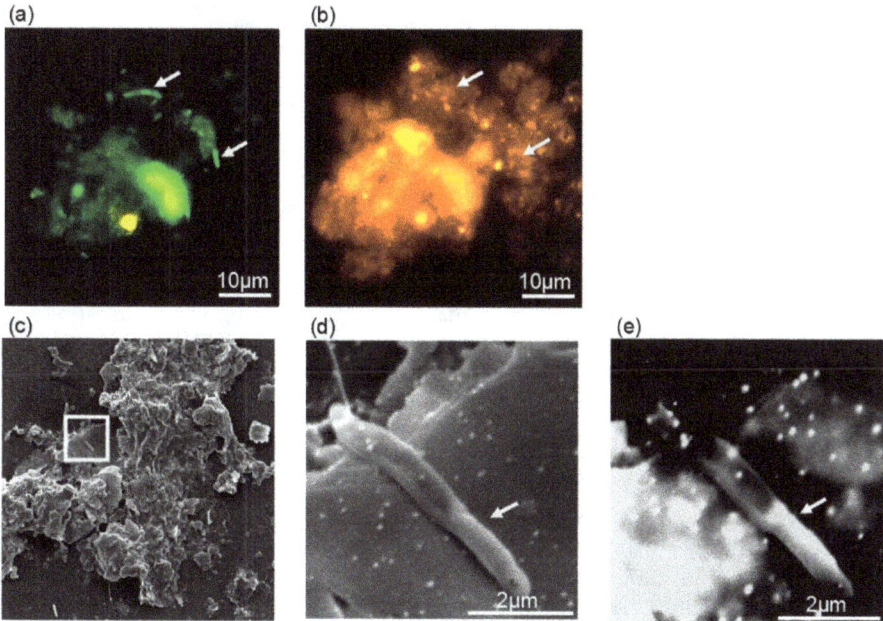

Fig. 4. Organic debris masks the probe signal in river water. *E. coli* cells expressing GFP were inoculated into river water samples, and subjected to FISH with Cy3 labeled probe ES445 (a, b) or SEM-ISH with biotin labeled probe ES445 (c, d, e). *E. coli* cells expressing GFP became attached to organic debris in the river water sample and delineated under blue excitation (a), but probe signals were masked under green excitation (b). High vacuum SEM imaging of *E. coli* cells attached to organic debris in the river water sample (c) was magnified (d). The topographic information was obtained with SE images (c and d), and cells hybridized with ES445 probe were detected with the BSE image (e). Arrow indicates introduced *E. coli* cell.

Certain bacterial cells happen to be buried in the surface of materials such as sediments or soil particles, and cell boundaries were unclear under SEM. In this case, it was difficult to distinguish individual cells within the particle structure using only SE signals. BSE images provided the probe signals from the hybridized cells in the same microscopic fields and clarified the existence of buried cells. We applied SEM-ISH with rRNA-targeted probes to examined bacteria communities on surface of sediment samples (Kenzaka et al., 2005a). SEM-ISH revealed the significant abundance of the *Cytophaga–Flavobacterium* cluster (detectable by probe CF319) on the surface of sediment particles and confirmed a wide distribution over the particle surface. When observed at high magnification, certain bacterial cells were found to be buried in the particles (Fig. 5a and 5b). SEM-ISH with rRNA-targeted probes identified the buried cells based on rRNA sequence.

(a)   (b)

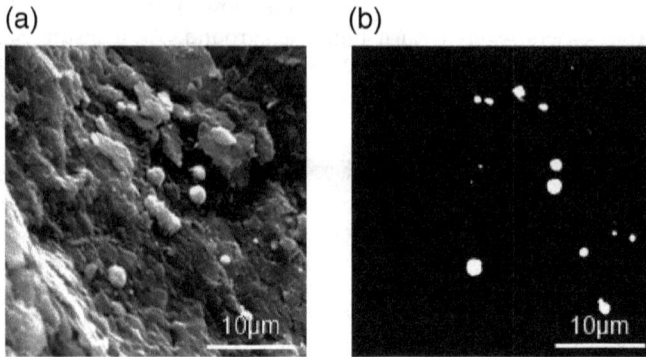

Fig. 5. High-vacuum SEM images of bacteria attached on surface of river sediment particles detected by ISH with the probe CF319 (targeted for *Cytophaga-Flavobacterium* phylum). The same microscopic fields are shown with SE image (a), BSE image (b). The topographic information was obtained with SE images (a), and cells hybridized with the probe CF319 were detected with the BSE image (b).

7. Access to yet-to-be-cultured bacteria with ISH

In the environment, more than 90% of the bacterial communities cannot be cultured by standard techniques, and the yet-to-be-cultured fraction includes diverse microorganisms that are only distantly related to the cultured ones. Culture-independent methods are essential to understand the genetic diversity, population structure, and ecological roles of the bacterial communities. The use of PCR-based clone libraries or metagenomics of an assemblage of microorganisms has great potential for the exploring novel species and sequences and for the understanding the composition and function of microbial communities and their dynamics in the environment (Handelsman, 2004).

Once novel RNA or DNA sequences are obtained by such approaches, ISH with newly designed probes targeting the novel sequences would be a valuable tool for the investigation of their distribution and abundance in the given environment. To validate probe specificity, ISH of clones with target sequences as inserts into plasmids (clone-ISH) would play important roles (Schramm et al., 2002).

8. Conclusion

SEM has been widely used in environmental microbiology to study cell and biofilm morphology. Combining morphological study with SEM and ISH with RNA- or DNA-targeted probes has demonstrated great potential in visualization of cells attached to materials without interference by autofluorescence and requiring no ultrathin sectioning of materials. SEM-ISH addresses some of the limitations of FISH alone and enhances the reliability of monitoring target cells in environmental samples in which the application of fluorescence-based methods is limited. The concepts of ISH in electron microscopic studies can lead to a new understanding of the spatial distribution of target cells as well as of the extent of cell heterogeneity on plant, metal, alloy, bioreactor, or the three-dimensional structures of the attachment matrix.

9. References

Amann, R.I., Binder, B.J., Olson, R.J., Chisholm, S.W., Devereux, R. & Stahl, D.A. 1990a. Combination of 16S rRNA-targeted oligonucleotide probes with flow cytometry for analyzing mixed microbial populations. *Applied and Environmental Microbiology,* Vol.56, No.6, pp. 1919-1925.

Amann, R.I., Krumholz, L. & Stahl, D.A. (1990b). Fluorescent-oligonucleotide probing of whole cells for determinative, phylogenetic, and environmental studies in microbiology. *Journal of Bacteriology,* Vol.172, No.2, pp. 762–770.

Amann, R.I., Ludwig, W. & Schleifer, K.H. 1995. Phylogenetic identification and in situ detection of individual microbial cells without cultivation. *Microbiological Reviews,* Vol.59, No.1, pp. 143-169.

Bouvier, T. & del Giorgio, P.A. (2003). Factors infuencing the detection of bacterial cells using fuorescence in situ hybridization (FISH): A quantitative review of published reports. *FEMS Microbiology Ecology,* Vol.44, No.1, pp. 3-15.

Brabec, V. & Leng, M. (1993). DNA interstrand cross-links of trans-diamminedichloroplatinum(II) are preferentially formed between guanine and complementary cytosine residues. *Proceedings of the National Academy of Sciences of the United States of America.* Vol.90, No.11, pp. 5345-5349.

Dalbiès, R., Payet, D. & Leng, M. (1994). DNA double helix promotes a linkageisomerization reaction in trans-diamminedichloroplatinum(II)-modified DNA. *Proceedings of the National Academy of Sciences of the United States of America.* Vol. 91, No.17. pp. 8147–8151.

DeLong, E.F., Taylor, L.T., Marsh, T.L. & Preston, C.M. (1999). Visualization and enumeration of marine planktonic archaea and bacteria by using polyribonucleotide probes and fluorescent in situ hybridization. *Applied and Environmental Microbiology,* Vol.65, No.12, pp. 5554-5563.

Fuchs, B.M., Wallner, G., Beisker, W., Schwippl, I., Ludwig, W. & Amann, R. (1998). Flow cytometric analysis of the in situ accessibility of *Escherichia coli* 16S rRNA for fluorescently labeled oligonucleotide probes. *Applied and Environmental Microbiology,* Vol.64, No.12, pp. 4973–4982.

Hacker, G.W. (1998). High performance Nanogold-silver in situ hybridisation. *European journal of histochemistry,* Vol.42, No.2, pp. 111-120.

Handelsman J. (2004). Metagenomics: application of genomics to uncultured microorganisms. *Microbiology and Molecular Biology Reviews,* Vol.68, No.4, pp. 669-685.

Hodson, R.E., Dustman, W.A., Garg, R.P. & Moran, M.A. (1995). In situ PCR for visualization of microscale distribution of specific genes and gene products in prokaryotic communities. *Applied and Environmental Microbiology,* Vol.61, No.11, pp. 4074-4082.

Kawakami, S. Kubota K., Imachi, H., Yamaguchi, T. Harada, H. & Ohashi, A. (2010). Detection of single copy genes by two-pass tyramide signal amplification fluorescence in situ hybridization (Two-Pass TSA-FISH) with single oligonucleotide probes. *Microbes and Environments,* Vol.25, No.1, pp. 15–21.

Kempf, V.A.J., Trebesius, K. & Autenrieth, I.B. (2000). Fluorescent in situ hybridization allows rapid identification of microorganisms in blood cultures. *Journal of Clinical Microbiology,* Vol.38, No.2, pp. 830-838.

Kenzaka, T., Yamaguchi, N., Tani, K. & Nasu, M. (1998). rRNA-targeted fluorescent in situ hybridization analysis of bacterial community structure in river water. *Microbiology,* Vol.144, No.8, pp. 2085-2093.

Kenzaka, T., Yamaguchi, N., Prapagdee, B., Mikami, E. & Nasu, M. (2001). Bacterial community composition and activity in urban rivers in Thailand and Malaysia. *Journal of Health Science,* Vol.47, No.4, pp. 353-361.

Kenzaka, T., Ishidoshiro, A., Yamaguchi, N., Tani, K. & Nasu, M. (2005a). rRNA sequence-based scanning electron microscopic detection of bacteria. *Applied and Environmental Microbiology,*Vol.71, No.9, pp. 5523-5531.

Kenzaka, T., Tamaki, S., Yamaguchi, N., Tani, K. & Nasu, M. (2005b). Recognition of individual genes in diverse microorganisms by cycling primed in situ amplification. *Applied and Environmental Microbiology,* Vol.71, No.11, pp. 7236-7244.

Kenzaka, T., Ishidoshiro, A., Tani, K. & Nasu, M. (2009). Scanning electron microscope imaging of bacteria based on DNA sequence. *Letters in Applied Microbiology,* Vol.49, No.6, pp. 796-799.

Kubota, K., Ohashi, A. Imachi, H. & Harada, H. (2006). Visualization of mcr mRNA in a methanogen by fluorescence in situ hybridization with an oligonucleotide probe and two-pass tyramide signal amplification (two-pass TSA-FISH). *Journal of Microbiological Methods*, Vol.66, No.3, pp. 521–528.

Lanoil, B.D. & Giovannoni, S.J. (1997). Identification of bacterial cells by chromosomal painting. *Applied and Environmental Microbiology,* Vol.63, No.3, pp. 1118-1123.

Lee, N., Nielsen, P.H. Andreasen, K.H., Juretschko, S., Nielsen, J.L., Schleifer, K.H. & Wagner, M. (1999). Combination of fluorescent in situ hybridization and microautoradiography-a new tool for structure-function analyses in microbial ecology. *Applied and Environmental Microbiology,* Vol.65, No.3, pp. 1289-1297.

Manz, W., Amann, R., Ludwig, W., Vancanneyt, M. & Schleifer, K.H. (1996). Application of a suite of 16S rRNA-specific oligonucleotide probes designed to investigate bacteria of the phylum Cytophaga-Flavobacter-Bacteroides in the natural environment. *Microbiology,* Vol.142, No.5, pp. 1097-1106.

Maruyama, F., Kenzaka, T., Yamaguchi, N., Tani, K. & Nasu, M. (2003). Detection of bacteria carrying the *stx*2 gene by in situ loop-mediated isothermal amplification. *Applied and Environmental Microbiology,* Vol.69, No.8, pp. 5023-5028.

Maruyama, F., Kenzaka, T., Yamaguchi, N., Tani, K. & Nasu, M. (2005). Visualization and enumeration of bacteria carrying a specific gene sequence by in situ rolling circle amplification. *Applied and Environmental Microbiology,* Vol.71, No.12, pp. 7933-7940.

Moraru, C., Lam, P., Fuchs, B.M., Kuypers, M.M.M. & Amann, R. (2010). GeneFISH – an *in situ* technique for linking gene presence and cell identity in environmental microorganisms. *Environmental Microbiology*, Vol.12, No.11, pp. 3057–3073.

Nielsen, P.E., Egholm, M, Berg, R.H. & Buchardt, O. (1991). Sequence-selective recognition of DNA by strand displacement with a thymine-substituted polyamide *Science,* Vol.254, No.5037, pp. 1497-1500.

Niki, H. & Hiraga, S. (1998). Polar localization of the replication origin and terminus in *Escherichia coli* nucleoids during chromosome partitioning. *Genes and Development*, Vol.12, No.7, pp. 1036-1045.

Obika, S. Nanbu, D., Hari, Y., Morio, K., In, Y., Ishida, T. & Imanishi, T. (1997). Synthesis of 2'-O,4'-C-methyleneuridine and -cytidine. Novel bicyclic nucleosides having a fixed C3'-endo sugar puckering. *Tetrahedron Letters*. Vol.38, No.50, pp. 8735-8738.

Pernthaler, A., Pernthaler, J. & Amann, R. (2002a). Fluorescence in situ hybridization and catalyzed reporter deposition for the identification of marine bacteria. *Applied and Environmental Microbiology*, Vol.68, No.6, pp. 3094-3101.

Pernthaler, A., Pernthaler, J., Schattenhofer, M. & Amann. R. (2002b). Identification of DNA-synthesizing bacterial cells in coastal North Sea plankton. *Applied and Environmental Microbiology*, Vol.68, No.11, pp. 5728-5736.

Pernthaler, A. & Amann, R. (2004). Simultaneous fluorescence in situ hybridization of mRNA and rRNA in environmental bacteria. *Applied and Environmental Microbiology*, Vol.70, No.9, pp.5426-5433.

Schönhuber, W., Fuchs, B., Juretschko, S. & Amann, R. (1997). Improved sensitivity of whole-cell hybridization by the combination of horseradish peroxidase-labeled oligonucleotides and tyramide signal amplification. *Applied and Environmental Microbiology*, Vol.63, No.8, pp. 3268-3273.

Schramm, A., Fuchs, B.M., Nielsen, J.L., Tonolla. M. & Stahl, D.A. (2002). Fluorescence in situ hybridization of 16S rRNA gene clones (Clone-FISH) for probe validation and screening of clone libraries. *Environmental Microbiology*, Vol.4. No.11, pp. 713-720.

Sekar, R., Pernthaler, A., Pernthaler, J., Warnecke, F., Posch, T. & Amann, R. (2003). An improved protocol for the quantification of freshwater actinobacteria by fluorescence in situ hybridization. *Applied and Environmental Microbiology*, Vol.69, No.5, pp. 2928-2935.

Sekiguchi, Y., Kamagata, Y., Nakamura, K., Ohashi, A. & Harada, H. (1999). Fluorescence in situ hybridization using 16S rRNA-targeted oligonucleotides reveals localization of methanogens and selected uncultured bacteria in mesophilic and thermophilic sludge granules. *Applied and Environmental Microbiology*, Vol.65, No.3, pp. 1280-1288.

Silahtaroglu, A.N., Tommerup, N. & Vissing, H. (2003). FISHing with locked nucleic acids (LNA): Evaluation of different LNA/DNA mixmers. *Molecular and Cellular Probes*, Vol.17, No.4, pp. 165–169.

Stender., H. (2003). PNA FISH: an intelligent stain for rapid diagnosis of infectious diseases. *Expert Review of Molecular Diagnostics*. Vol.3, No.5, pp. 649-655.

Tani, K., Kurokawa, K. & Nasu, M. (1998). Development of a direct in situ PCR method for detection of specific bacteria in natural environments. *Applied and Environmental Microbiology*, Vol.64, No.4, pp. 1536-1540.

Waar, K., Degener, J.E., van Luyn, M.J. & Harmsen, H.J.M. (2005) Fluorescent in situ hybridization with specific DNA probes offers adequate detection of *Enterococcus faecalis* and *Enterococcus faecium* in clinical samples. *Journal of Medical Microbiology*, Vol.54, No.10, pp. 937-944.

Wallner, G., Amann, R. & Beisker, W. (1993). Optimizing fluorescent in situ hybridization with rRNA-targeted oligonucleotide probes for flow cytometric identification of microorganisms. *Cytometry*, Vol.14, No.2, pp. 136–143.

Yamaguchi, N., Inaoka, S., Tani, K., Kenzaka, T. & Nasu, M. (1996). Detection of specific bacterial cells with 2-hydroxy-3-naphthoic acid-2'-phenylanilide phosphate and Fast Red TR in situ hybridization. *Applied and Environmental Microbiology,* Vol.62, No.1, pp. 275-278.

Yilmaz, L. S. & Noguera, D.R. (2004). Mechanistic approach to the problem of hybridization efficiency in fluorescent in situ hybridization. *Applied and Environmental Microbiology,* Vol.70, No.12, pp. 7126-7139.

Zwirglmaier, K., Ludwig, W. & Schleifer, K.H. (2004). Improved method for polynucleotide probe-based cell sorting, using DNA-coated microplates. *Applied and Environmental Microbiology,* Vol.70, No.1, pp. 494-497.

13

Study of Helminth Parasites of Amphibians by Scanning Electron Microscopy

Cynthya Elizabeth González[1], Monika Inés Hamann[1] and Cristina Salgado[2]
[1]Centro de Ecología Aplicada del Litoral,
Consejo Nacional de Investigaciones Científicas y Técnicas,
[2]Servicio de Microscopia Electrónica de Barrido,
Universidad Nacional del Nordeste (UNNE), Corrientes,
Argentina

1. Introduction

Amphibians, like all other animals, are subject to a variety of parasites and diseases, including viral, bacterial and fungal infections as well as some forms of cancer and tuberculosis (Hoff et al., 1984). Various viruses and bacteria such as *Pseudomonas* or *Salmonella*, and fungi such as *Candida*, are recorded as common infectious agents in amphibians, but currently the focus of studies are the fungi of genus *Batrachochytrium*, agents of the disease known as chytridiomycosis, which is considered as one of the factors responsible for the decline of amphibian populations in many parts of the world (Berger and Speare, 1998). In addition, protozoans of the genera *Opalina* and *Entamoeba* in the digestive tract, and trypanosomes in the circulatory system, as well as coccidian protozoa have been recorded in amphibians (Duellman and Trueb, 1986; Duszynski et al., 2007).

However, helminths are the most common invertebrate parasites of amphibians. One well known example among trematodes is the monogenean genus *Polystoma*, which infects the urinary bladder of adult amphibians around the world. Parasitic digenean trematodes include both larval stages (metacercariae) and adults. Cestodes are not common parasites in the gastrointestinal tract of amphibians, but when present may persist for a long time. Adult acanthocephalans adhere to the mucosa of the stomach or intestine. Finally, nematodes are particularly abundant in the digestive tract, lungs and blood vessels of these vertebrates (Pough et al., 2001). Amphibians are also hosts to other groups of less common parasitic invertebrates, such as annelids, pentastomids and arthropods (copepods, ticks, insects) (Tinsley, 1995).

Of these, analyses using scanning electron microscopy techniques have been mainly applied to digenean trematodes (flatworms), nematodes (roundworms) and acanthocephalans (thorny or spiny headed), particularly to their adult stages (Fig. 1-3).

The study of parasitic nematodes and trematodes by scanning electron microscopy began in the 1970s and involved mainly those organisms that produced diseases in humans and livestock, as well as parasites of crops (Halton, 2004). In particular, studies in amphibian hosts were first made by Nollen and Nadakavukaren (1974) and Nadakavukaren and

Nollen (1975) who provided details of the tegument of the trematodes *Megalodiscus temperatus* (Paramphistomatidae) and *Gorgoderina attenuata* (Gorgoderidae) from *Rana pipiens*. Regarding nematodes, Navarro et al. (1988) provided details of the cuticle of the species *Cosmocerca ornata*, *Oxysomatium brevicaudatum* (Cosmocercidae) and *Seuratascaris numidica* (Ascarididae) collected in ranid hosts from different geographical areas of the Iberian Peninsula.

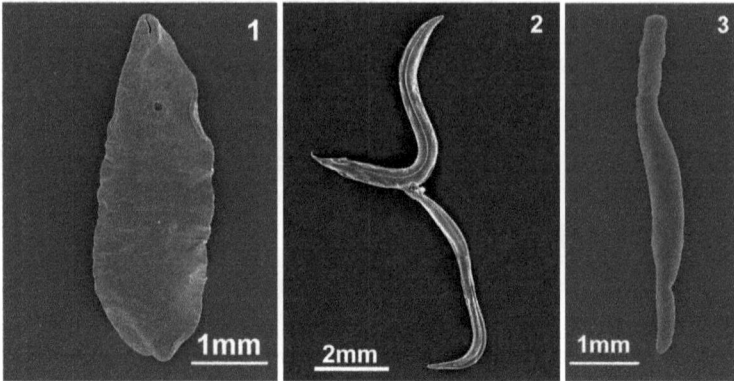

Fig. 1-3. Helminth parasites found in amphibian hosts. 1. Flatworms (Trematoda), general view. 2. Roundworms (Nematoda) male and female mating. 3. Thorny or spine headed (Acanthocephala), general view.

This chapter presents scanning electron micrographs taken during diverse studies undertaken to determine the helminth fauna of Argentinean anurans, especially those living in Northeastern Argentina. The survey includes the classes Trematoda (specifically subclass Digenea) and Nematoda (specifically subclass Secernentea) and the phyllum Acanthocephala. A total of five families of amphibians (Bufonidae, Cycloramphidae, Hylidae, Leptodactylidae, Leiuperidae) were analyzed, both at larval (tadpole) and adult stages, to study their helminth parasites. At the end of the chapter we present a summary of the present-day advances in this topic, including new contributions presented in this work; finally, we discuss possible future lines of research in this field of Parasitology.

The classification of helminths follows Anderson et al. (2009) and Gibbons (2010) for class Nematoda; Gibson et al. (2002) and Jones et al. (2005) for class Trematoda and Amin (1985) for Acanthocephala.

2. Preparation of helminth parasites of amphibians for observation by scanning electron microscopy

2.1 Collection of hosts and obtaining of helminth parasites

Adult frogs were hand captured, mainly at night, using the sampling technique defined as visual encounter survey (Crump and Scott, 1994). The individuals were transported live to the laboratory and killed in a chloroform solution ($CHCl_3$). The abdominal cavity of each frog was opened and the oesophagus, stomach, gut, lungs, liver, urinary bladder, kidneys, body cavity, musculature, integument and brain examined for parasites under a dissecting microscope (Fig. 4). Tadpoles were captured with a 45-cm-diameter dip net and kept alive in the laboratory until their dissection. They were killed using a chloroform solution, and subsequently all organs, musculature and body cavity were examined for parasites.

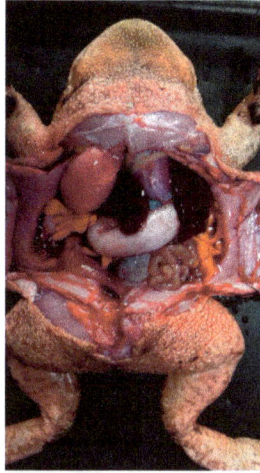

Fig. 4. Ventral view of adult amphibian showing all organ systems.

The analized amphibian species were: *Rhinella bergi, R. fernandezae, R. granulosa, R. schneideri* (Bufonidae), *Odontophrynus americanus* (Cycloramphidae), *Dendropsophus nanus, D. sanborni, Hypsiboas raniceps, Pseudis limellum, Phyllomedusa hypochondrialis, Scinax acuminatus, S. nasicus, Trachycephalus venulosus, Pseudis paradoxa* (Hylidae), *Physalaemus albonotatus, P. santafecinus, Pseudopaludicola boliviana, P. falcipes* (Leiuperidae), *Leptodactylus bufonius, L. chaquensis, L. elenae, L. latinasus, L. latrans, L. podicipinus* (Leptodactylidae), *Lepidobatrachus laevis* (Ceratophryidae). Anurans were identified using different guides and keys (Faivovich et al., 2005; Frost et al., 2004).

2.2 Procedure applied to helminth. Complications

In 1972, Allison et al. proposed a simplified four-step procedure that resulted in excellent preservation, support in the high vacuum and dissipation of surface charging of nematodes. The procedure involved: fixation, dehydration, treatment with an antistatic agent and gold-palladium coating. These authors obtained best results using fixation with 4% paraformaldehyde (phosphate-buffered) and AFA (acetic acid-formalin-alcohol); specimens were dehydrated in an ascending series of ethanol solutions to 70%, then transferred to 5% glycerine-95% ethanol from which the alcohol was allowed to evaporate, and cleared in 96.6% glycerol-0.05% potassium chloride-3.35% distilled water, 24 to 48 hours prior to examination. Subsequently, specimens were mounted on metal specimen stubs with Duco cement, outgassed in a vacuum evaporator for 1 hour or more and rotary-coated with gold-palladium.

In this study we basically followed the aforementioned procedure, with some modifications. Helminths were observed *in vivo*, counted, fixed and stored. Adult nematodes and trematodes were fixed in hot 4% formaldehyde; larval nematodes and trematodes were removed from cysts with the aid of preparation needlees and fixed in hot 4% formaldehyde; acanthocephalan larvae were placed in distilled water for 24 hours at 4° C for the proboscis to evert, and then fixed in hot 4% formaldehyde.

Another technique used for the study of nematodes and acanthocephalans by SEM consists of transferring specimens for 2 hours into 1% osmium tetroxide and dehydrate them in an ethanol series for 2 hours in each bath (Mafra and Lanfredi, 1998). In the case of trematodes, fixation can be made with paraformaldehyde, glutaraldehyde or a formaldehyde-glutaraldehyde mix such as Karnovsky's fixative in phosphate or cacodylate buffer. Postfixation is usually done for 2 to 3 hours at 4°C with cacodylate– or phosphate- buffered 1% osmium tetroxide (Karnovsky, 1965).

For processing of helminths, it should be taken into account that a hypertonic solution will cause shrinkage and almost complete disappearance of inflation in specimens, whereas a hypotonic solution may produce artificial inflations. In some cases, in spite of careful processing of samples for SEM study, good results are not achieved. Very frequently, the samples contain bacteria or debris, or tears of the cuticle of nematodes or tegument of trematodes (Fig. 5-9).

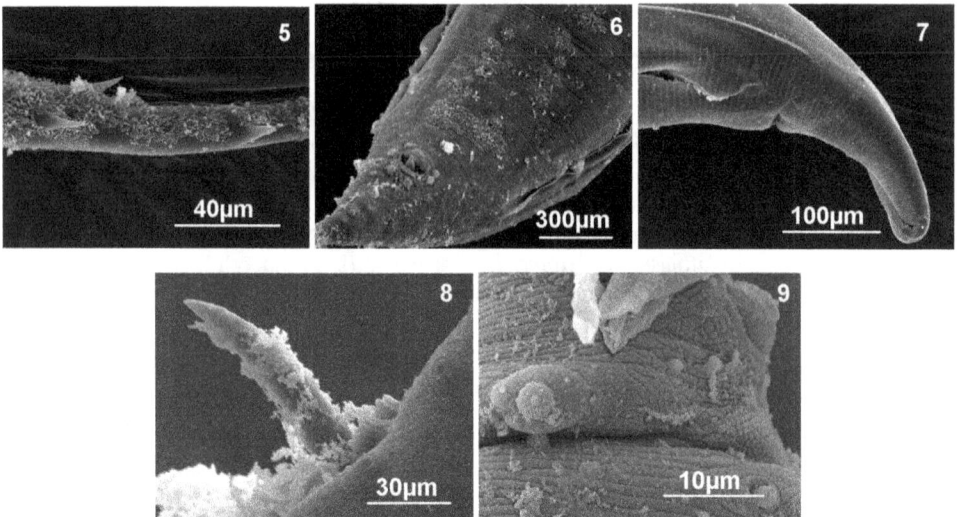

Fig. 5-9. Some complications in samples for SEM studies. 5. 6. Bacteria in posterior end of *Skrjabinodon* sp. (5) and in ventral surface of *Cosmocerca parva* (6). 7. Broken cuticle in posterior end of female of *Aplectana* sp. 8. 9. Debris in spicules and adanal papillae of *Cosmocerca podicipinus* (8) and in posterior end of *Falcaustra* sp. (9).

2.3 Characteristic of the scanning electron microscope used for this study

The Microscopy department at Universidad Nacional del Nordeste possesses a Jeol 5800LV scanning electron microscopy. Specimens are critical-point dried using a Denton Vacuum DCP-1 critical point drying apparatus, and sputter-coating is made using a Denton Vacuum Desk II sputter-coating unit.

For examination by scanning electron microscopy (SEM), specimens were dehydrated through an alcohol 70° and acetone series (70%, 85% and 100%; 15 minutes in each solution) and then subjected to critical point drying using CO_2; in the case of larval digenean trematodes, critical point drying time is shortened because these individuals are more fragile than adults. Samples were mounted on a metal sheet of copper or aluminum using double-sided tape. Then the specimens were sputter-coated with gold or gold-palladium for three minutes. Helminth parasites were observed in high vacuum in all cases.

Adequate fixation contributes to specimen preservation and stability within the microscope; dehydration allows outgassing and drying of specimens in the vaccum evaporator prior to coating without subsequent "bubbling" or shrinkage artifacts; glyceration provides an antistatic surface coating over all surface irregularities, even those which might not be adequately coated with 200 a of gold-palladium. The thin metal coating does not obscure fine structural detail but, in combination with the glycerol-KCl coating, ensures that the specimen can be exposed for extended periods of time to the electron beam without surface changes (Allison et al., 1972).

3. What can SEM studies tell us about helminth parasites of amphibians?

3.1 Nematoda

The body surface of nematodes is covered by a truly inert cuticle of extracellular material in the form of cross-linked collagens and insoluble proteins that are synthesised and secreted by the underlying epidermis (= hypodermis). In addition to proteins and collagen, the cuticle possesses glycoproteins, fibrin and keratin. It consists of three main layers and the epicuticle. The external layer is divided into internal and external cortex; the middle layer varies from having a granular uniform structure to presenting skeletal rods, fibers or channels; the internal basal layer can be laminated or grooved. The thin epicuticle may have a coating of quinone (Lee, 2002).

The above discussion refers specifically to the cuticle of the external body surface. In other body parts, such as the buccal cavity, excretory pore, vagina, cloaca and rectum, as well as the eggs and spicules in males, the cuticle is of a different nature.

The particular composition of the cuticle along the whole body of nematodes, and especially the ornamentations that they present in the anterior and posterior extremity, are the focus of primary study of these helminths by scanning electron microscopy.

The nomenclature for the structures detailed here is based on Chitwood and Chitwood (1975), Gibbons (1986) and Willmott (2009). We describe the modifications of the cuticle along the body surface, at the anterior and posterior body ends, and finally the cuticle of the eggs, vulva and spicules of these helminths.

3.1.1 Cuticle of body

The features analyzed along the body surface are the morphology and disposition of somatic papillae as well as the striation of the cuticle -longitudinal, transverse or oblique-, and the presence of annulations, punctuations, longitudinal ridges, alae -lateral, cervical or caudal-, inflation and spination.

There are several types of gross cuticular markings, namely, *transverse, longitudinal* and *oblique* markings. Of these, *transverse* and *longitudinal* markings are very common in nematode parasites of amphibians.

Tranverse markings: the superficial markings are grooves or ridges. *Striations* are defined as fine transverse grooves occurring at regular intervals; the distance between two striae is the intestrial region (*Aplectana hylambatis, Cosmocerca* spp.) (Figs. 10, 11). Deep striae are very commonly and are known as *annulations*; the distances between them are termed *annules* (*Aplectana delirae, Falcaustra* sp.) (Figs. 12, 13). In nematode parasites of the family Pharyngodonidae, annulations are much broader and more prominent (Figs. 14, 15, 34).

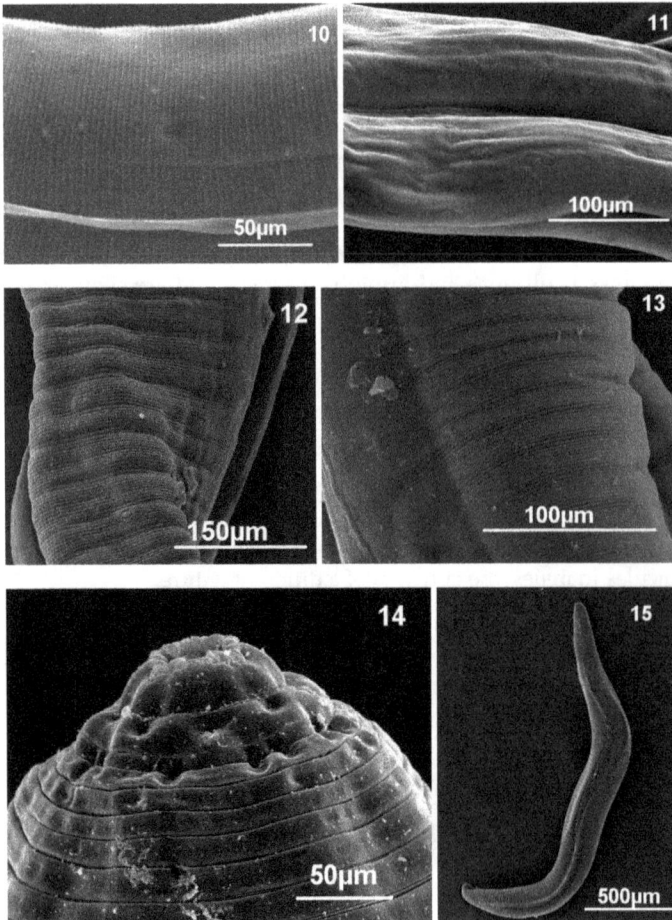

Fig. 10-15. Transverse marking in cuticle of nematode parasites of amphibians. 10. 11. Striations (*Aplectana hylambatis, Cosmocerca podicipinus*, respectively). 12. 13. Striations and annules (*A. delirae, Falcaustra* sp., respectively). 14. 15. Annulations (*Skrjabinodon* sp.).

Striations can be distributed uniformly along the whole body of the parasite (*Cosmocerca* spp.), or they may become less evident (*A. delirae*) or wider (*A. hylambatis*) at the body ends, or more marked, as for example, in the posterior body (*Skrjabinodon*) (Figs. 10-15).

Longitudinal markings: these may take the form of ridges or alae, or they may be merely the result of gaps in transverse markings.

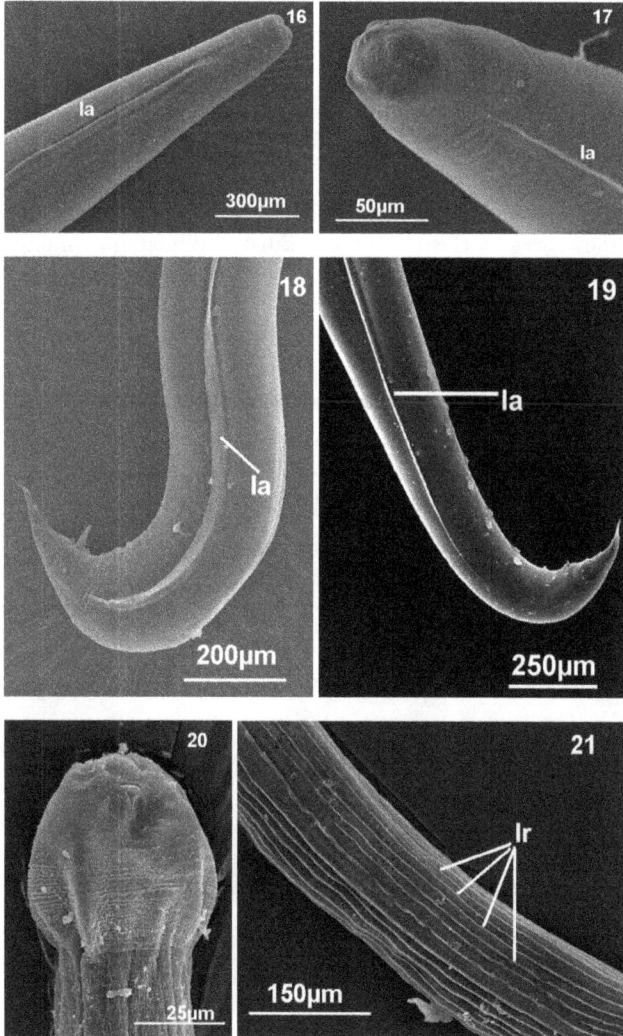

Fig. 16-21. Longitudinal markings in nematode parasites of amphibians. 16. 17. Lateral alae in anterior end of body in *Paraoxyascaris* sp. (16) and in *Porrocaecum* sp. (larvae) (17). 18. 19. Lateral alae in posterior end of body in *Cosmocerca* spp. 20. Cephalic cuticular vesicle in *Oswaldocruzia* sp. 21. Longitudinal ridges in *Oswaldocruzia* sp. la: lateral alae; lr: longitudinal ridges.

Alae: these are usually lateral or sublateral cuticular thickenings or projections. There are three types of alae: longitudinal, cervical and caudal.

Longitudinal alae: these are usually lateral or sublateral and occur in both sexes; extending along the length of the body. They occur in the families Cosmocercidae (*Aplectana, Cosmocerca, Cosmocercella, Paraoxyascaris*) (Figs. 16, 18, 19), Ascarididae (*Ortleppascaris, Porrocaecum*) (Fig. 17); in *Gyrinicola* sp. (Family Pharyngodonidae) lateral alae are present in males only.

Cervical alae: these structures are confined to the anterior part of the body. This modification of the cuticle does not occur in nematode species that parasitize amphibians.

In some parasitic nematodes, the cervical alae is modified as a *cephalic cuticular vesicle* as in the genus *Oswaldocruzia*; in some cases this cuticular vesicle is divided into a larger anterior part and a smaller posterior part (Fig. 20).

Caudal alae: these alae are confined to the caudal region of the body and limited to the males only; apparently they serve as clasping organs during copulation. Among nematodes that parasitize amphibians, they occur in the family Pharyngodonidae (*Parapharyngodon*). On the other hand, modified caudal alae occur in the males of some families such as the Molineidae, where they are called *bursa copulatrix* or *copulatory bursa* (see below).

Longitudinal ridges: these are raised areas that extend along the length of the body and are present on the submedian as well as on the lateral surfaces. *Oswaldocruzia* (Family: Molineidae) has longitudinal ridges throughout the body length that can disappear or appear along its body (Fig. 21). The system of longitudinal cuticular ridges or *synlophe* in this genus is very important as a taxonomic character.

Inflation: when the cuticle is swollen in a blister-like manner, this condition may be termed inflation (eg. *Rhabdias*; Figs. 22-24). Besides being inflated, the cuticle can be also striated. Inflation of the cuticle may be generalized over the whole body surface instead of restricted to certain areas.

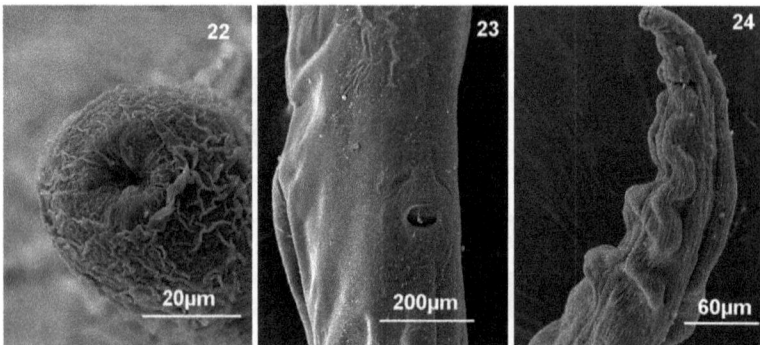

Fig. 22-24. Inflation in the cuticle of nematode parasites of genus *Rhabdias*. 22. Cephalic end. 23. Vulvar region. 24. Anterior end of body.

Papillae: these structures are nerve endings, some of which have a tactile function and others are chemoreceptors; they appear as cuticular elevations of different shapes and sizes (Figs. 25-27). According to their position, they are divided into labial, cephalic, cervical and genital papillae. In this section, we deal with the distribution and structure of somatic papillae, i.e. those that are distributed throughout the body of the nematode. These papillae are generally arranged in two subdorsal and two subventral rows along the body of the nematode.

Finally, in some genera of nematode parasites of amphibians, the excretory pore has conspicuous cuticularized walls, surrounded by a rough cuticular area (Pharyngodonidae); in others, the excretory pore is located in a depression of the cuticula (*Falcaustra*) (Figs. 28, 29).

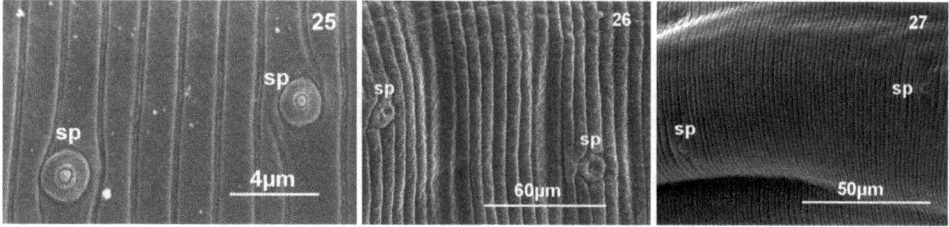

Fig. 25-27. Somatic papillae in nematode parasites of amphibians. 25. *Aplectana hylambatis*. 26. *Aplectana delirae*. 27. *Cosmocerca podicipinus*. sp: somatic papillae.

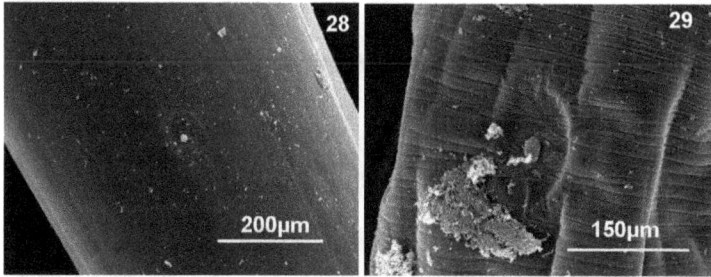

Fig. 28-29. Excretory pore in nematode parasites of amphibian. 28. *Falcaustra* sp. 29. *Skrjabinodon* sp.

Punctuations: this type of marking is frequent and appears as minute round areas of the cuticle which are arranged in definite patterns for each species. In nematode parasites of amphibians, this type of marking appears in males of the family Cosmocercidae, specifically in the genus *Cosmocerca* as part of the rosette papillae (see below; Figs. 38-44) and in the genus Cosmocercella (Figs. 46, 48).

3.1.2 Cuticle of anterior end of body

The cuticular modifications that occur in the anterior end of the body of nematode parasites of amphibians are: head papillae - cephalic papillae, externo-labial papillae, interno-labial papillae-, interlabia, deirids and amphids. These two latter structures are not modifications of the cuticle itself, but their opening in the cuticle can present diverse shape and structure.

Head papillae: these are tactile sensory organs usually located on the lips or labial region, including two circles of six labial papillae and one circle of four cephalic papillae. This arrangement has been proposed for ancestral nematodes considering a radial type of symmetry (De Coninck, 1965). The head papillae are divided into: *cephalic papillae*: outer circle of four head papillae (or latero-ventral and latero-dorsal papillae); *externo-labial papillae*: median circle of six head papillae, and *interno-labial papillae*: inner circle of six head papillae.

The basic structure proposed by De Coninck (1965) shows modifications in the nematodes that parasitize amphibians. For example, cosmocercid nematodes present three lips. The genus *Aplectana* has a circle of four internal labial papillae, 1 on each subventral lip and 2 on the dorsal lip, and a circle of six external labial papillae; the amphids are large, one on each subventral lip. The anterior end of the oesophagus presents three tooth-like projections covered with a thick cuticle, also called cuticular flap (Fig. 30). The genus *Cosmocerca* presents the same arrangement in the internal and external labial papillae and, as in the previous genus, the anterior end of the oesophagus bears three tooth-like projections (cuticular flap); the amphids are very prominent in some cases (Fig. 31). In the nematodes of family Atractidae, specifically in the genus *Schrankiana*, each lip has a cuticular flange overhanging the mouth opening (Fig. 32).

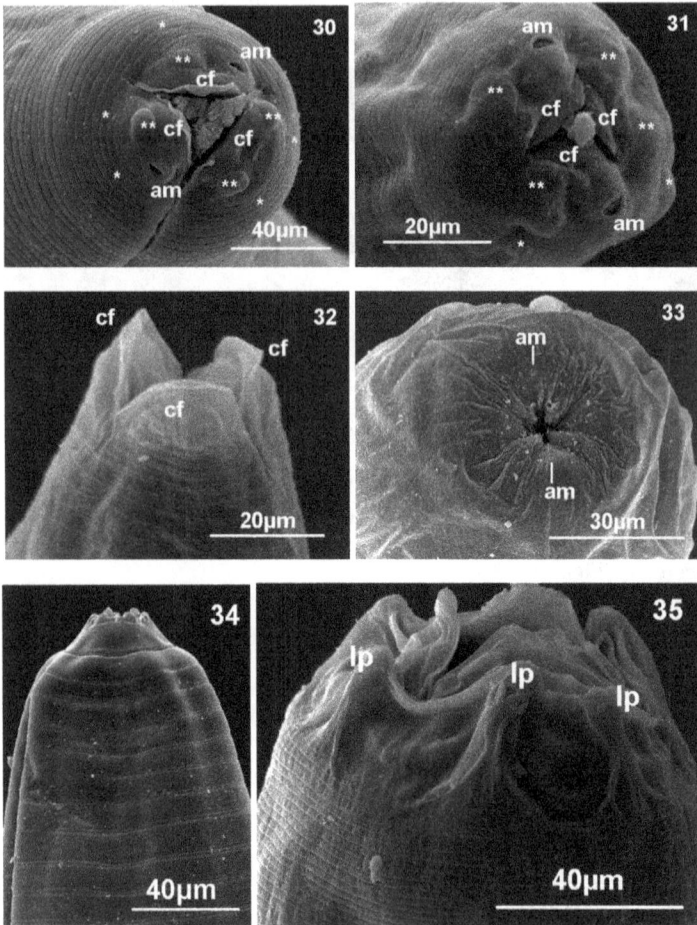

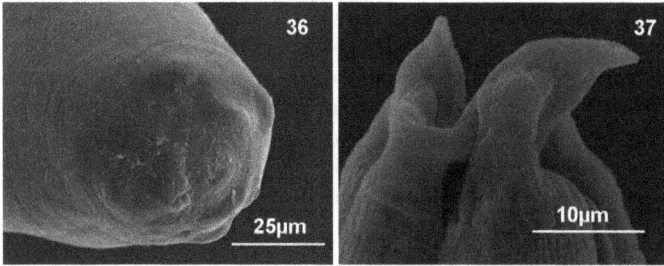

Fig. 30-37. Modifications of the cuticle in the anterior end of body. 30. *Aplectana* sp. 31. *Cosmocerca* sp. 32. *Schrankiana* sp. 33. *Rhabdias* sp. 34. *Pharyngodon* sp. 35. *Falcaustra* sp. 36. *Porrocaecum* sp. (larvae). 37. *Brevimulticaecum* sp. (larvae). am: amphids; cf: cuticular flap; lp: labial papillae; one asterisk: external labial papillae; two asterisk: internal labial papillae.

In the lung nematodes of genus *Rhabdias*, which have an inflated cuticle, the cephalic structures are in most cases very difficult to observe (Fig. 33). In oxyurid nematodes, eg. *Parapharyngodon*, there are three lips, each one bilobed with one small papillae (Fig. 34). In genus *Falcaustra* the mouth is surrounded by 3 large lips, each with 2 forked papillae, and one amphid on each ventrolateral lip (Fig. 35)

On the other hand, in nematode parasites found in larval stage, the cephalic structures of adult forms are generally not present. For example, *Porrocaecum*, a genus found in the liver of the host, has a very little developed lip anlagen (Fig. 36); *Brevimulticaecum*, found encapsulated in various organs of the hosts, presents two toothlike prominences, 1 dorsal and 1 ventral (Fig. 37).

Interlabia: these are cuticular outgrowths (neoformations) originating at the base of the lips or pseudolabia and extending between them, occurring in some ascarids and spirurids. These modifications of the cuticle are present in the larval stage of genus *Physaloptera*; this genus is characterized by the presence of four teeth on the internolateral face of each pseudolabium, an internal group of three teeth, two sublateral and one lateral, and a single externolateral tooth.

Deirids: a pair of sensory organs found laterally in the cervical region and usually protruding above the surface of the cuticle. That may be simple or claw-like and forked, and are situated laterally in the vicinity of the nerve ring in most species (Rhabdochonidae).

Amphids: a pair of glandular sensory organs situated laterally in the cephalic region and opening through the cuticle; according to some authors, they are chemoreceptors. They have various shapes and sizes but usually occur as two lateral pores (Figs. 30, 31).

3.1.3 Cuticle of posterior end of body

In the case of nematode parasites of amphibians, the special modifications of the cuticle in the posterior end of males, which are generally associated with the copula, are particularly important. These comprise: plectanes, rosettes papillae, vesiculated rosette, papillae, bursa, phasmids, spines suckers and caudal lateral alae. Phasmids are not modifications of the cuticle such as the deirids and the amphids, but their opening onto the cuticle can present different shapes and structures.

Plectanes: these are cross striated cuticular plates functioning as supports for the genital papillae in some males (Fig. 38). This structure is very common in the genus *Cosmocerca*, which is widely distributed in amphibians. In some species, these supports are very marked, as in *C. podicipinus*, in this case, the plectanes of each row are even fused to each other (Fig. 39). In other species, this support is not so developed (eg. *C. parva*); on the other hand, the same species in different hosts can present different degree of development of these structures (Figs. 40-43); indeed, in some immature specimens, the plectanes are imperceptible (Fig. 44).

These structures are arranged in two longitudinal rows on the ventral surface of the males; the number of pairs of plectanes varies among species (*C. podicipinus*, *C. cruzi*, *C. travassossi*: 5 pairs; *C. chilensis* and *C. rara*: 6 pairs; *C. longispicula*, *C. uruguayensis*, *C. vrcibradici*: 7 pairs). Furthermore, some species show wide intraspecific variation regarding the number of pairs of plectanes, even in the same host species (*C. parva*: 5-7 pairs; *C. brasiliensis*: 8-11 pairs; *C. paraguayensis*: 4-5 pairs) (Figs. 18, 19) (González and Hamann, 2010b).

Table 1 shows characteristics of the posterior end of males for species of *Cosmocerca* that parasitize amphibians, observed under SEM.

Cosmocerca spp.	Plectanes + rosette papillae	Adanal papillae*	Caudal papillae	Puncatations of rosette papillae	References
C. ornata	5 pairs	Not stablished	Not stablished	Not stablished	Navarro et al. (1988)
C. ornata	5 pairs	Not stablished	Not stablished	6-7 pad-like protuberances around the posterior border only	Grabda-Kazubska and Tenora (1991)
C. commutata	7 pairs	Not stablished	Several pairs	Interior and exterior rosette: 15	Grabda-Kazubska and Tenora (1991)
C. parva	5-7 pairs + 1 unpaired	1 + 2-4	3 pairs	Interior and exterior rosette: 12-16	Mordeglia and Digiani (1998)
C. podicipinus	5 pairs	3 pairs	Not stablished	Interior rosette: 11-12; exterior rosette: 12-15	González and Hamann (2010b)
C. parva	5-7 pairs	1 + 3 pairs	Not stablished	Interior rosette: 10-11; exterior rosette: 12-14	González and Hamann (2010b)
C. parva	4-5 pairs	1 + 3 pairs	Not stablished	Interior rosette: 12-15; exterior rosette: 12-15	González and Hamann (2008)

Table 1. Characteristics of the caudal region of males of *Cosmocerca* spp. that parasitize amphibian hosts, studied under SEM. *Arrangement of adanal papillae: unpaired papillae anterior to anus + pairs of adanal papillae.

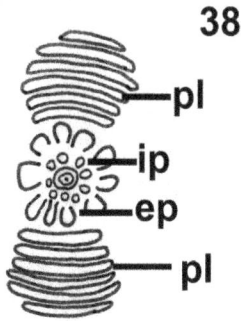

Fig. 38. Plectanes and rosette papillae in nematode parasites of genus *Cosmocerca*. pl: plectanes; ip: internal circle of punctations; ep: external circle of punctations.

Rosette papillae: these structures consist of papillae surrounded by punctuations. This modification is present in the genus *Cosmocerca*. In this genus, the plectanes are located outside the rosette papillae and are directed toward the anterior and posterior end (Fig. 38).

These rosette papillae are formed by two circles of punctuations, one internal and one external. The number of punctuations in each circle varies among species and within the same species for individuals collected from different hosts (González and Hamann, 2008; 2010b) (Figs. 38-44). In the genus *Cosmocercoides*, the caudal rosette papillae are not raised above the cuticular surface.

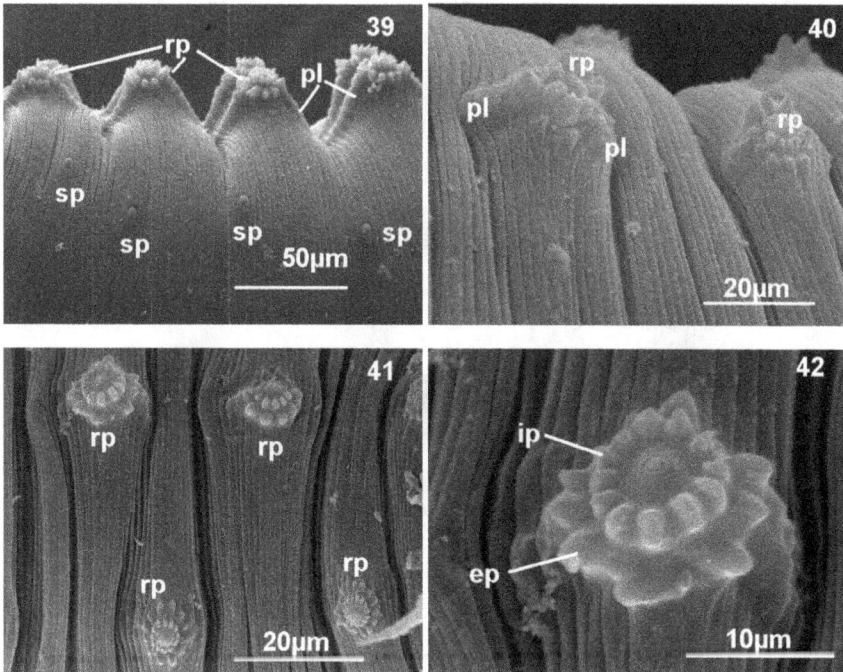

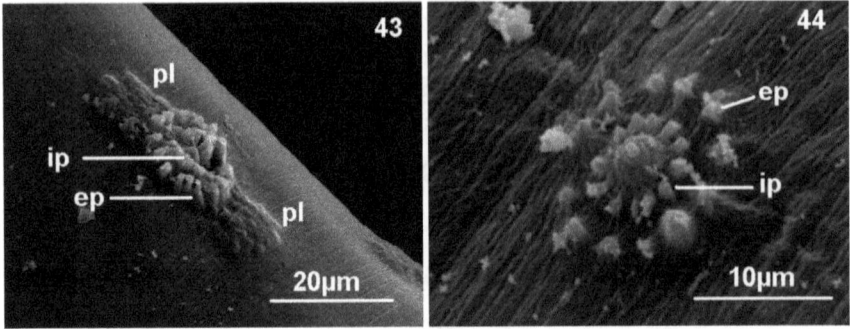

Fig. 39-44. Plectanes and rosette papillae in nematode parasites of genus *Cosmocerca*.
39. *Cosmocerca podicipinus* collected in *Pseudopaludicola falcipes* with fused plectanes.
40. *Cosmocerca parva* collected in *Rhinella granulosa* with very marked plectanes.
41. 42. *Cosmocerca parva* collected in *Rhinella schneideri*. 43. *Cosmocerca parva* collected in
Leptodactylus bufonius. 44. *Cosmocerca parva*, immature specimen, collected in *Rhinella fernandezae* with imperceptible plectanes. sp: somatic papillae; pl: plectanes; rs: rosette papillae; ip: internal circle of punctations; ep: external circle of punctations.

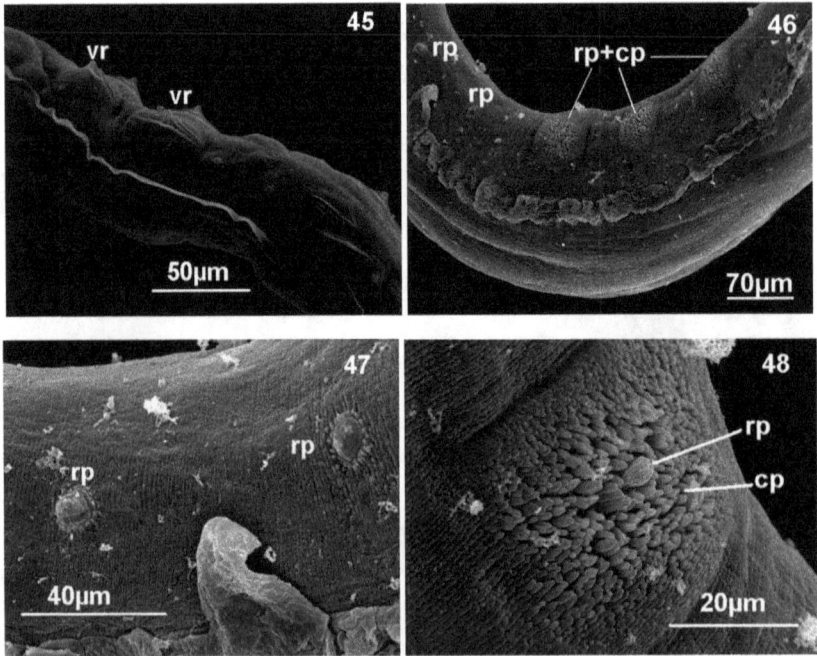

Fig. 45-48. Modifications of the cuticle in the posterior end of males of the genus
Cosmocercella. 45. Vesiculated rosette in *C. minor*. 46. Combination of different structures in
C. phyllomedusae. 47. Detail of small rosette papillae. 48. Detail of rosette papillae surrounded
by wide areas of cuticular punctations. vr: vesiculated rosette; rp: rosette papillae; cp:
cuticular punctation; rp+cp: rosette papillae surrounded by areas of cuticular punctations.

Vesiculated rosette: caudal rosette papillae raised on the surface of a clear vesicle. These modifications are present in the genus *Cosmocercella*. In *C. minor*, for example, there are 4 pairs of vesiculated papillae (Fig. 45) and in *C. phyllomedusae* there is a combination of different structures; this species has small rosette papillae in the preanal subventral surface (Figs. 46, 47), rosette papillae surrounded by wide areas of cuticular punctuations (Fig. 46, 48), and large unpaired vesiculated papillae extending almost to the level of the oesophagus.

Papillae: some genera of nematode parasites, such as *Aplectana, Raillietnema, Falcaustra* or *Paraoxyascaris* do not possess conspicuous structures in the posterior end such as plectanes, rosette papillae or vesiculated papillae; these genera have simple papillae with variable number and arrangement (Figs. 49-56). These structures can be divided into caudal papillae (located in the tail, posteriorly to the anus) and cloacal papillae (surrounding the cloaca; these can be precloacal, postcloacal and adcloacal). These papillae may be sessile (*Aplectana, Schrankiana, Falcaustra*) (Figs. 49-52, 54-56) or pedunculate (*Parapharyngodon*) (Fig. 53). On the other hand, other genera such as *Cosmocercella* and *Cosmocerca* have this type of papillae in addition to plectanes, rosettes and vesiculated rosette. These papillae are commonly surrounded by one or two small rosettes of punctuations (Fig. 51). In some cases they protrude above the surface of the cuticle (Fig. 52).

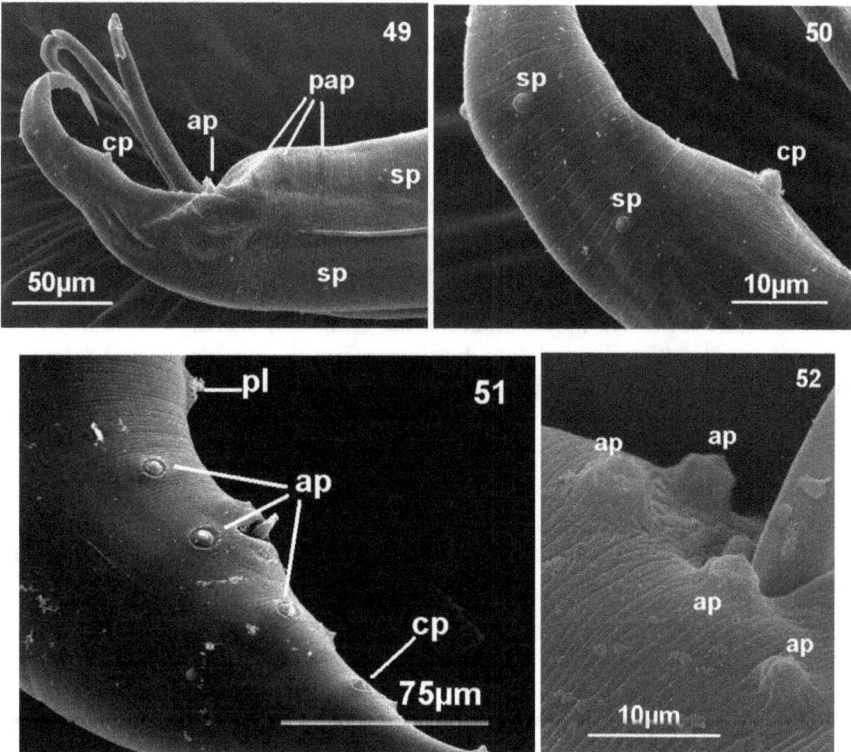

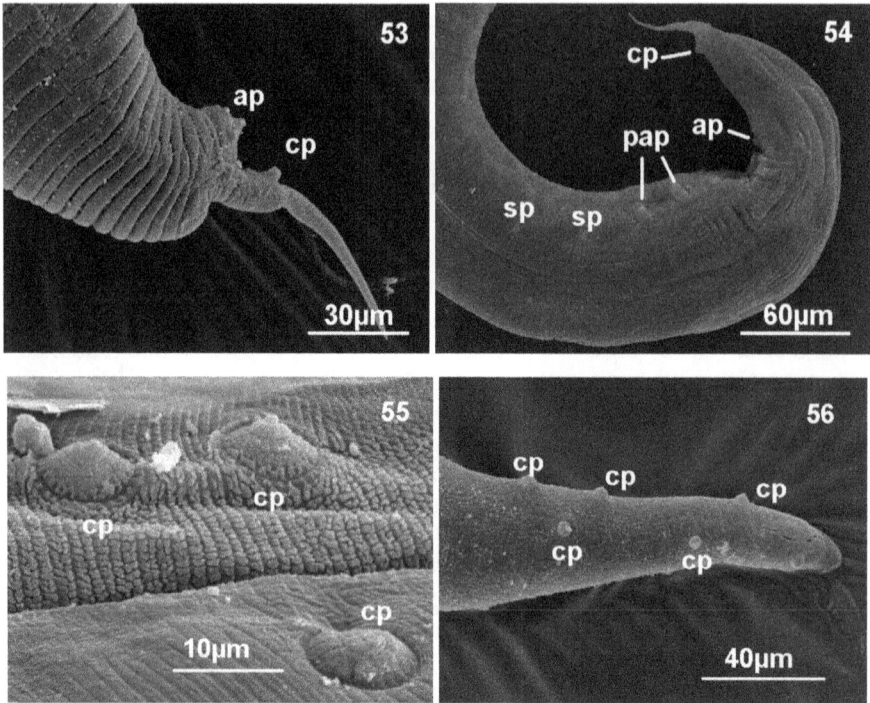

Fig. 49-56. Papillae in the posterior end of body of males nematode parasites of amphibians. 49. 50. Arrangement of caudal papillae in *Aplectana hylambatis*. 51. Arrangement of precloacal, adcloacal and postcloacal papillae in genus *Cosmocerca*. 52. Detail of precloacal, adcloacal and postcloacal papillae in *Cosmocerca*. 53. Pedunculate papillae in tail of *Parapharyngodon*. 54-56. Sessile papillae in *Schrankiana* (54), *Falcaustra* (55) and *Cosmocerca* (56). sp: somatic papillae; pl: plectane; pap: preanal papillae; ap: adanal papillae; cp: caudal papillae.

Bursa: this structure is present in nematodes of the genus *Oswaldocruzia*. This structure may be circular or oval, often divided into two symmetrical or asymmetrical lateral lobes, separated by a dorsal lobe and supported by rays or papillae. The rays of the bursa are visualized well under light microscope because they are not a part of the cuticle but embedded in the lobes.

Caudal lateral alae: these are sublateral or lateral longitudinal wings of the cuticle that occur on the male tail. Among nematode parasites of amphibians, they occur in the genus *Physaloptera*; however, these caudal alae develop in the adult stage, while it is typically the larval stage of *Physaloptera* that occurs as a parasite of amphibians.

Spines: some male and female nematode parasites, such as those of the genus *Skrjabinodon*, have cuticular spines on the tail filament (Fig. 57).

Suckers: this is a sucker-like pre-cloacal structure. A series of stages in sucker development occurs in some kathlaniids, indicating that there is first a concentration of copulatory muscles in this area which later becomes a sucker through modification of the cuticle. This modification of the cuticle is found in some species of genus *Falcaustra*.

Phasmids: these are paired glandular sensory organs situated laterally in the caudal region and opening to the surface by a slit or pore (Fig. 58).

As in the previous case, the structures and modifications of the cuticle that are observed in the posterior body end of adult specimens do not occur in larvae (Fig. 59).

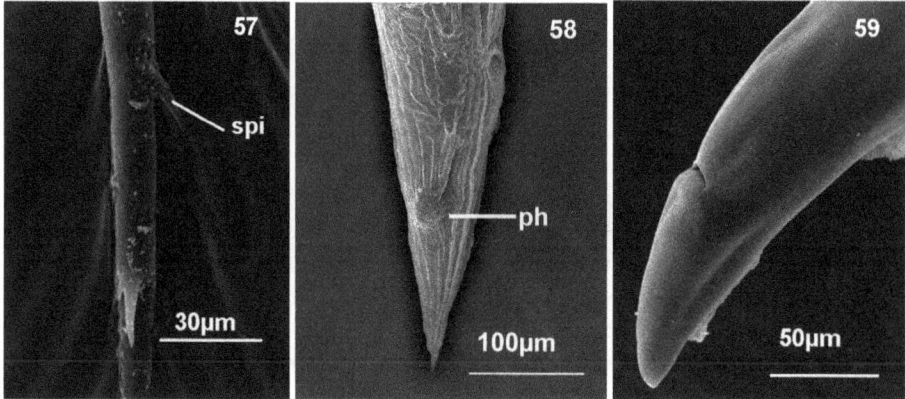

Fig. 57-59. Modifications of the cuticle in posterior end of body of nematode parasites. 57. Spines in the tail of *Skrjabinodon* sp. 58. Phasmid in *Rhabdias* sp. 59. Posterior end of *Physaloptera* sp. spi: spines; ph: phasmid.

3.1.4 Eggs

The eggs are variable in size, shape and structure; they usually have a many-layered shell with either smooth or rough, sometimes sculptured, external surface, and their poles may bear a characteristic operculum or plug. Among the nematode parasites of amphibians, eggs may present punctuations (Pharyngodonidae) (Figs. 60, 61), an operculum (*Gyrinicola*) or a thin membrane that has no special features, as in the Cosmocercidae.

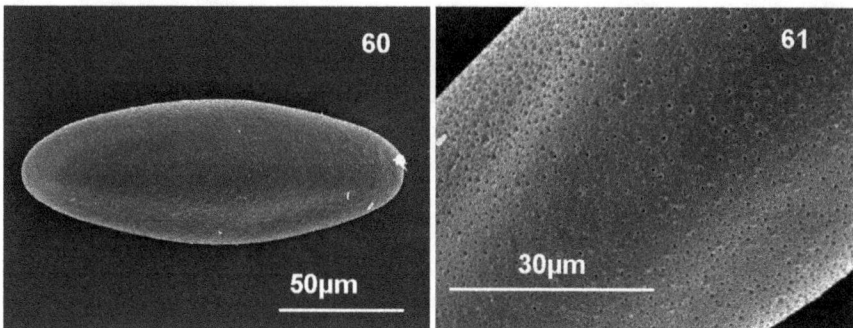

Fig. 60-61. Egg cuticle of nematode parasites of amphibians. 60. General view. 61. Detail of punctations.

3.1.5 Vulva

The area of the body inmediately anterior and posterior to the vulvar opening is called vulvar region. In most females of nematode parasites of amphibians, this may be simply an opening transversal to the longitudinal body axis without any special striation, as in the genus *Rhabdias* (Fig. 62), or with a striation that differs slightly from that of the rest of the body as in the genus *Cosmocercella* (Fig. 63); or it may present more complex structures as in *Aplectana* (Fig. 64); in this latter case, the cuticle forms an extension in the anterior side of the vulvar opening that can be observed as a vulvar flap (Gibbons, 1986).

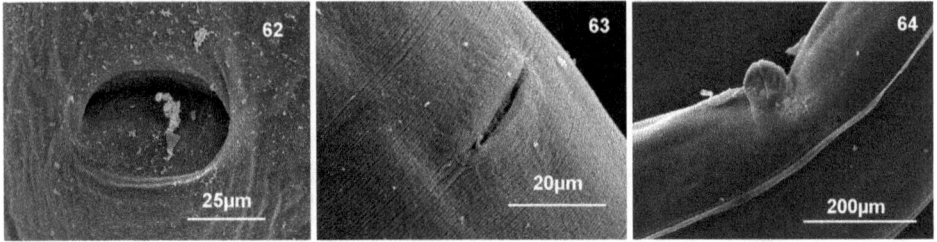

Fig. 62-64. Vulvar cuticle in nematode parasites of amphibians. 62. *Rhabdias* sp. 63. *Cosmocercella* sp. 64. *Aplectana* sp.

3.1.6 Spicules

Nematodes usually have two spicules; each one is essentially a tube covered by a sclerotized cuticle and containing a central protoplasmic core. In terms of the taxonomy of nematode parasites of amphibians, the importance of the spicules lies in their morphology and size, and not in the presence of ornamentation on the cuticle of these structures. In this case the spicules can be studied with SEM only when they are outside the individual, i.e., when protruding from the cloaca (Figs. 65-67).

Fig. 65-67. Spicules in nematode parasites of amphibians. 65. *Aplectana hylambatis*. 66. *Cosmocerca* sp. 67. *Falcaustra* sp.

3.2 Trematoda

The tegument of trematodes is syncytial and consists of a tegumental outer membrane (trilaminate), a matrix (with discoid bodies, membranous bodies and usually mitochondria) and a basal tegumental membrane. The tegument is variously interrupted by cytoplasmatic projections of gland cells and by openings of excretory pores. The tegumental surface often contains ornamentations such as spines between the outer and basal membranes; these are often present in different areas of the body; there are also numerous sensory papillae, pits and ridges of various configurations (Fried, 1997; Schmidt and Roberts, 2000). The surface topography of the cirrus of digenetic trematodes also shows spine-shape protrusions and papillae (Bušta and Našincová, 1987).

3.2.1 Tegument of suckers

The rim of the oral and ventral suckers in some species of digenean parasites of amphibians (both larval and adult) shows sensory papillae with variable morphology and distribution. Thus, papillae may appear as button-like structures and can be distributed as single and double papillae on the oral and ventral surfaces of the sucker (Hamann and González, 2009; Mata-López, 2006; Nadakavukaren and Nollen, 1975). In other digeneans, the oral sucker has a spongy surface with numerous pores (Whitehouse, 2002). Figures 68-70 show some characteristics of the tegument of the oral and ventral suckers of digenean trematodes found in Argentinean amphibians.

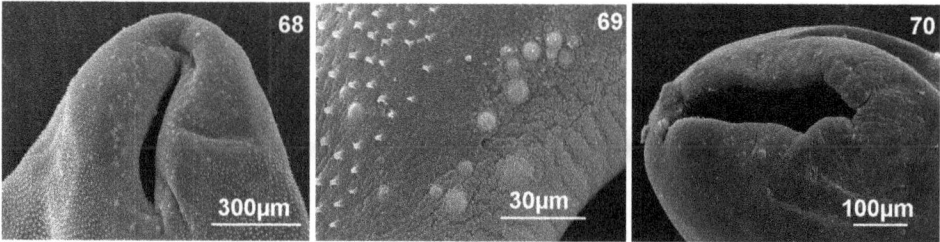

Fig. 68-70. Tegument of suckers of digenean parasites of amphibians. 68. Papillae on oral sucker of Macroderoididae. 69. Detail of papillae on oral sucker. 70. Detail of papillae on ventral sucker of Paramphistomatidae.

3.2.2 Tegument of the ventral surface

The tegument of digenean trematodes (larval and adults) that occur in amphibian hosts shows spines with varied morphology (e.g. scale-like spines) and variable distribution; they generally extend from the anterior end to variable levels of the posterior region (Hamann and González, 2009; Razo-Mendivil et al., 2006). In other digeneans, the surface of the tegument possesses regular ridges and interspersed protuberances (Nadakavukaren and Nollen, 1975). Figures 71-72 show the shape and distribution of spines found in the tegument of some digenean parasites of Argentinean amphibian.

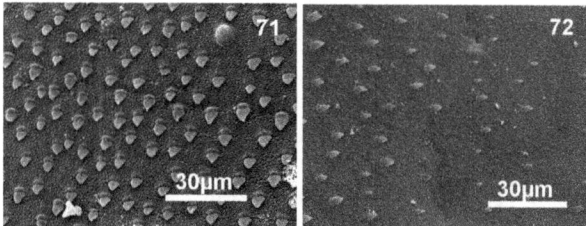

Fig. 71-72. Tegument of the ventral surface of Macroderoididae. 71. Spines of the anterior third of the body. 72. Spines of the posterior third of body.

3.3 Acanthocephalan

The body surface of acanthocephalans has 5 layers. The outermost layer is the epicuticle, followed by the cuticle which is composed mainly of lipoproteins; the third layer has a homogeneous nature; the fourth layer possesses fibrous bands besides mitochondria, bladders and lacunar channels, and the fifth layer contains scarce fibres but larger and more abundant lacunar channels than in the previous layer (Olsen, 1974).

Regarding this group of helminth parasites, most SEM studies are focused on the hooks that they possess in the proboscis, as well as their body spines. Likewise, they present sensory structures with diverse ornamentations in the posterior part of the bursa.

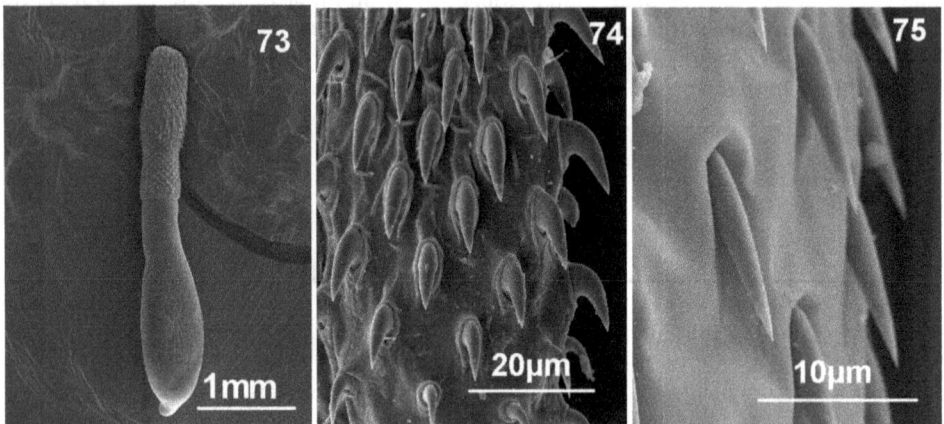

Fig. 73-75. Acanthocephalan parasites of amphibians. *Centrorhynchus* sp. 73. General view. 74. Detail of hooks of the proboscis. 75. Detail of spines of the proboscis.

In our study we found larval stages belonging to the genus *Centrorhynchus*. This genus is characterized for possessing an unarmed trunk, the proboscis divided into two regions (an anterior portion with hooks and a posterior portion with spines), and for having subterminal genital pores. Because the genital complex was not fully developed, specific determination was not possible. The proboscis presented 28 to 30 longitudinal rows of 20 to 23 hooks, 8 to 11 rooted hooks and 10 to 13 rootless spines (Figs. 73-75).

4. Helminth parasites of Argentinean amphibian studied with SEM: Synthesis and new contributions. Research perspectives

Up to the present, studies performed with scanning electron microscopy techniques on helminth parasites of Argentinean amphibians have included three families of nematodes: Rhabdiasidae, *Rhabdias füelleborni* (González and Hamann, 2008), Cosmocercidae, *Cosmocerca parva, C. podicipinus, Aplectana hylambatis, A. adaechevarriae* (González and Hamann, 2010b; Mordeglia and Digiani, 1998; Ramallo et al., 2008) and Physalopteridae, *Physaloptera* sp. (González and Hamann, 2010b), and one family of trematodes: Diplostomidae, *Lophosicyadiplostomum* aff. *nephrocystis* and *Bursotrema tetracotyloides* (Hamann and González, 2009).

New contributions presented in this work include, for the Class Nematoda, the families Molineoidae (*Oswaldocruzia* spp.), Pharyngodonidae (*Parapharyngodon, Pharyngodon, Skrjabinodon*), Cosmocercidae (*Aplectana* spp., *Cosmocerca* spp., *Cosmocercella* spp., *Paraoxyascaris* sp.), Kathlaniidae (*Falcaustra* sp.), Atractidae (*Schkrankiana* sp.) and Ascarididae (*Porrocaecum* sp., *Brevimulticaecum* sp.), and for the Phyllum Acanthocephala, family Centrorhynchidae (*Centrorhynchus* sp.).

Future research in this topic should focus on extending the geographical areas studied while at the same time, expanding the examination to other possible amphibian hosts.

Reports about helminth parasites of Argentinean amphibians studied under SEM refer mainly to specimens collected in host from the Northeast region, specifically Corrientes province, and the Northwest region, with only one record for Salta province so; thus, there is still a vast portion of the Argentinean territory that has not yet been studied (Fig. 76).

Fig. 76. Reports of helminth parasites of amphibians studied using SEM in Argentina. ¶: González and Hamann (2008); *: González and Hamann (2010b); +: Hamann and González (2009); £: Mordeglia and Digiani (1998); §: Ramallo et al. (2008).

Lavilla et al. (2000) reported a total of 271 amphibian species for Argentina (167 anurans and 4 gymnophions); of these, only 32 (11.8%) have been cited as hosts for helminth parasites (González and Hamann, 2004, 2005, 2006a, 2006b, 2007a, 2007b, 2008, 2009, 2010a, 2010b, 2011; Hamann and Pérez, 1999; Hamann and González, 2009; Hamann et al., 2006a, 2006b, 2009a, 2009b, 2010; Lajmanovich and Martinez de Ferrato, 1995; Lunaschi and Drago, 2007; Ramallo et al., 2007a, 2007b, 2008). Of all the anuran families, the most studied for helminth parasites are Hylidae, Bufonidae, Leiuperidae and Leptodactylidae. Nine species of hylids have been studied for helminth parasites, but only one of these studies included SEM: *Scinax nasicus* (González and Hamann, 2008; Hamann and González, 2009); similarly, seven species of bufonids have been studied for helminth parasites, but SEM was employed in only two cases: *Rhinella schneideri* and *R. granulosa* (González and Hamann, 2008; Mordeglia and Digiani, 1998; Ramallo et al., 2008), finally, six species of leiuperids and leptodactylids have been studied for helminth parasites, but only one of them was analyzed with SEM, the leiuperid *Physalaemus santafecinus* (González and Hamann, 2010b).

The SEM study of the tegument of helminth species (e.g. morphology of spines) collected in different localities could detect possible intraspecific variation related to geographical location; this phenomenon has been highly documented in both trematode (Grabda-Kazubska and Combes, 1981; Kennedy 1980a) and nematode (Chitwood, 1957) parasites of amphibians. Similarly, variations related to occurrence in a wide range of phylogenetically unrelated hosts, i.e. the cases of generalist helminths could detect possible intraspecific variation related with the host age or diet, previous exposure to the parasite, presence of another parasite and number of specimens present (Chitwood, 1957; Haley, 1962; Kennedy, 1980b; Watertor, 1967).

5. Importance of the use of scanning electron microscope for the study of helminth parasites of amphibians

In helminth parasites, all morphological aspects must be studied under light microscope, because these structures are very important in the context of their systematic classification. Some examples of these traits include, in the case of parasitic nematodes, the type of esophagus (oxyuroid, rhabditoid, strongyloid), presence of ventriculus and its shape, and the caecum and its shape; in the females, the arrangement of ovaries (prodelphic, amphidelphic or opisthodelphic), number of uteri (monodelphic or didelphic), structure of the ovoyector and the vagina and, in the males, the structure and measurements of the gubernaculum. In the case of trematodes, the distribution of vitelline follicles, the position of the testes and ovary, size of the eggs, position of oral and ventral suckers, reproductive structures, among others, are characteristics of taxonomic importance. Finally, regarding the internal anatomy of acanthocephalans, some particularly relevant structures are the proboscis receptacle, lemnisci, retractor muscle, testis, seminal vesicle, cement gland, Saefftigen's pouch, etc. Thus, the scanning electron microscope represents an additional tool for the study of this group of organisms. The importance of SEM lies in its ability to provide three-dimensional images with high magnification that allow understanding the spatial relationships among surface structures. It could be used to separate species that appear morphologically identical when examined under light microscope, validate species and demonstrate differences between populations or races (Gibbons, 1986; Hirschmann, 1983).

6. Acknowledgments

We thank Secretaría General de Ciencia y Técnica of Universidad Nacional del Nordeste, Corrientes, Argentina, for supporting partially this work.

We are grateful Dr. Graciela T. Navone, Dr. Julia I. Díaz, Dr. María del Rosario Robles at the Centro de Estudios Parasitológicos y de Vectores, La Plata, Argentina, Licentiate Rodrigo Cajade at Centro de Ecología Aplicada del Litoral, Corrientes, Argentina, Dr. Lorena Sereno at Centro de Energia Nuclear na Agricultura, Universidade de São Paulo, Brazil, Dr. Viviane Gularte Tavares dos Santos at Instituto de Biociências, Universidade Federal do Rio Grande do Sul, Brazil, for helping with literature search.

We are grateful Dr. Marta I. Duré and Dr. Eduardo F. Schaefer at Centro de Ecología Aplicada del Litoral for the photographs of the host and for helping with the edition of the map.

We thank to Graphic Designer Cecilia Rios Encina for help in photograph edition.

7. References

Anderson, R.; Chabaud, A. & Willmont, S. (2009). *Keys to the Nematode Parasites of Vertebrates. Archival Volumen.* CAB International, ISBN 978-1-84593-572-6, Wallingford, United Kingdom.

Allison, V.; Ubelaker, J.; Webster Jr., R. & Riddle, J. (1972). Preparations of Helminths for Scanning Electron Microscopy. *Journal of Parasitology*, Vol. 58, No. 2, (April 1972), pp. 414-416, ISSN 0022-3395.

Amin, O. (1985). Classification, In: *Biology of the Acanthocephala*, D. Crompton and B. Nickol (Ed.), 27-72, Cambridge University Press, ISBN 0-521-24674-1, Cambridge, United Kingdom.

Berger, L. & Speare, R. (1998). *Chytridiomycosis*: a New Disease of Wild and Captive Amphibians. Australian & New Zealand Council for the Care of Animal in Research and Teaching. *Newsletter*, 11 (4), (December 1998), pp. 1-3.

Bušta, J. & Našincová, V. (1987). Ultrastructure of the Surface of External Genitals of six Species of Digenetic Trematodes Studied by Scanning Electron Microscopy. *Folia Parasitologica*, Vol. 34, No. 2, pp. 137-143, ISSN 0015-5683.

Chitwood, M. (1957). Intraespecific Variation in Parasitic Nematodes. *Systematic Zoology*, Vol. 6, No. 1, (March 1957), pp. 19-23, ISSN: 0039-7989

Chitwood, B. & Chitwood, M. (1975). *Introduction to Nematology* (Second Revised Edition). University Park Press, ISBN 0-8391-0697-1. Maryland, United State of America.

Crump, M. & Scott Jr, N. (1994). Visual Encounters Surveys. In: *Measuring and Monitoring Biological Diversity - standard methods for amphibians*, W. Heyer, M. Donnelly, R. McDiarmid, L. Hayek and M. Foster (Ed.), 84-91, Smithsonian Institution Press, ISBN-10 1560982845, Washington, United State of America.

De Coninck, L. (1965). Classe de Nématodes. In: *Traité de Zoologie. Anatomie, Systématique, Biologie*, P. Grassé (Ed.), pp. 1-217, Vol 4(2) Némathelminthes, pp. 1-217. Masson et Cie, ISBN : 9782225585104, Paris, France.

Duellman, W. & Trueb, L. (1986). *Biology of Amphibians*. The Johns Hopkins University Press, ISBN 0-8018-4780-X, London, United Kingdom.

Duszynski, D.; Bolek, M. & Upton, S. (2007). Coccidia (Apicomplexa: Eimeriidae) of Amphibians of the World. *Zootaxa*, Vol. 1667, (December 2007), pp. 1-77, ISBN 978-1-86977-183-6.

Faivovich, J.; Haddad, C.; García, P.; Frost, D.; Campbell, J. & Wheeler, W. (2005). Systematic Review of the Frog Family Hylidae, with Special Reference to Hylinae: Phylogenetic Analysis and Taxonomic Revision. *Bulletin of the American Museum of Natural History*, Vol. 294, (June 2005), pp. 1-240, ISSN 0003-0090.

Fried, B. (1997). An Overview of the Biology of Trematodes. In: *Advances in Trematode Biology*. Fried, B. and Graczyk T. (Ed.). CRC Press, ISBN 0-8493-2645-1, New York, United State of America.

Frost, D.; Grant, T.; Faivovich, J.; Bain, R.; Haas, A.; Haddad, C.; De Sá, R.; Channing, A.; Wilkinson, M.; Donnellan, S.; Raxworthy, C.; Campbell, J.; Blotto, B.; Moler, P.; Drewes, R.; Nussbaum, R.; Lynch, J.; Green, D. & Wheeler, W. (2004). The Amphibian Tree of Life. *Bulletin of the American Museum of Natural History*, Vol. 297, pp. 1-370, ISSN 0003-0090.

Gibbons, L. (1986). *SEM Guide to the Morphology of Nematode Parasites of Vertebrates*. CAB International, ISBN 085198-569-6, Oxu, United Kingdom.

Gibbons, L. (2010). *Keys to the Nematode Parasites of Vertebrates. Supplementary Volume*. CAB International/The Natural History Museum, ISBN 978-1-84593-571-9, Wallingford, United Kingdom.

Gibson, D.; Jones, A. & Bray, R. (2002). *Keys to the Trematoda. Vol 1*. CAB International/The Natural History Museum, ISBN 0-85199-547-0, London, United Kingdom.

González, C. & Hamann, M. (2004). Primer Registro de *Cosmocerca podicipinus* Baker y Vaucher, 1984 (Nematoda, Cosmocercidae) en *Pseudopaludicola falcipes* (Hensel, 1867) (Amphibia, Leptodactylidae) en Argentina. *FACENA*, Vol. 20, (December 2004), pp. 65-72, ISSN 0325-4216.

González, C. & Hamann, M. 2005. *Gyrinicola chabaudi* Araujo & Artigas, 1982 (Nematoda: Pharyngodonidae) in Tadpoles of *Scinax nasicus* (Cope, 1862) (Anura: Hylidae) from Corrientes, Argentina. *FACENA*, Vol. 21, (December 2005), pp. 145-148, ISSN 0325-4216.

González, C. & Hamann, M. (2006a). Nematodes Parásitos de *Chaunus granulosus major* (Müller & Hellmich, 1936) (Anura: Bufonidae) en Corrientes, Argentina. *Cuadernos de Herpetología*, Vol. 20, No. 1, (September 2006), pp. 43-49, ISSN 0326-551X.

González, C. & Hamann, M. (2006b). Helmintos Parásitos de *Leptodactylus bufonius* Boulenger, 1894 (Anura: Leptodactylidae) de Corrientes, Argentina. *Revista Española de Herpetología*, Vol. 20, pp. 39-46, ISSN 0213-6686.

González, C. & Hamann, M. (2007a). Nematode Parasites of two Species of *Chaunus* (Anura: Bufonidae) from Corrientes, Argentina. *Zootaxa*, Vol. 1393, (January 2007), pp. 27-34, ISSN 1175-5326

González, C. & Hamann, M. (2007b). The First Record of Amphibians as Paratenic Host of *Serpinema* larvae (Nematoda: Camallanidae). *Brazilian Journal of Biology*, Vol. 67, No. 3, (August 2007), pp. 579-580, ISSN 1519-6984.

González, C. & Hamann, M. (2008). Nematode Parasites of two Anurans Species, *Rhinella schneideri* (Bufonidae) and *Scinax acuminatus* (Hylidae), from Corrientes, Argentina. *Revista de Biología Tropical*, Vol. 56, No. 4, (December 2008), pp. 2147-2161, ISSN 0034-7744.

González, C. & Hamann, M. (2009). First Report of Nematodes in the Common Lesser Escuerzo *Odontophrynus americanus* (Duméril and Bibron, 1841) (Amphibia: Cycloramphidae) from Corrientes, Argentina. *Comparative Parasitology*, Vol. 76, No. 1, (January 2009), pp. 122-126, ISSN 1525-2647.

González, C. & Hamann, M. (2010a). Larval Nematodes Found in Amphibians from Northeastern Argentina. *Brazilian Journal of Biology*, Vol. 70, No. 4, (November 2010), pp. 1089-1092, ISSN 1519-6984.

González, C. & Hamann, M. (2010b). First Report of Nematode Parasites of *Physalaemus santafecinus* (Anura: Leiuperidae) from Corrientes, Argentina. *Revista Mexicana de Biodiversidad*, Vol. 81, No. 3, (December 2010), pp. 677-687, ISSN 1870-3453.

González, C. & Hamann M. (2011). Cosmocercid Nematodes of Three Species of Frogs (Anura: Hylidae) from Corrientes, Argentina. *Comparative Parasitology*, Vol. 78, No. 1, (January 2011), pp. 212-216, ISSN 1525-2647.

Grabda-Kazubska, B. & Combes, C. (1981). Morphological Variabily of *Haplometra cylindracea* (Zeder, 1800) (Trematoda, Plagiorchiidae) in Populational and Geografic Aspects. *Acta Parasitologica Polonica*, Vol. 28, No. 5, pp. 39-65, ISSN 1230- 2821.

Grabda-Kazubska, B. & Tenora, F. (1991). SEM Study on *Cosmocerca ornata* (Dujardin, 1845) and *C. commutata* (Diesing, 1851) (Nematoda, Cosmocercidae. *Acta Parasitologica Polonica*, Vol. 36, No. 1, pp. 45-50, ISSN 0065-1478.

Haley, A. (1962). Role of Host Relationship in the Systematics of Helminth Parasites. *Journal of Parasitology*, Vol. 48, No. 5, (October 1962), pp. 671-678, ISSN 0022-3395.

Halton, D. (2004). Microscopy and the Helminth Parasite. *Micron*, Vol. 35, No. 5, (July 2004), pp. 361-390, ISSN 0968-4328.

Hamann, M. & González, C. (2009). Larval Digenetic Trematodes in Tadpoles of six Amphibian Species from Northeastern Argentina. *Journal of Parasitology*, Vol. 95, No. 3, (June 2009), pp. 623-628, ISSN 0022-3395.

Hamann, M. & Pérez, V. (1999). Presencia de *Haematoloechus longiplexus* Stafford, 1902 (Trematoda, Haematoloechidae) en Anfibios de Argentina. *FACENA*, Vol. 15, No. (December 1999), pp. 157-162, ISSN 0325-4216.

Hamann, M., González, C. & Kehr, A. (2006a). Helminth Community Structure of the Oven Frog *Leptodactylus latinasus* (Anura, Leptodactylidae) from Corrientes, Argentina. *Acta Parasitologica*, Vol. 51, No. 4, (December 2006), pp. 294-299, ISSN 1230-2821.

Hamann, M.; Kehr, A. & González, C. (2006b). Species Affinity and Infracommunity Ordination of Helminths of *Leptodactylus chaquensis* (Anura: Leptodactylidae) in two Contrasting Environments from Northeastern Argentina. *Journal of Parasitology*, Vol. 92, No. 6, (April 2006), pp. 1171-1179, ISSN 0022-3395.

Hamann, M.; Kehr, A.; González, C.; Duré, M. & Schaefer, E. (2009a). Parasite and Reproductive Features of *Scinax nasicus* (Anura: Hylidae) from a South American Subtropical Area. *Interciencia*, Vol. 34, No. 3, (March 2009), pp. 214-218, ISSN 0378-1844.

Hamann, M.; Kehr, A. & González, C. (2009b). Niche Specificity of two *Glypthelmins* (Trematoda) Congeners Infecting *Leptodactylus chaquensis* (Anura: Leptodactylidae) from Argentina. *Journal of Parasitology*, Vol. 95, No. 4, (August 2009), pp. 817-822, ISSN 0022-3395.

Hamann, M.; Kehr, A. & González, C. (2010). Helminth Community Structure of *Scinax nasicus* (Anura: Hylidae) from a South American Subtropical Area. *Diseases of Aquatic Organisms*, Vol. 93, No. 1, (December 2010), pp. 71-82, ISSN 0177-5103.

Hirschmann, H. (1983). Scanning Electron Microscopy as a Tool in Nematode Taxonomy, In: *Concepts in Nematode Systematics*, A. Stone, H. Platt and L. Khalil (Ed.), 95-111, Academic Press, New York, United State of America and London, United Kingdom.

Hoff, G.; Frye, F. & Jacobson, E. (1984). *Diseases of Amphibians and Reptiles*. Plenum Press, ISBN 978-0306417115. New York, United State of America.

Jones, A.; Bray, R. & Gibson, D. (2005). *Keys to the Trematoda. Vol 2*. CAB International/The Natural History Museum, ISBN 0-85199-547-0, London, United Kingdom.

Karnovsky, M. (1965). A Formaldehyde-glutaraldehyde Fixative of High Osmolality for use in Electron Microscopy. *The Journal of Cell Biology*, Vol. 27, pp. 137A-138A (Abstr.), ISSN 0021-9525.

Kennedy, J. (1980a). Geographical Variation in Some Representatives of *Haematoloechus* Looss, 1899 (Trematoda: Haematoloechidae) from Canada and the United State. *Canadian Journal of Zoology*, Vol. 58, No. 6, (June 1980), pp. 1151-1167, ISSN 0008-4301.

Kennedy, J. (1980b). Host-induced Variations en *Haematoloechus buttensis* (Trematoda: Haematoloechidae). *Canadian Journal of Zoology*, Vol. 58, No. 3, (March 1980), pp. 427-442, ISSN 0008-4301.

Lajmanovich, R. & Martinez de Ferrato, A. (1995). *Acanthocephalus lutzi* (Hamman 1891) Parasite de *Bufo arenarum* en el Río Paraná, Argentina. *Revista de la Asociación de Ciencias Naturales del Litoral*, Vol. 26, No. 1, pp. 19–23, ISSN 0325-2809.

Lavilla, E.; Richard, E. & Scrocchi, G. (2000). Categorización de los Anfibios y Reptiles de la República Argentina. Edición Especial Asociación Herpetológica Argentina. Argentina: 1-97, ISBN 987-98331-0-4.

Lee, D. (2002). Cuticle, moulting and exsheathment. In: *The Biology of Nematodes*, D.L. Lee, (Ed.), 171-209, Taylor and Francis, ISBN 0-415-27211-4, London, United Kingdom.

Lunaschi, L. & Drago, F. (2007). Checklist of Digenean Parasites of Amphibians and Reptiles from Argentina. *Zootaxa*, Vol. 1476, (May 2007), pp. 51–68, ISSN 1175-5326.

Mafra, A. & Lanfredi, R. (1998). Reevaluation of *Physaloptera bispiculata* (Nematoda: Spiruroidea) by Light and Scanning Electron Microscopy. *Journal of Parasitology*, Vol. 84, No. 3, (June 1998), pp. 582–588, ISSN 0022-3395.

Mata-López, R. (2006). A New Gorgoderid Species of the Urinary Bladder of *Rana zweifeli* from Michoacán, Mexico. *Revista Mexicana de Biodiversidad*, Vol. 77, No. 2, (December 2006), pp. 191-198, ISSN 1870-3453.

Mordeglia, C. & Digiani, M. (1998). *Cosmocerca parva* Travassos, 1925 (Nematoda: Cosmocercidae) in Toads from Argentina. *Memorias do Instituto Oswaldo Cruz*, Vol. 93, No. 6, (November/December 1998), pp. 737-738, ISSN 0074-0276.

Nadakavukaren, M. & Nollen, M. (1975). A Scanning Electron Microscopic Investigation of the Outer Surfaces of *Gorgoderina attenuata*. *International Journal for Parasitology*, Vol. 5, No. 6, (December 1975), pp. 591-595, ISSN 0020-7519.

Navarro, P.; Izquierdo, S.; Pérez-Soler, P.; Hornero, M. & Lluch, J. (1988). Contribución al Conocimiento de la Helmintofauna de los Herpetos Ibéricos. VIII. Nematoda Ascaridida Skrjabin et Schultz, 1940 de Rana *spp. Revista Ibérica de Parasitología*, Vol. 48, No. 2, (April, May, June 1988), pp. 167-173, ISSN 0034-9623.

Nollen, M. & Nadakavukaren, M. (1974). *Megalodiscus temperatus*: Scanning Electron Microscopy of the Tegumental Surfaces. *Experimental Parasitology*, Vol. 36, No. 1, (August 1974), pp. 123-130, ISSN 0014-4894.

Olsen, O. (1974). *Animal parasites, their life cycles and ecology*, Dover Publications, Inc., ISBN 0839106432, New York, United State of America.

Pough, F.; Andrews, R.; Cadle, J.; Crump, M.; Savitzky, A. & Wells, K. (2001). *Herpetology* (Second Edition), Prentice Hall, ISBN 0-13-030795-5, New Jersey, United State of America.

Ramallo, G; Bursey, C. & Goldberg, S. (2007a). Two New Species of Cosmocercids (Ascaridida) in the Toad *Chaunus arenarum* (Anura: Bufonidae) from Argentina. *Journal of Parasitology*, Vol. 93, No. 4, (August 2007), pp. 910-916, ISSN 0022-3395.

Ramallo, G; Bursey, C. & Goldberg, S. (2007b). Primer Registro de *Oswaldocruzia proencai* (Nematoda: Molineoidae), Parásito de *Rhinella schneideri* (Anura: Bufonidae) en Salta, Argentina. *Acta Zoológica Lilloana*, Vol. 51, No. 1, pp. 91–92, ISSN 0065-1729.

Ramallo, G; Bursey, C. & Goldberg, S. (2008). New Species of *Aplectana* (Ascaridida: Cosmocercidae) in the Toads, *Rhinella granulosa* and *Rhinella schneideri* (Anura: Bufonidae) from Northern Argentina. *Journal of Parasitology*, Vol. 94, No. 6, (December 2008), pp. 1357-1360, ISSN 0022-3395.

Razo-Mendivil, U.; León-Règagnon, V. & Pérez-Ponce de León, G. (2006). Monophyly and Systematic Position of *Glypthelmins* (Digenea), Based on Partial lsrDNA Sequences and Morphological Evidence. *Organisms, Diversity & Evolution*, Vol. 6, No. 4, (November 2006), pp. 308–320, ISSN 1439-6092.

Schmidt, G. & Roberts, L. (2000). *Foundations of Parasitology* (Sixth Edition), Mcraw-Hill, ISBN: 0071168966, New York, United State of America.

Tinsley, R. (1995). Parasitic Diseasse in Amphibians: Control by the Regulation of Worm Burdens. *Parasitology*, Vol. 111, No. Suplement 1, (January 1995): pp. 153-178, ISSN 0031-1820.

Watertor, J. (1967). Intraespecific Variation of Adult *Telorchis bonnerensis* (Trematoda: Telorchiidae) in Amphibian and Reptilian Hosts. *Journal of Parasitology*, Vol. 53, No. 5, (October 1967), pp. 962-968, ISSN 0022-3395.

Whitehouse, C. (2002). A Study of the Frog Lung Fluke *Haematoloechus* (Trematoda: Haematoloechidae) Collected from Areas of Kentucky and Indiana. *Proceedings of the Indiana Academy of Science*, Vol. 1, No. 1, pp. 67–76, ISSN 0073-6767.

Willmott, S. (2009). Glossary and Keys to Subclasse, In: *Keys to the Nematode Parasites of Vertebrates. Archival Volume*, R. Anderson, A. Chabaud and S. Willmott (Ed.), pp. 1-17. CAB International, ISBN: 978-1-84593-572-6, Wallingford, United Kingdom.

14

Ionizing Radiation Effect on Morphology of PLLA: PCL Blends and on Their Composite with Coconut Fiber

Yasko Kodama[1,*] and Claudia Giovedi[2]
[1]*Instituto de Pesquisas Energéticas e Nucleares – IPEN–CNEN/SP,*
[2]*Centro Tecnológico da Marinha em São Paulo – CTMSP,*
Brazil

1. Introduction

The problem of non-biodegradable plastic waste remains a challenge due to its negative environmental impact. In this sense, poly(L-lactic acid) (PLLA) and poly(ε-caprolactone) (PCL) have been receiving much attention lately due to their biodegradability in human body as well as in the soil, biocompatibility, environmentally friendly characteristics and non-toxicity (Tsuji & Ikada, 1996; Kammer & Kummerlowe, 1994; Dell'Erba et al., 2001; Yoshii et al., 2000; Zhang et al., 2005). The controlled degradation of polymers is sometimes desired for biomedical applications and environmental purposes (Michler, 2008).

PLLA is a poly(α-hydroxy acid) and PCL is a poly(ω-hydroxy acid) (Tsuji & Ikada, 1996). PLLA is a hard, transparent and crystalline polymer (Mochizuki & Hirami, 1997). On the other hand, PCL can be used as a polymeric plasticizer because of its ability to lower elastic modulus and to soften other polymers (Kammer & Kummerlowe, 1994). Both polymers, PLLA and PCL, can be used in biomedical applications, which require a proper sterilization process. Nowadays, the most suitable sterilization method is high energy irradiation. Ionizing radiation exposure induces to crosslinking or scission of polymer main chain (Broz et al., 2003), besides other chemical alterations. Nature of those alterations is affected by chemical structure of polymer, and also by gaseous compounds present, as oxygen. Irradiation in the presence of air or oxygen leads to oxidized products formation that normally are undesirable, being less thermally stable and decreasing crosslinking degree by reaction of polymeric radicals (Charlesby et al., 1991).

The market for biodegradable polymers had shown strong growth from 2001 up to 2005. A number of major plant expansions for commercial scale production had been planned. The major classes of biopolymers, polylactic acid and aliphatic-aromatic co-polyesters has been used in a wide variety of niche applications, particularly for manufacture of rigid and flexible packaging, bags and sacks and foodservice products. In 2005, starch-based materials were the largest class of biodegradable polymer and polylactic acid (PLA) was the second

---

* Corresponding Author

largest material class followed by synthetic aliphatic-aromatic co-polyesters. Product development and improvement has a crucial role to play in the further development of the biodegradable polymers market. Biodegradable polymers can be found in a wide range of end use markets. Continued progress in terms of product development and cost reduction will be required before they can effectively compete with conventional plastics for mainstream applications. The main markets for PLA are thermoformed trays and containers for food packaging and food service applications. In 2005, packaging was the largest sector with 39% of total biodegradable polymer market volumes. Loose-fill packaging was the second largest sector, followed by bags and sacks. Fibers or textiles is an important sector for PLA, and accounted for 9% of total market volumes. Others include a wide range of very small application areas, the most important of which are agriculture and fishing, medical devices, consumer products and hygiene products. While the cost of some biodegradable plastics is high compared to conventional polymers, from a marketing perspective, it is important not only to consider the material cost, but also all associated costs, including the costs of handling and disposal, which are of course lower for biodegradable plastics. Users of biodegradable plastics can differentiate themselves from the competition by demonstrating how innovative and proactive they are for the benefit of the environment (Platt, 2006).

Polylactic acid is a biodegradable polymer derived from lactic acid. It is a highly versatile material and is made from 100% renewable resources like corn, sugar beet, wheat and other starch rich products. The homopolymer of L-lactide is a semicrystaline polymer. Due to high costs, the focus was initially on the manufacture of medical grade sutures, implants and controlled drug release applications. PLA has many potential uses, including many applications in the textile and medical industries as well the packaging industry.

From a commercial point of view the most important synthetic biodegradable aliphatic polyester was traditionally polycaprolactone (PCL). The ring opening polymerization of ε-caprolactone yields a semicrystalline polymer, which is regarded as tissue compatible and was originally used in the medical field as a biodegradable suture in Europe. Polycaprolactone aliphatic polyesters have long been available commercially for use as adhesives, compatibilizers, modifiers and films, as well as medical applications. Caprolactone limits moisture sensitivity, boosts melt strength, and helps to plasticize the starch (Platt, 2006).

In order to improve some desirable properties two or more polymers can be mixed to form polymeric blends (Utracki, 1989). The original reasons for preparing polymer blends are to reduce costs by combining high-quality polymers with cheaper materials (although this approach is usually accompanied by a drastic worsening of the properties of the polymer) and to create a polymer that has a desired combination of the different properties of its components (Michler, 2008). However, according to Michler (2008), usually different polymers are incompatible. Improved properties can be only realized if the blend exhibits optimum morphology. According to Sawyer (2008), in polymer science, the term morphology generally refers to form and organization on a size scale above the atomic arrangement, but smaller than the size and shape of the whole sample. Thus, improving compatibility between the different polymers and optimizing the morphology are the main issues to address when producing polymer blends.

In general, the morphology results from the complex thermomechanical history experienced by the different constituents during processing. So, parameters like the composition, viscosity ratio, shear rate/shear stress, elasticity ratio and interfacial tension among the component polymers, as well as processing conditions such as time, temperature and type of mixing, rotation speed of rotor of mixing determine the final size, shape and distribution of dispersed phase during the melting process (Dell'Erba et al., 2001; Michler, 2008; Nakayama & Tanaka, 1997) and are very important to define the characteristics of the obtained material. One common feature of semicrystalline polymers is a hierarchical morphology. The scales of structural details within them range from nanometers to millimeters. Under certain conditions, macromolecules are able to form periodic structures that involve adjacent chains or chain segments. By repeated folding of a flexible polymer chain results in densely packed and highly ordered domains. Polymers normally are partially crystalline, highly ordered domains will coexist with regions of an amorphous phase and a crystalline fraction with identical chemical compositions but divergent physical properties (Michler, 2008). Moreover, the structural modifications induced by ionizing radiation may alter the morphology of the samples.

PLLA:PCL blends have attracted great interest as temporary absorbable implants in human body, but they suffer from poor mechanical properties due to macro phase separation of the two immiscible components, and to poor adhesion between phases (Dell'Erba et al., 2001). Chemical structure influences the biodegradation of solid polymers. Enzymatic and non enzymatic degradations occur easier in the amorphous region (Mochizuki & Hirami, 1997; Tsuji & Ishizaka, 2001). The crystallinity and the resulting morphology are usually controlled by using either different proportions of stereoisomers (enantiomers, e.g. L-lactide and D-lactide) of the same monomer or a defined ratio of comonomers of related polyesters (Michler, 2008).

The morphology of the blends affects also the biodegradation of the polymers. So the control of the morphology of an immiscible polymer processed by melting is important for the tailoring of the final properties of the product (Dell'Erba et al., 2001; Michler, 2008). Kikkawa et al. (2006) cited that one of the approaches used to generate biodegradable materials with a wide range of physical properties is blending, and miscibility of blends is one of the most important factors affecting the final polymer properties. In particular, surface structure and morphology of biodegradable polymer blends have a great impact on the enzymatic degradation behavior by enzymes. According to Michler (2008), a high degree of crystallinity yields low degradation rates since it has been shown that hydrolytic degradation (cleavage of ester bonds) preferentially occurs in the amorphous regions.

Plastic solid waste has become a serious problem recently concerning environmental impact. In this scenario, preparation of polymers and composites based on coconut husk fiber would lead to a reduction on the cost of the final product. Additionally, it will reduce the amount of agribusiness waste disposal in the environment. In Brazil, coconut production is around 1.5 billion fruits by year in a cultivated area of 2.7 million hectares, but the coconut husk fiber has not been used much for industrial applications. According to Sawyer (2008), composites can contain short or continuous fibers, inorganic or organic. Even though according to Michler (2008), the word "composite" should only be applied to polymers with inorganic components, in this chapter it will be applied to the mixture of polymeric blend and natural coconut fiber.

It is worthy of note that the use of natural fibers as reinforcement in polymeric composites is an important research field that has been growing in the last decades, (Kapulskis et al, 2006;

Martins et al., 2006; Tomczak et al., 2007). Thus, by incorporating fibers of low cost to the polymeric blend, it is possible to obtain an improvement of the mechanical properties without loss of the original characteristics of polymeric components.

Regarding the irradiation effects, vegetable fiber, like as coconut fiber, is composed by cellulose and lignin, which suffer chemical alteration by irradiation such as scission or cross-linking. In the case of natural polymers, such as cellulose, main chain scission occurs predominantly due to irradiation and as a result molecular weight decrease (Chmielewski, 2005).

The structures and morphologies of polymers have been under investigation for more than 60 years. Scanning electron microscopy (SEM) was introduced in the 1960´s, and has been used to investigate fracture of surfaces, phase separation in polymer blends and crystallization of spherulites. Several improvements have been made in electron generation and electron optics that have enhanced the resolution power. Also, computerized SEM led to the introduction of the digital scanning generator for digital image recording and processing. Consequently, SEM is at present the most popular of the microscopic techniques. This is because of the user-friendliness of the apparatus, the ease specimen preparation, and the general simplicity of image interpretation. The limitation is that only surface features are easily accessible. Furthermore, chemical analysis of different elements is usually possible with SEM (energy dispersive or wavelength dispersive analysis of X-rays, EDXA, WDXA) (Michler, 2008). According to Sawyer (2008), SEM is used to evaluate features as the size, distribution, and adhesion of the fibers or particles, their adhesion to the matrix play a major role in revealing the strength and toughness enhancement. Regarding to the studied material, coarse structures such as larger particles in a matrix can easily be studied at low temperature (brittle) fracture surfaces in the SEM, since the fracture path follows the shape of the particle (Michler, 2008) considering that from a practical point of view, a good indicator of the degree of compatibility is the heterogeneity of the polymer system (e.g how the sizes of the dispersed domains depend on the processing conditions).

The objective of this chapter is to present the effect of ionizing radiation on the morphology of PLLA:PCL blend and composites containing coconut fiber. SEM and field emission (FE) SEM micrographs were taken of non irradiated and irradiated samples with gamma rays and electron beam.

2. Experimental

Coconut fiber

Coconut fibers were from three different origins: Embrapa- Empresa Brasileira de Pesquisa Agropecuária, Paraipaba region, Ceará; Projeto Coco Verde – Quissamã, Rio de Janeiro; Poematec – Ananindeua, Pará.

Size reduction of the coconut fibers was carried out using helix mill Marconi – model MA 680, from Laboratório de Matéria-prima Particulados e Sólidos Não Metálicos – LMPSol, Departamento de Engenharia de Materiais of Escola Politécnica/USP.

The fiber size distribution was measured using sieves of the Tyler series 16, 20, 35 and 48, fiber sizes of 1.0mm, 0.84mm, 0.417mm, and 0.297mm, respectively. The 0.417mm fiber size was used for the assays. The triturated material was separated using a sieve shaker Produtest, for 1 min.

In order to remove lignin from coconut fiber surface, fibers were soaked with Na_2SO_3 2% aqueous solution for 2h using ultrasound. Fibers from Embrapa were washed several times with tap water and finally, tree times with deionized water, as described by Calado et al. (2000).

Fiber acetylation was performed as described by d'Almeida et al. (2005). As received fibers from Embrapa were soaked in a solution of acetic anhydrate and acetic acid (1.5:1.0, w:w). It was used as a catalyst, 20 drops of sulfuric acid in 500mL solution. Set were submitted to ultrasound for 3h, then for more 24h rest at the same solution. Fibers were washed using tap water and for more 24h rested in deionized water. Fibers were separated from water and washed with acetone, after that, were evaporated at room temperature.

Fiber residue of non irradiated and irradiated samples were obtained using a furnace at 600°C, air atmosphere, by 2 hours. Residues were analyzed in a Scanning Electron Microscope, SEM, Jeol, JXA-6400, at Centro Tecnológico da Marinha em São Paulo (CTMSP). It was used Energy Dispersion Spectroscopy, EDS, to analyze chemical elements present in the residues.

Preparation of blend sheets

PLLA pellets were dried in a vacuum oven at 90°C and PCL pellets were dried at 40°C overnight to avoid hydrolysis of polymers during the melt-processing. Sheets of PCL and PLLA homo-polymers and blends with PCL:PLLA weight ratio of 25:75, 50:50 and 75:25 were prepared using a twin screw extruder (Labo Plastomill Model 150C, Toyoseki, Japan) equipped with a T-die (60mm width and 1.05mm thickness). T-die temperature was set at 205°C for PLLA homo-polymer and its blends, and at 90°C for PCL. Extruded sheets were quenched using a water bath set at room temperature. The take up speed was selected at 0.35 m min⁻¹. As the take up speed was set at slightly higher than the extrusion out-put speed, finally obtained thickness of films was around 1 mm.

Preparation of composite pellets and sheets

PCL (pellets, $\bar{M}_W$=2.14 10^5 g mol⁻¹; $\bar{M}_W / \bar{M}_n$= 1.423), PLLA (pellets, $\bar{M}_W$=2.64 10^5 g mol⁻¹ $\bar{M}_W / \bar{M}_n$= 1.518 – Gel Permeation Chromatographic values) and dry coconut fiber (from Embrapa, Ceará, Brazil) were used to prepare blends and composites. A Labo Plastomil model 50C 150 of Toyoseiki twin screw extruder was used for pellets preparation. Pellets of PLLA:PCL 80:20 (w:w) blend and composites containing 5 and 10% of untreated and chemically treated coconut fiber were prepared at AIST.

Sheets (150mm x 150mm x 0.5mm) of PCL, PLLA, PLLA:PCL 80:20 (w:w) blend and composites containing 5 and 10% untreated and chemically treated coconut fiber were prepared using Ikeda hot press equipment of JAEA– Japan Atomic Energy Agency. Mixed pellets of samples were preheated at 195°C for 3 min and then pressed by under heating at the same temperature for another 3 min under pressure of 150 kgf cm⁻². The sample was then cooled in the cold press using water as a coolant for 3 min.

Gamma irradiation

Samples were irradiated at IPEN–CNEN/SP (Brazil) using a Co-60 irradiator Gammacell model 220, series 142 from Atomic Energy of Canada Limited. Doses of 25, 50, 75, 100 and

500 kGy were applied at a dose rate of 4.3 kGy h^{-1}. Samples were cut 10 × 100 cm^2, and irradiated at room temperature in air.

Electron beam irradiation

Hot pressed sheets were irradiated using a electron beam accelerator (E=2 MeV, 2 mA) applying radiation doses from 10 kGy to 500 kGy with a dose rate of 0.6 kGy s^{-1}, at JAEA, Takasaki, Japan.

Scanning Electron Microscopy (SEM)

Morphology of the fractured surfaces of the non-irradiated homopolymers and blends was examined by a scanning electron microscope (SEM) DS-720 TOPCON Co. The photomicrographs of the cryogenic fractured surface of the blends sheet were taken after 4-5 min gold coating.

SEM micrographs of the irradiated homopolymers and blends sheets; and coconut fibers surfaces from cryogenic fractured samples were obtained using a scanning electron microscope model JXA-6400 (JEOL) at Centro Tecnológico da Marinha em São Paulo.

Field Emission Scanning Electron Microscopy (FE-SEM)

Photomicrographs of the, cryogenic fractured, non-irradiated and irradiated samples were taken using a field emission scanning electron microscope, JEOL, JSM – 7401F, acceleration voltage 1.0 kV at Central Analítica IQ-USP.

3. Results and discussion

Figure 1 shows scanning electron micrographs of coconut fiber surface, as received samples from Embrapa.

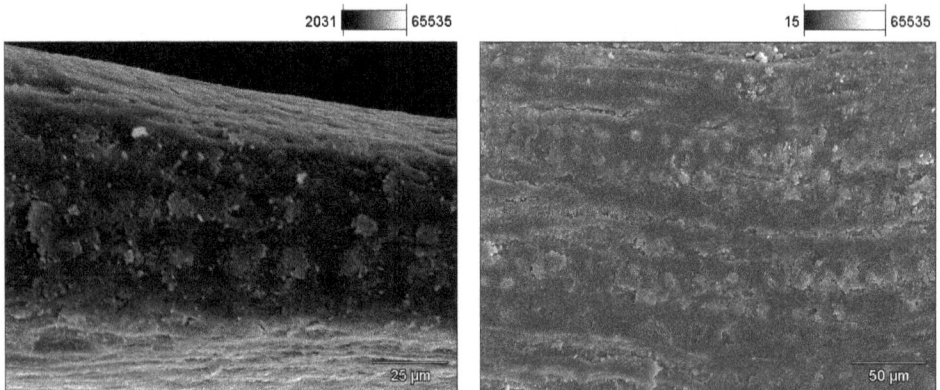

Acceleration voltage: 20.0 kV Magnification: 700×   Acceleration voltage: 20.0 kV Magnification: 350×

Fig. 1. Scanning electron micrographs of coconut fiber surface, as received from Embrapa.

There are some studies involving chemical treatment of vegetal fibers to improve its compatibility to polymers (Kapulskis et al, 2006; Abdul-Khalil & Rozman, 2000; Lee et al., 2004). Cell wall of a plant in the dry state consist mainly of carbohydrates combined to lignin, and few quantities of protein, starch and inorganic compounds, chemical composition varies from plant to plant and through different parts of the same plant (Rowell et al., 2000). The lignocellulosic fibers characteristics vary considerably with the place where they are produced, and possess different chemical compositions that affect their physical and chemical properties (Tomczak et al., 2007).

According to Calado et al. (2000), coconut coir fiber treated with Na_2SO_3 suffers lignin removal from surface and its roughness increases. This increase of surface roughness improves their adhesion due to promotion of mechanical interaction between fibers and polymeric matrix. In this sense, scanning electron micrographs of coconut fibers surface, chemically treated with Na_2SO_3 are shown in Figure 2.

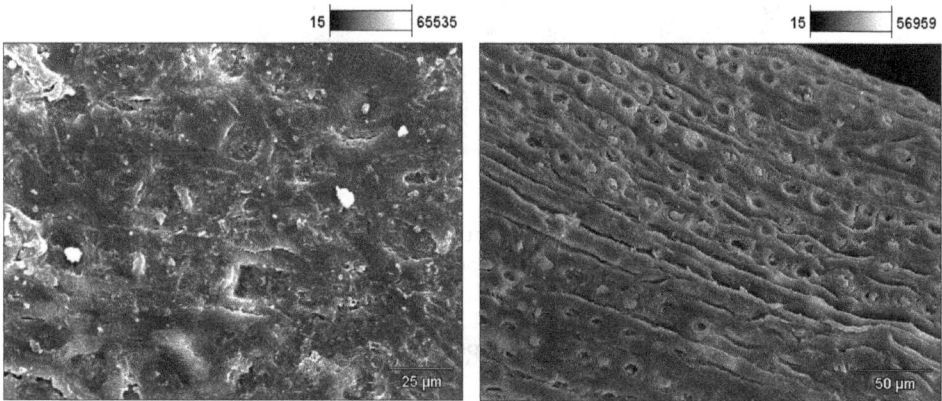

Acceleration voltage: 20.0 kV   Magnification 600×   Acceleration voltage: 20.0 kV Magnification: 350×

Fig. 2. Scanning electron micrographs of coconut fibers surface from Embrapa , chemically treated with Na_2SO_3.

Chemical treatment with anhydride acetic reduces the number of hydroxyl radicals, as shown in Figure 3. This treatment would reduce cellulose molecules polarity and then would improve the compatibility with thermo rigid matrix used in composites.

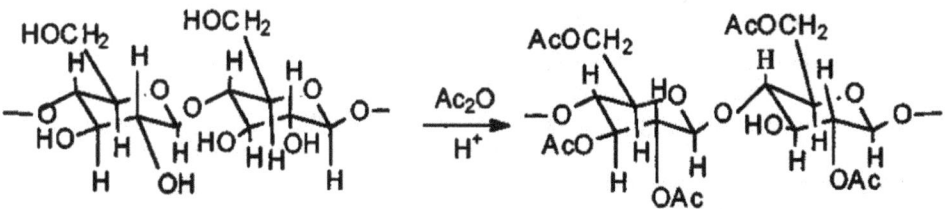

Fig. 3. Cellulose acetylation reaction (Calado et al., 2000).

Calado et al. (2000) observed clear difference on the morphologies of non treated and chemical treated surfaces of coir fibers. Same behavior was observed on the fibers studied in this chapter.

It was possible to observe roughness increase on the surface treated with Na_2SO_3 and anhydride acetic, as shown in Figure 4.

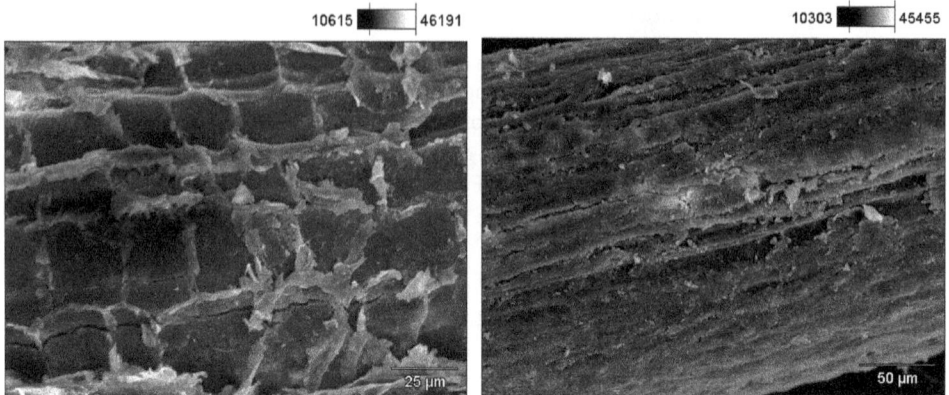

Acceleration voltage: 20.0 kV   Magnification: 600×   Acceleration voltage: 20.0 kV   Magnification: 350×

Fig. 4. Scanning electron micrographs of coconut fibers surface from Embrapa, chemically treated with Na_2SO_3 and anhydride acetic, respectively.

X rays spectra by EDS of Embrapa coconut fiber, non irradiated with 0.297 mm up to 0.417 mm size, are shown in Figure 5. It was possible to identify that chemical composition of the residue varied for the same fiber. It can be explained by the fact that chemical composition varies from plant to plant and also through different parts of the same plant (Rowell et al., 2000; Severiano et al., 2010).

In the residue of thermal decomposition in air of fibers from three different origins, most common elements found were K, Na, Si, Ca and, in some cases, Fe. In the literature, they found P, Mg and , in small concentrations, Cu, Zn, Mn (Rosa et al., 2001), in addition to Cl, S, Br e Rb (Mothé & Miranda, 2009) that were not observed in samples in this study. This variation can be attributed to soil where coconut tree was grown. Peak that had attributed to Zr, in fact is due to P, as they appear at the same channel of energy and, it is more probable to find P in higher quantity in soil than Zr. Peak that was attributed to As observed in some spectra was due to coating process used to allow image formation.

Scanning electron micrographs of surfaces of cryogenic fractured non irradiated as extruded blend samples were taken. As extruded PCL micrograph shows a homogeneous morphology, as shown in Figure 6. The as extruded PCL:PLLA 50:50 micrograph shows spheres with different sizes and shapes, as shown in Figure 7. According to Michler (2008), when a polymer is cooled down from the melt, the (primary) crystallization starts from initial points that are randomly distributed in the volume. Such starting points are either homogeneous or heterogeneous nuclei (i.e. nucleating agents, impurities, or filler particles). This radial growth results in a characteristic arrangement of lamellae. The superstructures (spherulites and sheaf-like boundless of lamellae) come in a variety of forms depending on the polymer and its crystalline structure. These superstructures generally form a texture consisting of one or more spherulite types with a characteristic spherulite size distribution.

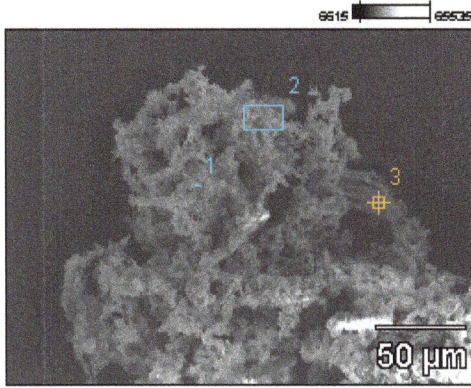

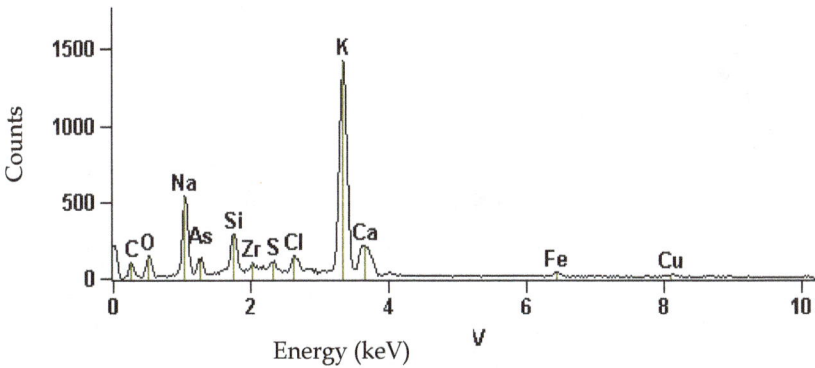

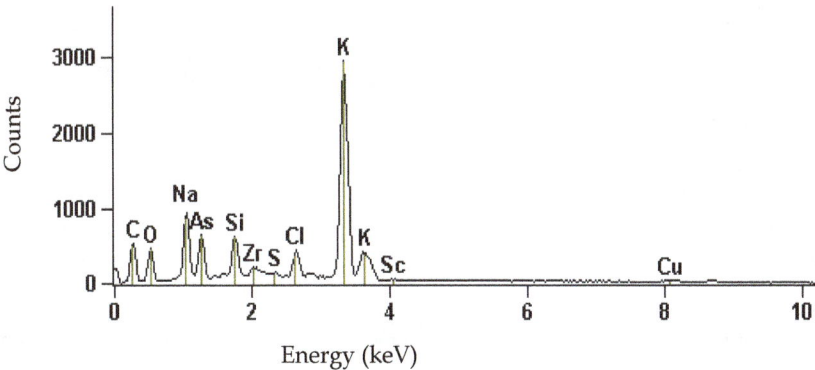

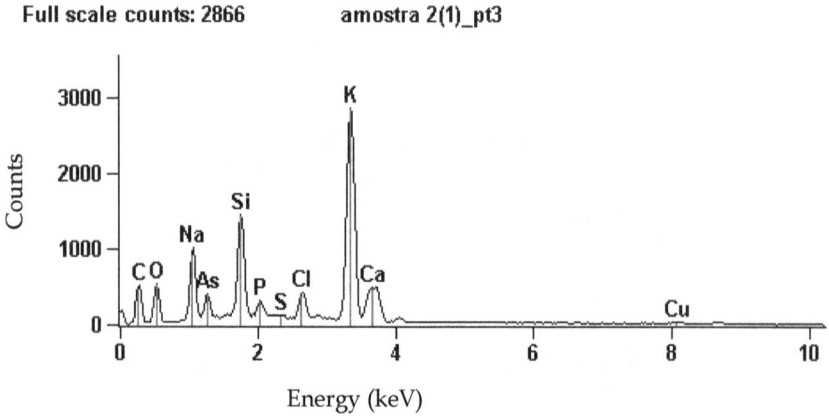

Fig. 5. Scanning electron micrograph of non irradiated Embrapa coconut fiber (0.297 and 0.417 mm) residue and X ray spectra by EDS (points 1, 2 and 3).

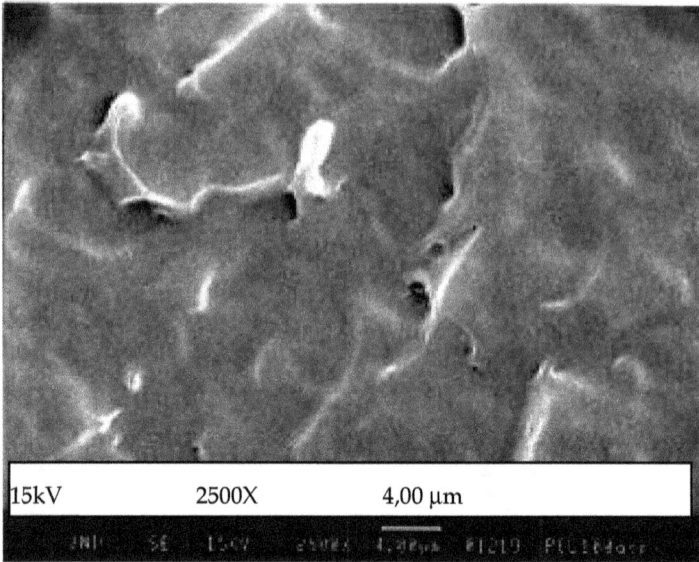

Fig. 6. Scanning electron micrograph of as extruded PCL, cryogenic fractured sample.

It is possible to observe the continuous PLLA-rich phase and the PCL-rich dispersed phase with a maximum domain size of 1.5 μm, as visualized before by Tsuji et al. (2001) in PCL:PLLA solution-cast blends. The PCL-rich phase is homogeneously dispersed in the PLLA matrix. One can observe some cracks or voids in the PLLA film probably caused by the temperature of processing, as shown in Figure 8. It has mentioned before that the degradation of aliphatic polyesters can occur because of melting at high temperatures (Yoshii et al., 2000; Kodama et al., 1997).

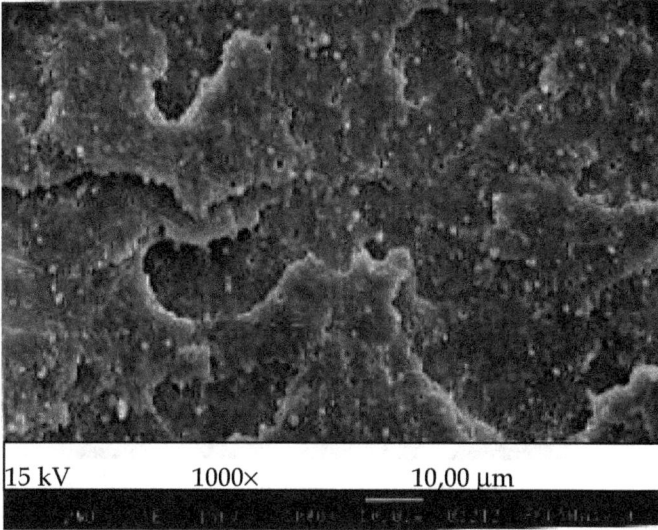

Fig. 7. Scanning electron micrograph of as extruded PCL:PLLA 50:50 (w:w), cryogenic fractured sample.

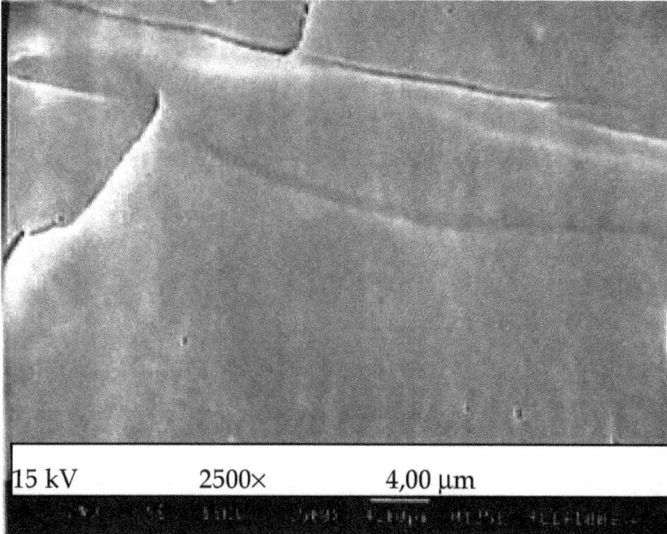

Fig. 8. Scanning electron micrograph of as extruded PLLA, cryogenic fractured sample.

Micrographs of annealed PLLA (Figure 9) and PCL:PLLA 50:50 (w:w) (Figure 10) shows changes in the morphology due to the crystallization of PLLA. Utracki (1989) explained that depending on the crystallization conditions various types of morphology can be obtained, which proceeds through melt, nucleation, lamellar growth and, spherulitic growth.

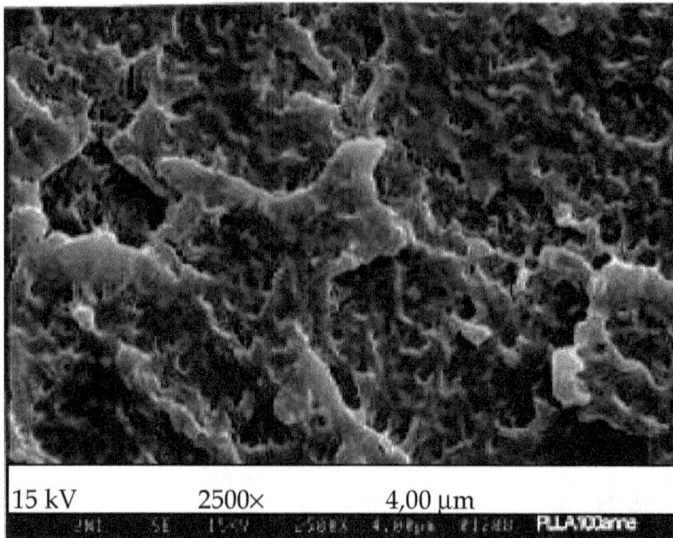

Fig. 9. Scanning electron micrograph of *annealed* PLLA, cryogenic fractured sample.

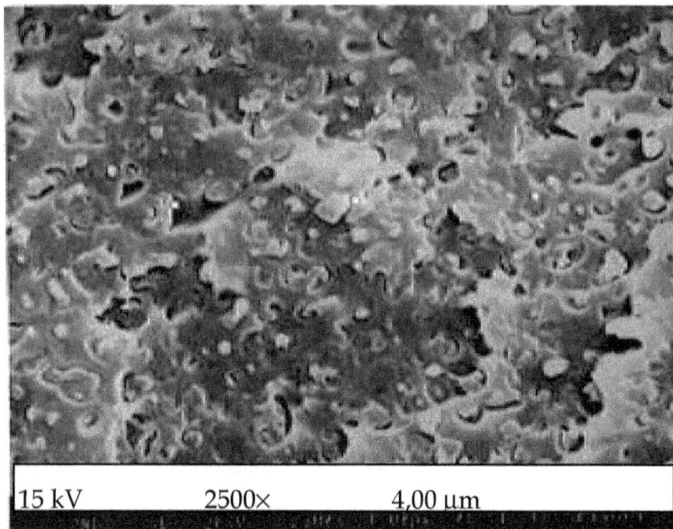

Fig. 10. Scanning electron micrograph of *annealed* PCL:PLLA 50:50 (w:w), cryogenic fractured sample.

Although the temperature of annealing was lower than the PLLA melting temperature (T_m), the thermal treatment allowed the crystallization of PLLA. Moreover, even though as extruded and annealed samples temperatures of processing were the same, one can observe the reduction of the cracks on the PLLA annealed sample. It is also possible to verify some morphological changes. The spheres are apart from the matrix, and in addition the matrix was changed due to the crystallization of PLLA. The blends are not miscible and the after extrusion cooling from the melt to room temperature causes the phase separation due to the difference between the melting temperatures of both blend components, PCL and PLLA.

In preliminary studies by differential scanning calorimetry (DSC) no change in the PLLA melting temperature was observed by increasing the PCL content in the blends. It was observed that, as PCL amount increased, PCL T_m peak increased in the region of the glass transition temperature of PLLA (Kodama et al., 1997). Although the immiscibility occurs, it is possible to observe by SEM some interfacial interaction, as the spheres seem to be covered by a thin layer of the polymeric matrix of the blends.

It should be noted that the blends were well mixed during extrusion as shown by the distribution and the size of the spherulites in the matrix. Dell'Erba et al. (2001) have found that it was reasonable to assume that low interfacial tensions were obtained in PLLA:PCL blends because of their similar chemical nature of the blends components, which allows interpolymer polar interactions across phase boundaries, thus favoring a well-dispersed morphology.

Preliminary studies have shown that although both are semi-crystalline polymers, only PCL crystallizes during extrusion. PLLA is amorphous and crystallizes after annealing, which was observed by x-ray diffraction of the non-irradiated samples (Broz et al., 2003). Also the orientation of crystallites in the blends was observed by x-ray diffraction, PLLA crystallizes in the α form with 10_3 helical conformation (Zhang et al., 2005). The thermal treatment increases the quantity of spheres, it was possible to notice some ellipsoids. This suggests that the new spherulites were formed due to the crystallization of amorphous PLLA, as observed previously (Broz et al., 2003). In this case, the segregation is also clear. It is possible to observe the separated spheres from the matrix and the cavities. It was discussed in literature (Dell'Erba et al., 2001) that the PLLA spherulites growth mechanism does not change when different amounts of PCL are present in the blend. Additionally, the presence of PCL enhances the PLLA crystallization rate, suggested to likely occur through the increase in the nucleation rate, it was observed that the presence of PCL domains in the PLLA matrix causes a small lowering in the half time of crystallization (Dell'Erba et al., 2001).

Furthermore, the results indicated that even though PLLA and PCL are immiscible, revealed by the presence of two glass transition temperatures for the blends very close to those found for pure PLLA and PCL, they are not highly incompatible (Dell'Erba et al., 2001). The binary mixture of (two) polymers is considered a compatible blend, when a homogeneous solid system is formed, without phase separations. It means a complete mutual solubility of the two polymers in molten state as well. This compatibility is reflected in, among other physical and mechanical properties, the fact that the system will have one single glass transition temperature (T_g) (IAEA-TECDOC-1420, 2004).

According to Michler (2008), γ or electron irradiation initiates pronounced crosslink in the amorphous parts of semi-crystalline polymers, whereas the structure inside the lamellae (the crystallinity) is not destroyed, so long as critical doses are not used. In this study, micrographs of PCL and PLLA homopolymers and PCL:PLLA 50:50 (w:w) blend irradiated with 100 kGy and 500 kGy, respectively, are shown in Figure 11 up to Figure 16.

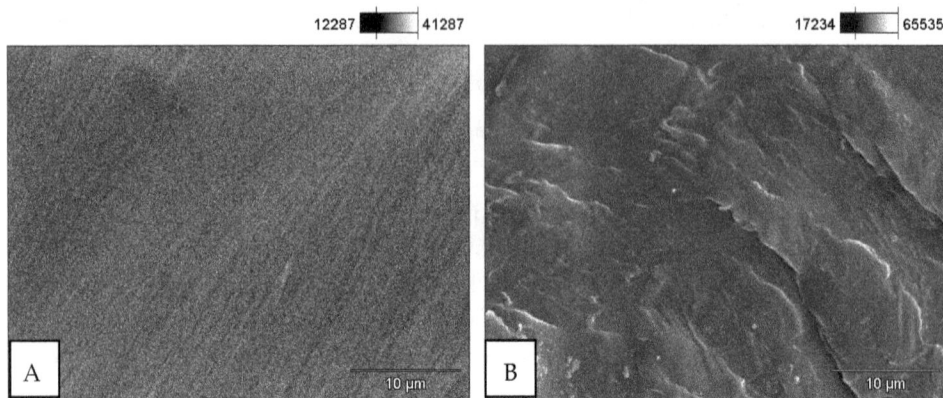

Acceleration voltage: 20.0 kV   Magnification: 2500

Fig. 11. Scanning electron micrographs of as extruded PCL, cryogenic fractured sample: A) non irradiated and B) irradiated with 100 kGy.

In Figure 12 it is possible to observe that lamellar structures increase on the fractured surface of PCL irradiated with 500 kGy, indicating that possibly significant crosslinking occurred.

Comparing the scanning electron micrographs, it is observed very few changes on the surface of ruptured samples. In earlier studies by size exclusion chromatography of electron beam irradiated PCL in air, it was observed a small increase followed by a decrease of crosslinking degree up to 5 kGy and after that an increase of gel-content of 15 wt %, indicating the enhance of crosslinking degree up to 200 kGy radiation dose (Södergard & Stolt, 2002). Even though PCL crosslinking predominates at radiation doses higher than 5 kGy and random chain-scission at lower doses (Södergard & Stolt, 2002; Ohrlander et al.,2000). In this chapter, only few changes could be seen by SEM for the irradiated PCL up to 100 kGy. However, the ruptured sample surface of irradiated PCL with 500 kGy became full of scales suggesting that the increase of crosslinking density induced by the ionizing radiation caused this alteration.

Some differences observed on micrographs A and B on Figure 13 probably are related to different regions analyzed. Apparently, an increase of granulation occurred on the polymeric matrix for 100 kGy irradiated sample.

Comparing Figures 13 and 14, some cracks appear, and polymeric surface seems to become smoother, probably due to significant scission of polymeric chains.

The surface of PLLA sample became rough. This fact is correlated to chain scissions promoted by gamma radiation. In the literature it was observed that PLA mainly undergoes chain-scissions at doses below 250 kGy. At higher doses of radiation, crosslinking reactions increase as a function of the increasing radiation dose. The reactions occur in the amorphous phase of the polymer (Södergard & Stolt, 2002). Samples submitted to doses in the range of 30 up to 100 kGy showed a marked depression in mechanical properties attributed to oxidative chain-scissions in amorphous region (Nugroho et al., 2001). Apparently no changes are visible by SEM on the irradiated PLLA with 500 kGy radiation dose, in contrast to other properties observed previously in the literature (Södergard & Stolt, 2002; Ohrlander et al., 2000; Nugroho et al., 2001).

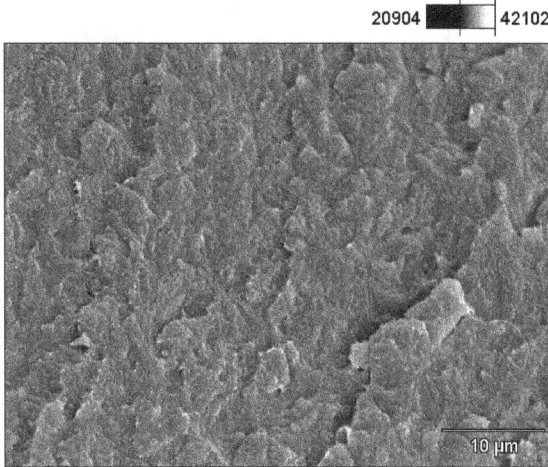

Acceleration voltage: 10.0 kV   Magnification: 2000×

Fig. 12. Scanning electron micrograph of as extruded PCL, cryogenic fractured sample irradiated with 500 kGy.

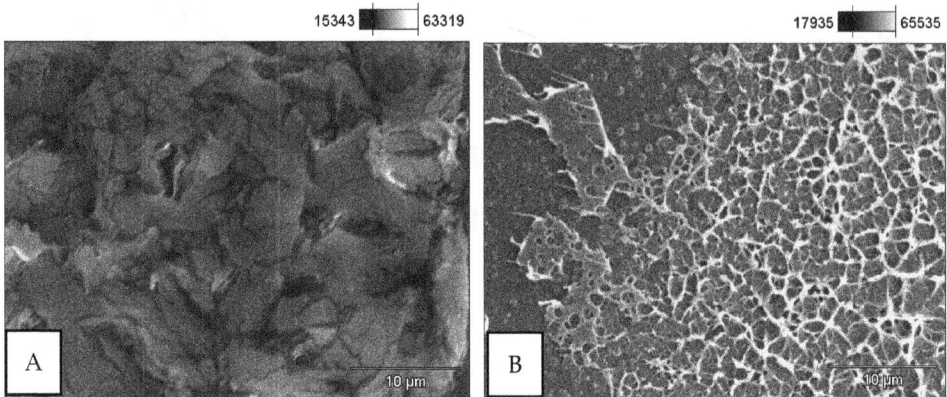

Acceleration voltage: 20.0 kV   Magnification: 2500×

Fig. 13. Scanning electron micrographs of as extruded PLLA, cryogenic fractured sample: A) non irradiated and B) irradiated with 100 kGy.

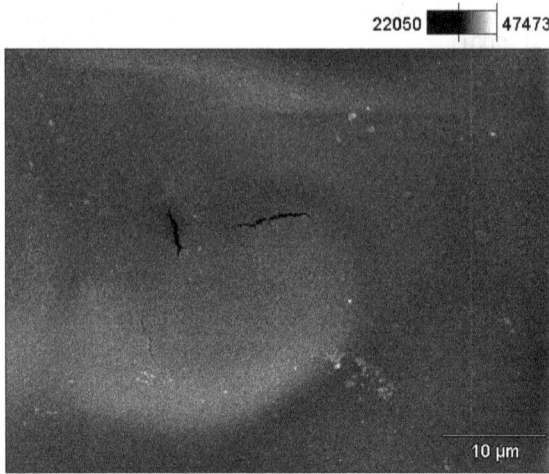

Acceleration voltage: 10.0 kV   Magnification: 2000×

Fig. 14. Scanning electron micrograph of as extruded PLLA, cryogenic fractured sample irradiated with 500 kGy.

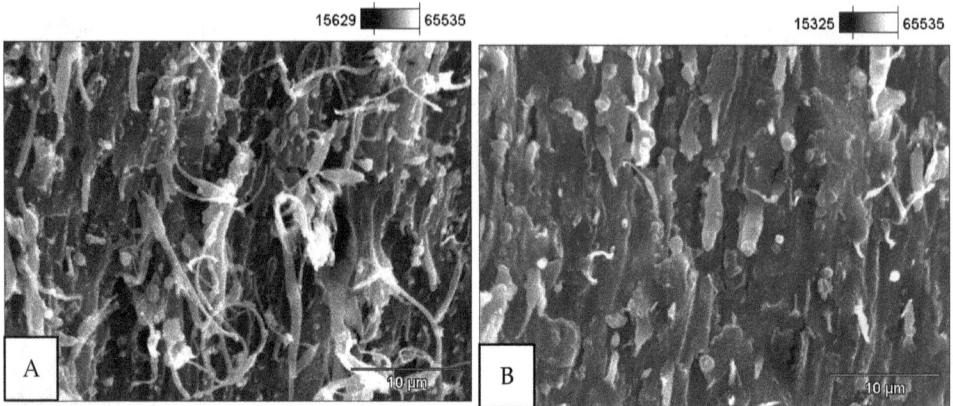

Acceleration voltage: 20.0 kV   Magnification: 2500

Fig. 15. Scanning electron micrographs of as extruded PCL:PLLA 50:50 (w:w), cryogenic fractured sample: A) non irradiated and B) irradiated with 100 kGy.

Broz et al. (2003) observed that microstructure of PLLA 0.4 in PCL was characterized by relatively wide quantity of spherical inclusions of PLLA on PCL matrix. Particles had sizes varying from 5 μm to 100 μm, apparently isolated on the matrix. Similar micrograph was observed in Figure 15A. Irradiated blends micrographs, shown in Figures 15B and 16, suffered scales formation similar to that observed for irradiated PCL. The interface between spherical inclusions and polymeric matrix seemed to be clean, suggesting that exists a weak adhesion between two phases, in consonance with absence of thermal transitions dislodgement observed by DSC. This lack of adhesion is unexpected since both polymers had been stated as miscible in the molten state.

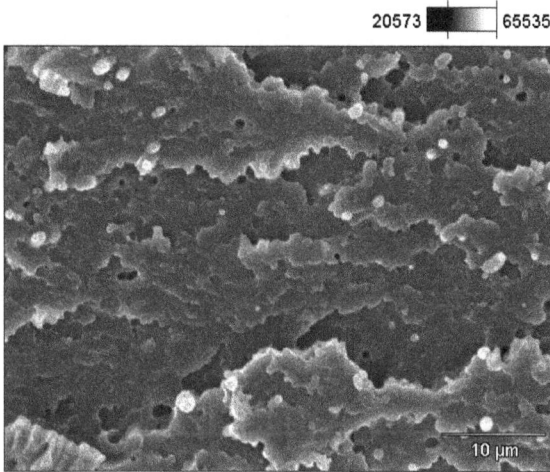

Acceleration voltage: 10.0 kV   Magnification: 2000××

Fig. 16. Scanning electron micrograph of as extruded PCL:PLLA 50:50 (w:w), cryogenic fractured sample irradiated with 500 kGy.

It seems that the ionizing radiation induced some shape alteration in the PCL dispersed phase in blends that were irradiated with 100 kGy. Likewise, samples submitted to irradiation processing up to dose of 500 kGy present the matrix with decreased PCL spherulites disperse phase, suggesting that some interaction between both polymeric phases had been promoted by the ionizing radiation. Previous studies demonstrated that gamma radiation does not affect significantly thermal properties of the blends when doses were kept bellow 75 kGy. A small decrease of PLLA T_m occurred probably due to PLLA main chain-scission. Thermal treatment induces PLLA T_m variation on irradiated blends with high concentration of PLLA. The crystallinity of PCL homopolymer and PLLA homopolymer as well as in the studied blends was not significantly affected by irradiation up to 100 kGy (Kodama et al., 2005). On the other hand, after irradiation with higher doses, PCL samples were more thermally stable than PLLA and blend (Nugroho et al., 2001). Thermal properties of PLLA were not affected by gamma radiation up to 100 kGy (Kodama et al., 2006a). Other results obtained previously by Kodama et al.(2006b) have shown that both, gamma and EB radiation, at doses up to 500 kGy, do not cause sample degradation to any significant extent to be detectable by FTIR (Fourier Transform Infrared Spectroscopy). As well, the miscibility of the polymeric blends was not affected by the irradiation process (Kodama et al., 2006b).

It can be observed in Figure 17A and B that fractured surface of non irradiated blend presented several needles like structures orthogonal to the surface, apparently related to PCL, due to the proportion o this component in the blend. Radiation dose seemed to induce these structures to diminish, probably related to PCL crosslinking that occurred with 100 kGy radiation dose.

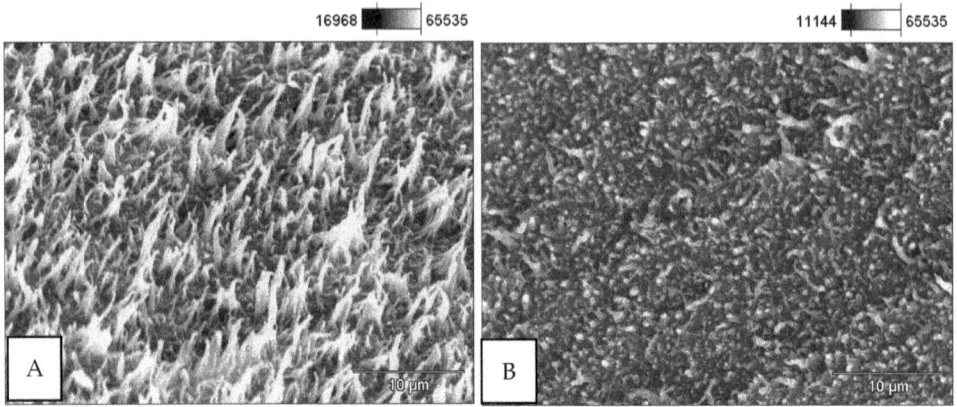

Acceleration voltage: 20.0 kV   Magnification: 2500×

Fig. 17. Scanning electron micrographs of as extruded PCL:PLLA 75:25 (w:w), cryogenic fractured sample: A) non irradiated and B) irradiated with 100 kGy.

Following micrographs were obtained using field emission scanning electron microscopy (FE-SEM) that dispense the use of Au⁰ coating and energy for image obtaining is lower. Micrograph of PCL:PLLA 20:80 (w:w), non irradiated is shown in Figure 18. It was observed a rough surface.

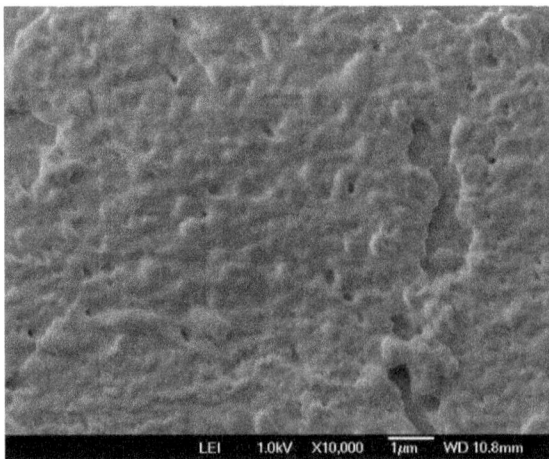

Fig. 18. Field emission scanning electron micrograph of as extruded PCL:PLLA 20:80 (w:w), cryogenic fractured sample, non irradiated.

In Figure 19 is observed surface micrographs of samples of PCL:PLLA 20:80 (w:w) prepared by hot press process , non irradiated and irradiated with 20 kGy radiation dose. This dose look as if do not affect significantly blend surface. Considering that conventional radiation dose for sterilization is 25 kGy, it is not expected that significant alteration occurs on the blend morphology sterilized by ionizing radiation.

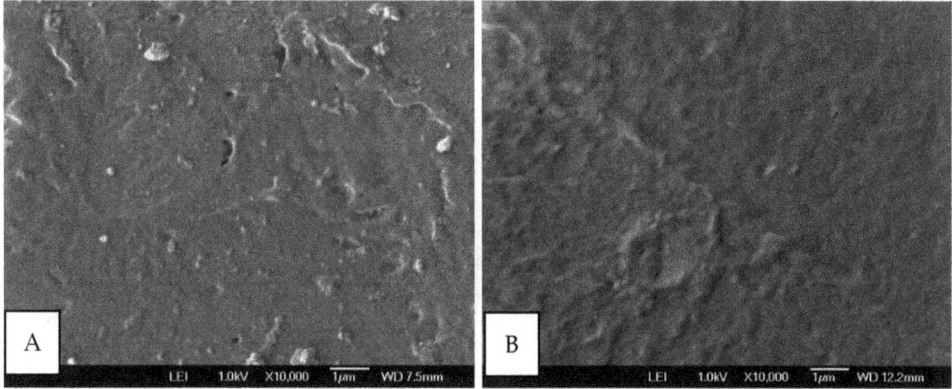

Fig. 19. Field emission scanning electron micrographs of hot pressed PCL:PLLA 20:80 (w:w) blends: A) non irradiated and B) irradiated with 20 kGy.

Arbelaiz et al. (2006) studied composites of linen fiber and PCL, and observed by SEM that fibers were clean and almost without adhesion points with PCL polymeric matrix, that indicated low wettability of fibers and lack of adhesion between phases. In Figure 20 it was no possible to observe in this sample regions containing fibers as observed by the authors mentioned before, probably it is related to its low concentration on the composite (5 or 10%) studied in this chapter. It was just observed what looks like a fragment of fiber.

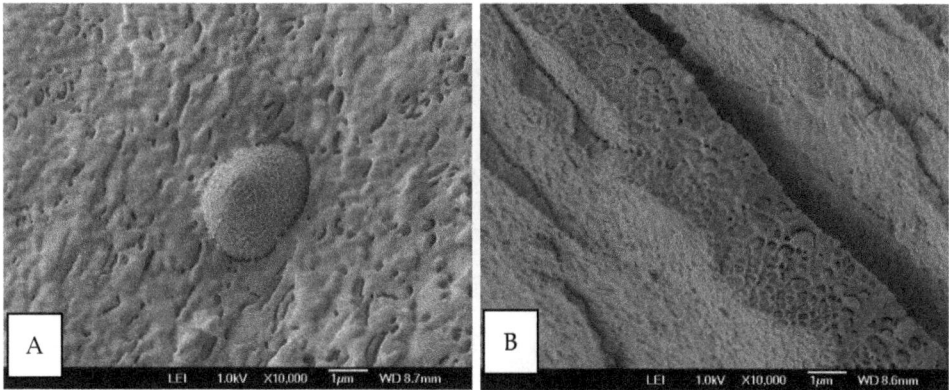

Fig. 20. Field emission scanning electron micrographs of composite with 10% non chemically treated coconut fiber: A) non irradiated; B) irradiated with 100 kGy, cryogenic fractured.

It was no possible to observe in Figure 21 significant alteration between structure existent and blend matrix. Apparently, acetylating process was not effective referring to the adhesion. However, it seems that ellipsoidal structures apart from polymeric matrix increased. Visually, structures suffered elongation and size reduction. Irradiated sample have smoother surface, probably related to scission process prevail of PLLA with doses above 100 kGy.

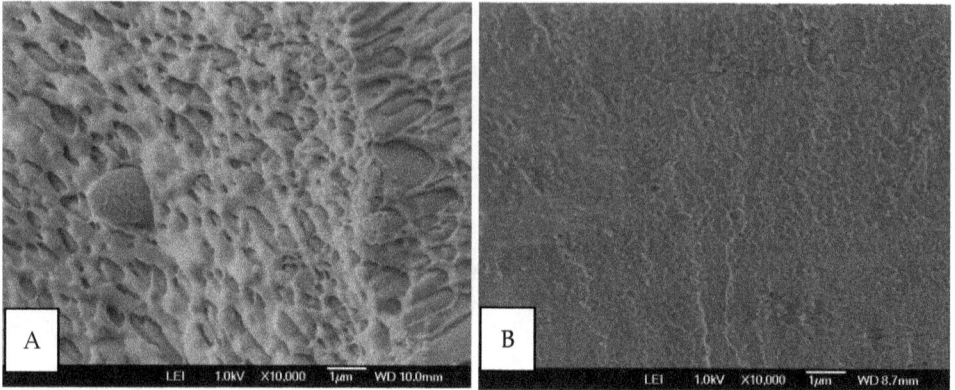

Fig. 21. Field emission scanning electron micrographs of composite with 10% acetylated coconut fiber: A) non irradiated; B) irradiated with 100 kGy, cryogenic fractured.

Micrographs of hot pressed sheets surface of composites containing 10% non chemically treated coconut fiber, non irradiated, and EB irradiated with 50 kGy and 100 kGy radiation doses, respectively, are shown in Figure 21. It was not possible to observe fiber presence on the surface analyzed. Neither any significant alteration on the irradiated surface, in the dose range studied. It suggests that cryogenic fractured surfaces allows the observation of irradiation effect on the polymeric bulk that otherwise could not be observed on sample surface. Probably it occurs due to the fact that species formed by energy deposition of radiation through polymeric matrix reacts mainly on the bulk than on the surface.

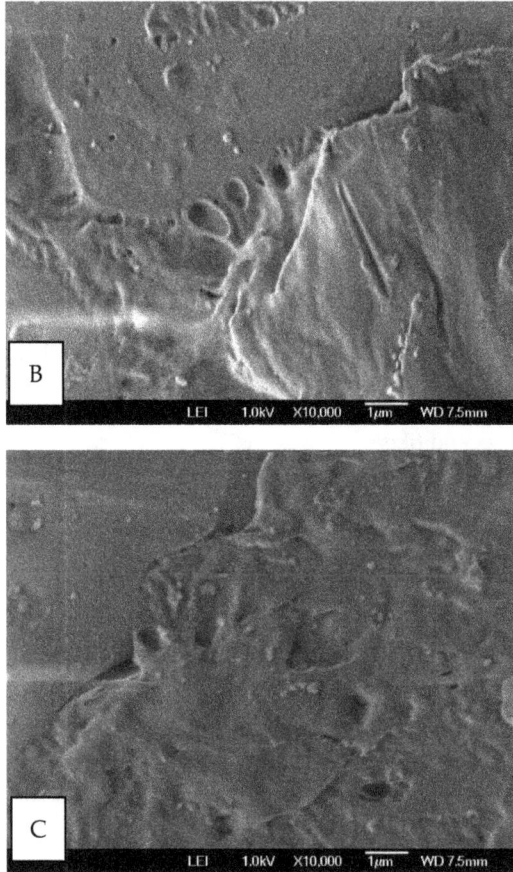

Fig. 22. Field emission scanning electron micrograph surface of hot pressed sheet of composite containing 10% non chemically treated coconut fiber: A) non irradiated; B) EB irradiated with 50 kGy, and C) EB irradiated with 100 kGy.

4. Conclusion

Due to coalescence effect it was possible to observe spherical inclusions of PLLA in PCL:PLLA blend. Increasing radiation dose induced elongation of inclusions, as well, lamellar structures increase in the PCL matrix. Radiation doses higher than 100 kGy altered morphologies of samples surfaces, that became smoother, attributed to the presence of smaller fractions of PLLA, as a result of long chain scission, and high crosslinking reaction in PCL phase. Radiation processing and chemical acetylating did not promote measurable interaction between fibers with polymeric matrix.

The SEM micrographs of the fractured homopolymers and blends have shown their immiscibility. The crystallization of PLLA could be observed on the annealed samples. Samples irradiated with 100 kGy presented little variation on the morphology, even supposing that the structural modifications induced by ionizing radiation may alter the morphology of the samples. It seems that some shape alteration in the PCL dispersed phase in blends occurred. Likewise, samples submitted to irradiation processing up to dose of 500 kGy presented the matrix with decreased PCL spherulites disperse phase, suggesting that some interaction between both polymeric phases had been promoted by the ionizing radiation. However, in PCL homopolymer and PCL:PLLA 50:50 irradiated with 500 kGy samples it was possible to observe significant alteration. The ruptured sample surface of irradiated PCL with 500 kGy became full of scales probably due to an increase of crosslinking density induced by the ionizing radiation. The surface of PLLA sample became rough with 100 kGy radiation dose correlated to chain scissions promoted by gamma radiation. On the other hand, apparently no changes are visible by SEM on the irradiated PLLA with 500 kGy radiation dose, in contrast to the observed previously in the literature. It was also studied blends and composites based on PCL, PLLA, and coconut fiber. Acetylation of fibers was not effective in order to induce any interaction between fibers and polymeric matrix, as expected. Ionizing radiation neither promoted detectable interaction between polymeric matrix and fibers.

5. Acknowledgements

We are grateful to the financial support from JICA and IAEA. Additionally, to Dr. Akihiro Oishi and Dr. Kazuo Nakayama, from National Institute of Advanced Industrial Science and Technology – AIST, Japan, for samples preparation and valuable discussion; to Dr. Naotsugu Nagasawa and Dr. Masao Tamada , from Japan Atomic Energy Agency – JAEA, Japan, for samples preparation and irradiation. We also would like to thank Dr. Morsyleide Freitas Rosa from Embrapa for providing coconut fiber; to Prof. Dr. Hélio Wiebeck, and Mr. Wilson Maia from Laboratório de Matérias-Primas Particuladas e Sólidos Não Metálicos – LMPSol, Departamento de Engenharia de Materiais, Escola Politécnica da USP (EPUSP) for coconut fiber size reduction and segregation; also to Eng. Elisabeth S.R. Somessari, Eng. Carlos G. da Silveira, and Mr. Paulo de Souza Santos, from IPEN, for blends and composites irradiation. In addition, to Centro Tecnológico da Marinha em São Paulo – CTMSP, for SEM and SEM EDS micrographs. We would like also to thank Dr. Luci Diva Brocardo Machado for helpful discussion.

6. References

Abdul Khalil, H.P.S., Rozman, H.D. (2000). Acetylated plant-fiber-reinforced polyester composites: a study of mechanical, hygrothermal, and aging characteristics, *Polymer-Plastics Technology and Engineering* Vol. 39 (No. 4) : 757-781.

Advances in Radiation Chemistry of Polymers, IAEA-TECDOC-1420, IAEA, Vienna, 2004.

Broz, M.E., VanderHart, D.L., Washburn, N.R. (2003). Structure and mechanical properties of poly(D,L-lactic acid)/poly(epsilon -caprolactone) blends, *Biomaterials* Vol. 24: 4181-4190.

Calado, V., Barreto, D.W., D'Almeida, J.R.M., (2000). The effect of a chemical treatment on the structure and morphology of coir fibers, *Journal of Materials Science Letters* Vol. 19: 2151-2153.

Charlesby, A., Clegg, D.W. & Collyer, A.A. (Ed.). (1991). *Irradiation Effects on Polymers*, Elsevier Applied Science, London and New York.

Chmielewski, A.G. New Trends in radiation processing of polymers, In: International Nuclear Atlantic Conference; Encontro Nacional de Aplicações Nucleares, 7th, ago. 28 - set. 2, 2005, Santos, SP. Anais... São Paulo: ABEN, 2005.

D'Almeida, A.L.F.S.; Calado, V.; Barreto, D.W. (2005). Acetilação da fibra de bucha (Luffa cylindrica), Polímeros: Ciência e Tecnologia Vol.15(No. 1): 59-62.

Dell'Erba, R., Groeninckx, G., Maglio, G., Malinconico, M., Migliozzi, A., (2001). Imiscible polymer blends of semicrystalline biocompatible components: thermal properties and phase morphology analysis of PLLA/PCL blends, Polymer Vol. 42: 7831-7840.

Kammer, H.W., Kummerlowe, C. (1994). Poly (ε-caprolactone) Comprising Blends - Phase Behavior and Thermal Properties, in Finlayson, K. (ed.) Advances in Polymer Blends and Alloys Technology, Technomicv, USA, 5, pp. 132-160.

Kantoğlu, Ö., Güven, O., (2002). Radiation induced crystallinity damage in poly(L-lactic acid), Nuclear Instruments and Methods in Physics Research B Vol. 197: 259-264.

Kapulskis, T.A., de Jesus, R.C., Innocentini-Mei L.H. Modificação química de fibras de coco visando melhorar suas interações interfaciais com matrizes poliméricas biodegradáveis. "XIII Congresso Interno de Iniciação Científica da UNICAMP – PIBC 2005, www.prp.unicamp.br/pibic/congressos/xiiicongresso/cdrom/html/area3.html. Accessed in 18/09/06

Kikkawa, Y., Suzuki, T., Tsuge, T., Kanesato, M., Doi, Y., Abe, H., (2006). Phase structure and enzymatic degradation of poly(L-lactide)/atactic poly(3-hydroxybutyrate) blends: an atomic force microscopy study, Biomacromolecules Vol. 7: 1921-1928.

Kodama, Y., Machado, L.D.B., Nakayama, K. (2005). Thermal Properties of Gamma Irradiated Blends Based on Aliphatic Polyesters. In: International Nuclear Atlantic Conference - INAC 2005, August 28 to September 2, 2005, Santos, SP, Brazil, Associação Brasileira de Energia Nuclear - ABEN ISBN 85-99141-01-5. 1CD-ROM.

Kodama, Y., Machado, L.D.B., Nakayama, K. Effect of Gamma Rays on Thermal Properties of Biodegradable Aliphatic Polyesters Blends, In: V Congresso Brasileiro de Análise Térmica e Calorimetria, April 02 to 05, 2006a, Poços de Caldas, MG, Brazil, Brazilian Association of thermal analysis and calorimetry – ABRATEC.

Kodama, Y., Machado, L.D.B., Giovedi, C., Nakayama, K. FTIR Investigation of Irradiated Biodegradable Blends. In: 17° Congresso Brasileiro de Engenharia e Ciência dos Materiais – CBECIMAT, 2006b, Foz do Iguaçu, PR, Brazil.

Lee, S.H., Ohkita, T., Kitagawa, K. (2004). Eco-composite from poly (L-lactic acid) and bamboo fiber, Holzforschung Vol. 58: 529-536.

Martins, M.A., Forato, L.A., Mattoso, L.H.C., Colnago, L.A. (2006). A solid state ^{13}C high resolution NMR study of raw and chemically treated sisal fibers, Carbohydrate Polymers Vol. 64: 127-133.

Michler, G.H. (2008). Electron Microscopy of Polymers, Springer-Verlag.

Mochizuki, M., Hirami, M., (1997). Structural effects on the biodegradation of aliphatic polyesters, Polymers for AdvancedTechnology Vol. 8: 203-209.

Mothé, C.G.; de Miranda, I.C. (2009) Characterization of sugarcane and coconut fibers by thermal analysis and FTIR, Journal of Thermal Analysis and Calorimetry Vol. 97: 661-665.

Nakayama K. & Tanaka, K. (1997). Effect of heat treatment on dynamic viscoelastic properties of immiscible polycarbonate-linear low density polyethylene blends, *Advanced Composite Materials* Vol. 6 (No. 4): 327-339.

Nugroho, P., Mitomo, H., Yoshii, F., Kume, T. (2001). Degradation of poly(l-lactic acid) by γ -irradiation, *Polymer Degradation and Stability* Vol. 72: 337-343.

Ohrlander, M., Erickson, R., Palmgren, R., Wisén, A., Albertsson, A.-C. (2000). The effect of electron beam irradiation on PCL and PDXO-X monitored by luminescence and electron spin resonance measurements, *Polymer* Vol. 41: 1277-1286.

Platt, D.K. (2006). *Biodegradable Polymers: Market Report*, Smithers Rapra Limited, United Kindom.

Rosa, M.F.; Santos, F.J.S.; Montenegro, A.A.T.; Abreu, F.A.P; Correia, D.; Araújo, F.B.S.; Norões, E.R.V. (2001). *Caracterização do pó da casca de coco verde usado como substrato agrícola*, Embrapa, Comunicado Técnico, Vol. 54: 1-6.

Sawyer, L.C., Grubb, D.T. & Meyers, G.F. (2008). *Polymer Microscopy 3rd ed*, Springer.

Severiano, L.C.; Lahr, F.A.R.; Bardi, M.A.G. Machado; L.D.B. (2010). Estudo do efeito da radiação gama sobre as propriedades térmicas de madeira usadas em patrimônios artísticos e culturais brasileiros. In: VII CONGRESSO DE ANÁLISE TÉRMICA E CALORIMETRIA, 2010, São Paulo, SP. *Anais...* São Paulo: 25 a 28 de abril. 1 CD-ROM.

Rowell, R.M., Han., J.S., Rowell, J.S., Characterization and Factors Effecting Fiber Properties, In: Frollini, E.; Leão, A.L.; Mattoso, L.H.C. (Ed.). *Natural Polymers and Agrofibers Composites: preparation, properties and applications*, São Carlos: USP-IQSC/Embrapa Instrumentação Agropecuária, Botucatu: UNESP, São Paulo, 2000.

Södergard, A. & Stolt, M. (2002). Properties of lactic acid based polymers and their correlation with composition, *Progress in Polymer Science* Vol. 27: 1123-1163.

Tomczak, F., Sydenstricker, T.H.D., Satyanaryana, K.G. (2007). Studies on lignocellulosic fibers of Brazil. Part II: Morphology and properties of Brazilian coconut fibers, *Composites: part A* Vol. 38: 1710-1721.

Tsuji, H. & Ikada, Y. (1996) Blends of aliphatic polyesters. I. Physical properties and morphologies of solution-cast blends from poly (DL-lactide) and poly(ε-caprolactone), *Journal of Applied Polymer Science* Vol. 60: 2367-2375.

Tsuji, H., Ishizaka, T., (2001). Blends of aliphatic polyesters. VI. Lipase-catalyzed hydrolysis and visualized phase strucuture of biodegradable blends from poly(e-caprolactone) and poly(L-lactide). International Journal of Biological Macromolecules Vol. 29: 83-89.

Utracki, L.A. (1989). *Polymer Alloys and Blends: Thermodinamics and Rheology*, Hanser, New York.

Yoshii, F., Darvis, D., Mitomo, H., Makuuchi, K. (2000). Crosslinking of poly (ε-caprolactone) by radiation technique and its biodegradability, *Radiation Physics and Chemistry* Vol. 57: 417-420.

Zhang, J., Duan, Y., Sato, H., Tsuji, H., Noda, I., Yan, S., Ozaki, Y. (2005). Crystal modifications and thermal behavior of poly (L-lactic acid) revealed by infrared spectroscopy, *Macromolecules* Vol. 38: 8012-8021.

15

Pathogenic Attributes of Non-*Candida albicans* *Candida* Species Revealed by SEM

Márcia Cristina Furlaneto[1], Célia Guadalupe Tardeli de Jesus Andrade[2],
Luciana Furlaneto-Maia[3], Emanuele Júlio Galvão de França[1]
and Alane Tatiana Pereira Moralez[1]

[1]*Department of Microbiology,*
[2]*Electronic Microscopy and Microanalysis Laboratory,*
State University of Londrina (UEL),
[3]*Technological Federal University of Paraná (UTFPR), Londrina-PR,*
Brazil

1. Introduction

The advent of microscopy provided an expressive progress on the knowledge of the biological world. Particularly important was the development of the electron microscopy at 1930s, making possible to find out a universe of unimaginable dimensions. Its great highlight is the much shorter wavelength of the electron that increases the resolution power of the equipment (Lee, 1993). Currently the electron microscopy is considered a specialized field of science (Bozzola and Russel, 1999). Although electron microscopy is useful to answer important questions about the ultrastructure of biological materials, it is also represent an additional tool that may be used as an ally in several research fields.

The scanning electron microscope (SEM) is useful to analyze microstructural features of solid bodies' surfaces, such as yeast cells. Besides, it leads to the formation of a three-dimensional image as a direct result of the great depth of field (Lee, 1993). Even samples observed by naked eye may be analyzed at low magnifications with great depth of focus, making possible to obtain images with a pronounced resolution using detectors of secondary electrons. Thus, the electron microscopy may contribute to reveal this nanometer´s world including fungal cell structure and the interaction between fungal cells and their microenviroment.

Most of the ultrastructural studies of *Candida* are based on the polymorphic species *Candida albicans*. In a pioneering study, the employment of SEM allowed the analyses of the surface features of different morphologies (budding yeast cells, germ tubes, hyphae and chlamydospores) of *C. albicans* (Barnes et al., 1971). Since then, many studies have applied SEM to elucidate several biological features of *Candida*. Recently, an updated useful review of the ultrastructural biological features of superficial candidiasis was presented (Jayatilake, 2011).

The evolution of intensive care medicine prolongs life expectancy leading populations to high susceptibility for candidal infection. Although ubiquitous in nature, *Candida* species can cause various infections that can vary from a relatively mild skin mycoses to life-threatening systemic disease. Over the past two decades, an increase in the number of cases caused by non-*Candida albicans Candida* (NCAC) species has been reported. In this context, *Candida parapsilosis* and *Candida tropicalis* are among the commonest species of *Candida* responsible for nosocomial blood infection worldwide (Krcmery & Barnes, 2002, Almirante et al., 2006, Colombo et al., 2006, Nucci & Colombo, 2007). Besides, *C. parapsilosis* has gained increasing recognition as the most common etiological agent causing *Candida* nail infections (reviewed in Trofa et al., 2008).

Yeasts belonging to genus *Candida* produce daughter cells by budding that readily separate at sites of septation. However other morphologies also occur, including pseudohyphal cells that also grow by budding and display distinct constrictions at septa, although they are more elongated and do not readily separate, and true hyphal cells formed as long thin tubes with parallel cell walls that lack septal constrictions.

With reference to morphological characteristics, *Candida* budding cells display oval, round, or cylindrical shapes. *C. parapsilosis* does not form true hyphae and exists in either a yeast phase or a pseudohyphal form. Differently, *C. tropicalis* can exist in multiple morphogenetic forms, including yeast phase, pseudohyphae and true hyphae, being considered a polymorphic fungi.

For pathogenic yeasts, it is widely accepted that yeast form cells are essential for efficient dissemination through the body, whereas the filamentous forms are required for tissue invasion. For *C. albicans*, strains lacking hyphal formation exhibited lower ability to invade tissue compared with wild-type strains (Jayatilake et al., 2006).

Most pathogenic *Candida* species have developed a wide range of putative virulence factors to assist in their ability to colonize host tissues, cause disease, and overcome host defenses. Despite intensive research to identify pathogenic factors in yeast, particularly in *C. albicans*, relatively little is known about the virulence attributes associated with NCAC species. Although non-*albicans* species seem to share common virulence determinants with *C. albicans* it is believed that they have a particular repertoire of specific virulence traits (Haynes, 2001). Currently, we still have much to learn about the virulence of NCAC species, particularly *C. parapsilosis* and *C. tropicalis*.

The purpose of this chapter is to sum up some of the recent ultrastructural findings of *C. parapsilosis* and *C. tropicalis* virulence-associated characteristics that are thought to contribute in their process of pathogenesis.

2. Ultrastructural features of pathogenic attributes

Multiple characteristics have been proposed to be putative virulence factors related to the pathogenesis of *C. parapsilosis* and *C. tropicalis*, including adherence to host surfaces (cells and tissues) and medical devices, formation of filamentous forms, biofilm formation, production of hydrolytic enzymes and phenotypic switching (reviewed in Trofa et al., 2008, reviewed in van Asbeck et al., 2009, reviewed in Silva et al., 2011).

Recently, França et al. (2010) described a correlation between *in vitro* haemolytic and proteinase activities in clinical isolates of *C. parapsilosis* and *C. tropicalis* and site of fungal isolation. According to these authors, anatomical sites of isolation seem to be correlated with these activities, particularly for *C. parapsilosis* isolates.

Currently, we are employing the SEM to evaluate many events related to pathogenicity of clinical strains of *C. parapasilosis* and *C. tropicalis*, including adherence to biotic substrates, agar invasion capability, switching morphogenesis and morphological alterations of *Candida* cells by compounds from natural resources.

Precise imaging of yeasts depends on the adequate preservation. To SEM analysis, yeast samples are well fixed by immersion on glutaraldehyde in phosphate buffer at proper concentrations (Hayat, 2000). The use of buffered osmium tetroxide at 1%, for 1 hour, is recommended as post fixation to preserve cellular content and surfaces. In our studies, samples of planktonic cells or biofilms are critical point dried after ethanolic dehydration. In order to preserve the colonies organization some steps were optimized, such as omission of osmium tetroxide, ethanolic dehydration and critical point dried. As a routine in our laboratory colonies are freeze-dried to avoid distortions and to maintain their architecture. Besides, sputtering was performed using a thick layer (50 nm) of gold.

2.1 Adherence patterns *in vitro*

Adherence is essential for members of the genus *Candida* to develop their pathogenic potential since it triggers the process that leads to colonization and allows their persistence in the host. Furthermore, different intra-species adherence ability has been reported for *Candida* species (reviewed in Silva et al., 2011). For instance, *C. tropicalis* exhibited higher ability to colonize reconstituted human epithelium (RHE) than did *C. parapsilosis* and *Candida glabrata* (Jayatilake et al., 2006). Adherence of *C. albicans* to epithelial cells is greater than that of *C. parapsilosis* (Lima-Neto et al., 2011).

For *C. parapsilosis* its emergence as a major opportunistic and nosocomial pathogen may relate to an ability to colonize the skin and adhere to inert polymeric surfaces and forms biofilms on these surfaces, such as catheters, prosthetic valves, artificial dentures and others (Douglas, 2003). According to Panagoda et al. (2001) the initial adherence of *C. parapsilosis* to surfaces is associated with cell surface hydrophobicity.

As cited previously, *C. parapsilosis* is a common etiological agent causing *Candida* nail infections. In Brazil, *C. parapsilosis* is the first or second most common cause of onychomycosis lesions (Figueiredo et al., 2007, Martins et al. 2007). Candidal onychomycosis, infection affecting nails, is increasingly found especially in immunocompromised patients. Far more than being a simple esthetics problem, infected nail serves as a reservoir of infections of the skin and mucous membrane. Multiple virulence mechanisms of *Candida* are involved in the pathogenesis of nail infections (reviewed in Jayatilake et al., 2009).

Therefore, ultrastructural investigations of the interface of *C. parapsilosis* and the keratinised substrates from human source reveal important features, which may help to clarify the pathogenesis of superficial candidiasis.

We have recently initiated experiments to verify the *in vitro* adherence pattern of *C. parapsilosis sensu stricto* (formerly *C. parapsilosis* group I) isolates obtained from candidal onychomycosis with keratinous substrates from human source. In a recent work, SEM was employed to verify the capability of *C. parapsilosis* cells to adhere and grow as biofilm on human natural substrates (nail and hair). In addition, the adherence pattern of isolates exhibiting distinct colonies phenotypes (smooth and crepe) was compared (Oliveira et al. 2010). This analysis allowed us to observe, for the first time, extracellular material and biofilm formation by *C. parapsilsois* on keratinised substrates.

In the present study, we compared ultrastructural features related to adhesion of *C. parapsilosis* cells of isolates obtained from distinct clinical sources (nail infection and tracheal secretion), on soft keratin (cutaneous stratum corneum - the outermost layer of skin) and hard keratin (nail and hair) substrates, following growth on these substrates as sole nitrogen source. In general, the surfaces of the budding cells (blastoconidia) and pseudohyphae were generally smooth except for occasional bud scars. Based on the SEM images, a different pattern of adhesion was observed for the isolates tested (Figure 1). For the onychomycosis (finger nail lesion) isolate the cell population attached to keratin substrates consisted mainly of cells in the budding-yeast phase of growth (blastoconidia) (Figure 1 A1, B1 and C1). Besides, on stratum corneum keratin short hyphal form was observed (Fig. 1A1). Differently, the isolate obtained from tracheal secretion (colonization site) the cellular population consisted mostly of pseudohyphae, particularly on stratum corneum (Figure 1 A2), a pattern that could indicate that this situation favors cellular morphologies with capacity for tissue invasion. Furthermore, SEM analysis also revealed that the tracheal secretion isolate presented different morphological pattern according to the substrate that they were in contact. For instance, cells adhered to hair keratin, consisted mainly of blastoconidia (Figure 1 C2). Overall, there was a loose association between yeast cells and keratinous substrates. However, on stratum corneum extracellular material was seen evolving cells from the onychomycosis isolate by forming a biofilm-like structure (Fig. 1A1). This feature was not observed on the other two sources of human keratin (nail and hair). These results extend our knowledge about the course of adhesion of *C. parapsilosis* on keratinized substrates which may help to clarify the pathogenesis of superficial candidiasis.

2.2 Invasion capability

Morphogenesis between yeast and hyphal growth, which facilitates fungal tissue invasion and enables the fungus to evade the defense system of the host is generally accepted as virulence traits of *C. albicans*. For instance, Jayatilake et al. (2008, 2009) have demonstrated that multiple host–fungal interactions such as cavitations, thigmotropism, and morphogenesis take place during candidal tissue invasion.

For non-*albicans Candida* species it is suggested that filamentous forms (hyphae and/or pseudohyphae) also assist in the invasive penetration of physical barriers (reviewed in Jayatilake, 2011). However, relatively little is known about the invasive potential regarding NCAC species. According to the literature, the invasiveness of non-*albicans Candida* species is variable among species. For instance, in RHE model *C. tropicalis* exhibited higher ability to invade this tissue than did *C. parapsilosis* and *C. glabrata* (Jayatilake et al., 2006).

Although it is well established that *C. tropicalis* is a polymorphic fungus few studies have analyzed the importance of its morphology on virulence. Recently, Silva et al. (2010) demonstrated that only filamentous forms of *C. tropicalis* were able to invade an oral epithelium.

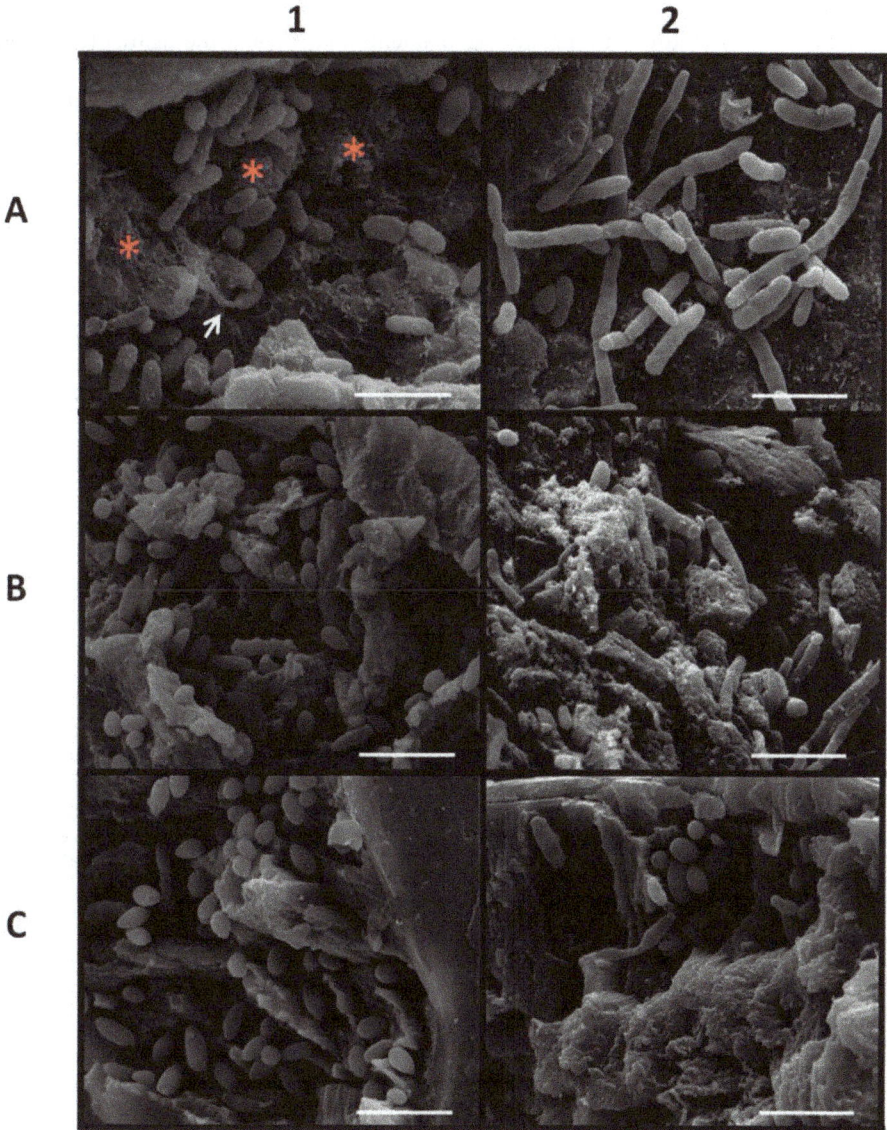

Fig. 1. These figures illustrate the *in vitro* adherence pattern of *Candida parapsilosis* isolates recovered from onychomycoses (1) and tracheal secretion (2) to human keratinous substrates. (A) Stratum corneum. On this substrate the cellular shape and the extend of extracellular material (*) were isolate-dependent. Note that cells from the onychomycosis isolate exhibits an oval shape while that cells from the tracheal secretion isolate displays pseudohyphae forms. (B) Nail. The pattern of cell morphology on nail keratin was also isolate-dependent. (C) Hair. The cellular features were independent to the site of yeast isolation. Short hyphal form (arrow). Scale bars = 100μm.

According to Brown et al. (1999), *C. albicans* cells respond to growth in contact with agar (semi solid matrix) medium by producing filaments that invade the agar. Production of invasive hyphae during growth in synthetic medium may occur by the same mechanism that is involved in production of invasive lesions during candidiasis. Recently, it has been showed that invasive filamentation of *C. albicans* into agar medium is promoted by a cell wall-linked protein (Zucchi et al., 2011). Thus, agar invasion tests enable sorting strains by their degree of invasiveness.

In this study, we employed the agar invasion assay to determine the invasive potential of a switch variant strain, exhibiting a crepe morphotype (Figure 2A), obtained from a clinical *C. tropicalis* isolate recovered from tracheal secretion. For this, cells were grown on the surface of YPD medium for 4 days a 37°C. After colony were washed off the agar surface, with a stream of water and gentle rubbing, cells that had invaded the agar remained as macroscopically visible microcolonies on the surface of the washed plate (Fig. 2B). The parental isolate also invade, although not to the same extent (data not shown). This data suggests that colonies of *C. tropicalis* exhibiting distinct morphologies differ in their capabilities to invade agar. Similar data were observed by *C. parapsilosis* (Laffey and Bluter, 2005) using the agar invasion assay.

Further we analyzed the invaded agar at ultrastructural level, by the employment of fracture technique. Fracture is valuable to reveal internal surfaces and it is performed by dipping the fixed samples in liquid nitrogen and breaking with a sharp scalpel. Growth in YPD shows filamentous forms invading the agar at different planes and angles, as well as yeast cell forms (Figure 3). Thus, SEM may be useful for the detailed analysis of extend and pattern of yeast cells in the course of the invasion process. This is the first report of the employment of SEM to examine the pattern of agar invasion by *Candida* cells.

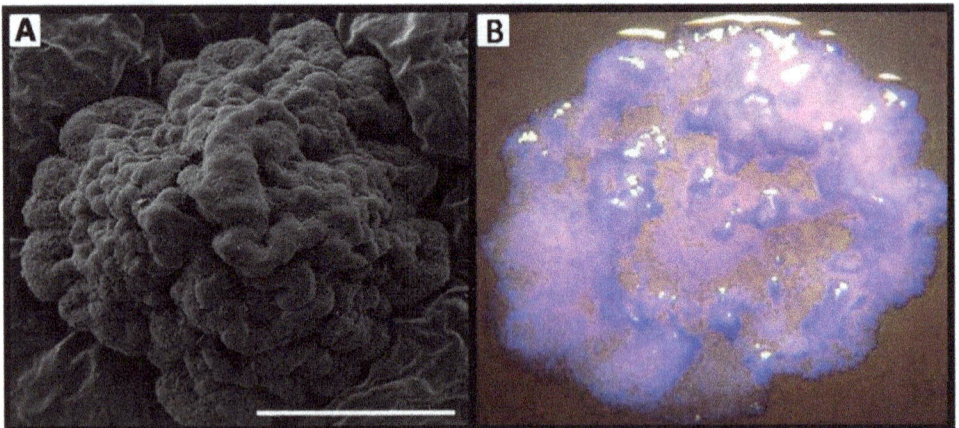

Fig. 2. (A) Scanning electron micrograph shows *C. tropicalis* morphotype crepe following 96 h growth on YPD agar. (B) Photomicrograph of footprint of attached cells after washing out the colony. Scale bar = 1mm (A). 2.5X (B).

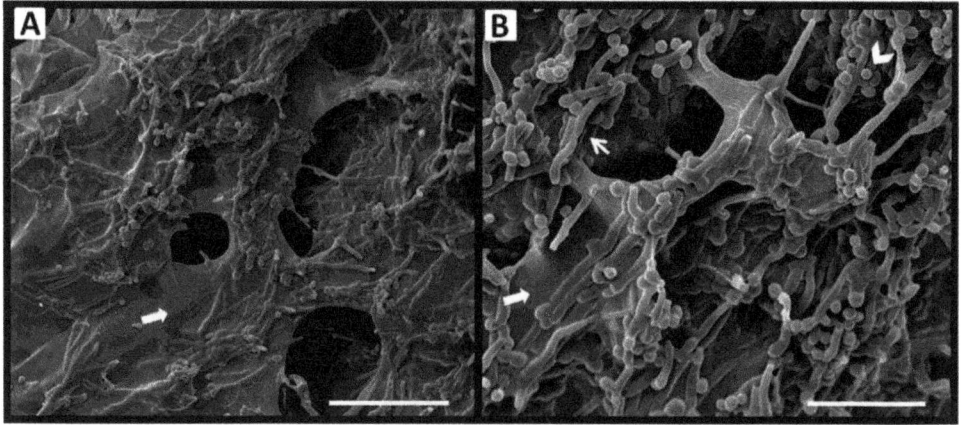

Fig. 3. Scanning electron micrographs of agar invasion by *C. tropicalis* crepe after cells were washed off the agar surface. Cells includes on agar substrate (large arrow) appear like filamentous form (thin arrow) and blastoconidia (head arrow) Scale bars = 100µm (A) and 50µm (B).

2.3 Switching morphotypes: Ultrastructure and morphological types

Phenotypic switching represents an epigenetic state that occurs in a small fraction of the population, is random and reversible. This biological phenomenon is related to the occurrence of spontaneous emergence of colonies with different morphologies that enables the microorganism to undergo rapid microevolution and to adapt to different environments, including various anatomical sites in the human body (reviewed in Soll, 1992). Thus, the switching phenotype event has also being considered a candidal virulence factor (Segal, 2004).

Furthermore, switching has been demonstrated to regulate virulence-associated characteristics in *C. albicans*, such as adhesion, expression of cell surface hydrophobicity and biofilm formation (Kennedy et al., 1988, Antony et al., 2009, Lohse and Johnson, 2009). Concerning NCAC species, *C. parapsilosis* distinct switch phenotypes exhibited differential ability to form biofilm on polystyrene surfaces (Laffey and Butler, 2005). For *C. tropicalis*, França et al. (2010) also found a correlation in phenotypic switching and biofilm formation.

For fungi this event is defined as the reversible change manifested as altered colony morphology at a rate higher than the somatic mutation rate (reviewed in Soll, 1992).

In yeast, phenotypic switching was originally described in *C. albicans* strains (Soll, 1992), but is also known to exist in other *Candida* species, such as *C. tropicalis* (Soll et al., 1988, França et al., 2010).

Ultrastructural investigations revealed a relationship between *C. albicans* switched variant colonies and microstructure (Radford et al., 1994, 1997). In a pioneer study we report the presence of extracellular material, resembling a biofim-like colony, throughout the development of *C. tropicalis* switch colonies, suggesting that its presence is correlated with the complex architecture of colonies (França et al. 2010).

SEM was successfully employed for the analyses of whole *Candida* colonies architecture (França et al. 2010, Furlaneto et al., 2012). Additional studies on switching event in clinical isolates of *C. tropicalis* are in progress. For instance, different architectures exhibited by *C. tropicalis* colonies morphotypes are shown in Figure 4. The smooth phenotype colony (Figure 4A) showed a hemispherical shape character, while the rough phenotype exhibited more complex architecture and was characterized by the presence of deep central and peripheral depressions areas (Figure 4B). The irregular wrinkled colony was characterized by a highly wrinkled centre and an irregular periphery (Figure 4C). Crepe colony was characterized by the presence of aerial hyphae on the colony surface (Figure 4D). The preparation of colonies by a freeze-drying technique allowed their architecture preservation with maintenance of the phenotypes observed at lower magnitude (data not shown).

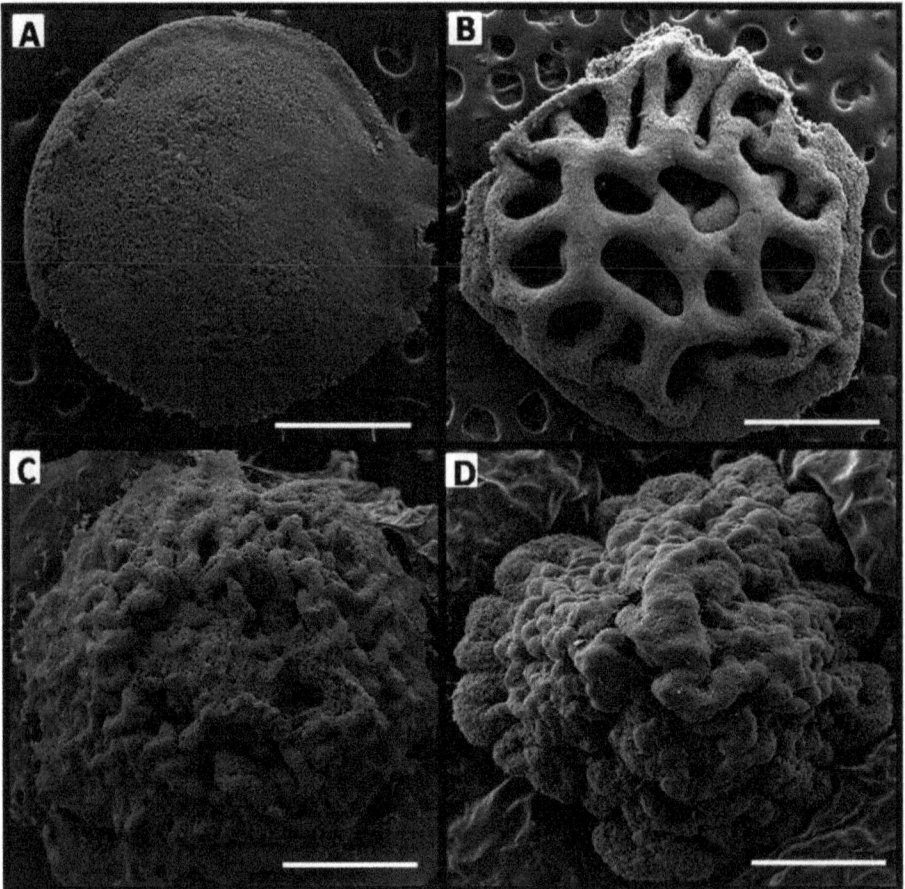

Fig. 4. Different architectures exhibited by *C. tropicalis* colonies morphotypes following 96 h incubation on YPD agar. (A) Smooth colony. (B) Rough colony, shows cells in a tridimensional disposition. (C) Cells are establishing an irregular wrinkled colony and (D) crepe colony shows the homogeneous substance coating its surface, besides filamentous forms. Scale bars = 1mm.

The ultrastructural analysis allowed the observation of the arrangement of individual cells within the colonies. After 4 days of colony development, the whole smooth and irregular wrinkled colonies consisted entirely of yeast cells (not shown). The crepe colony phenotype also comprised mainly yeast cells as observed at depressions areas (Figure 5A). Most interesting was the presence of extracellular material forming a biofilm-like colony where many of the cells were almost hidden by this material. It was observed as fibrils, with enlarged structures, connecting neighbouring cells (Figure 5B).

A *C. tropicalis* variant exhibiting a myceliated phenotype is shown in Figure 6. The whole colony surface is formed by aerial mycelia with a prominent centre (Figure 6A). The aerial hyphae showed a compact nature that is composed by hyphae and blastoconidia (Figure 6B).

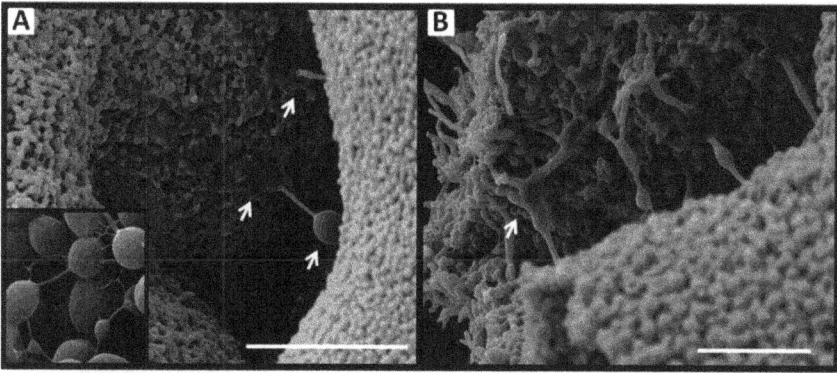

Fig. 5. *C. tropicalis* crepe morphotype following 96 h incubation on YPD agar.
(A) Extracellular material (arrows) is seeing forming a biofilm-like colony. (B) Fractured colonies reveal details of extracellular material recovering and connecting cells. Scale bar = 100μm (A), 50μm (B). Inset shows fibrilar extracellular material connecting cells (5000x)

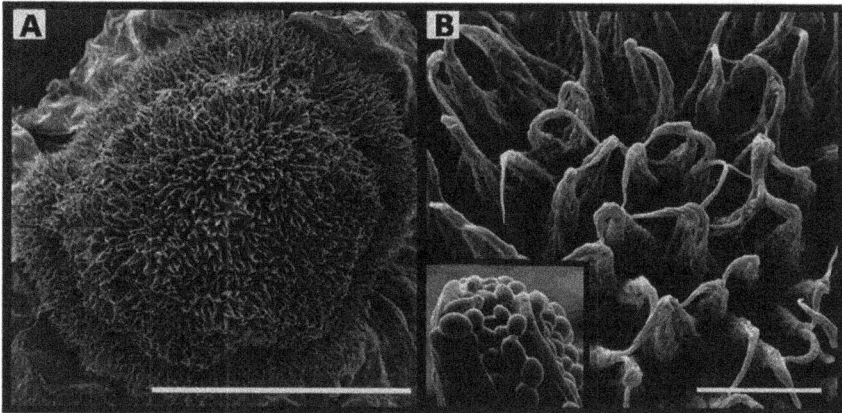

Fig. 6. Electron micrographs of the *C. tropicalis* myceliated morphotype following 96 h incubation on YPD agar. (A) Whole myceliated colony, (B) In higher magnification aerial hyphae showing compact nature. Scale bars = 1mm (A), 100μm (B). Inset shows detail of a distal end of an aerial hyphae that is composed by hyphae and blastoconidia (5000x)

2.4 Effect of antifungal compounds on yeast morphology

The therapy of deep fungal infections, particularly those caused by opportunistic pathogens, including *Candida* species, remains a difficult medical problem. Besides, compared with antibiotics, the development of antifungal agents has been relatively limited. Widespread use of antifungal agents could be an explanation for the emergence of the more resistant non-*albicans* species of *Candida* (Pfaller & Diekema, 2004). Fluconazole is a systemic antifungal drug effective against most of the *Candida* species; however, the emergence of fluconazole resistance has been reported in NCAC species, particularly *C. tropicalis* and *C. parapsilosis* (Yang et al., 2004, Pereira et al., 2010, Oxman et al., 2010, Bruder-Nascimento et al., 2010). The emergence of yeast species with decreased susceptibility to contemporary antifungal regimens demonstrates the need for new antifungal agents.

Many studies have addressed the search for natural compounds with antifungal activity. As an example, Duarte et al. (2005) screened 35 medicinal plants commonly used in Brazil for anti-*Candida albicans* activity. In this context, the flavonoid baicalein, originally isolated from the roots of *Scutellaria baicalensis* Georgi (a Chinese herb) has been tested against *C. albicans* (Cao et al., 2008, Huang et al., 2008). According to these authors, antifungal activity was observed on free cells as well as on biofilm.

Ultrastructural investigations of the effect of natural compounds on morphology are limited. However, SEM analysis allowed the observation of irregular budding patterns and pseudohyphae formation in *C. albicans* type strain treated with compounds isolated from pomegranate peels (Endo et al., 2010). In contrast, the exposure of cells from the same type strain to berberine (alkaloid found in medicinal herbs) did not affect cell morphology (Iwazaki et al., 2010).

We employed the scanning electron microscopy to evaluate the effect of baicalein alone and in combination with fluconazole on the morphology of *C. parapsilosis* and *C. tropicalis*. For *C. parapsilosis*, SEM analyses showed control cells (untreated cells) with a normal budding profile where no extracellular material was seen (not shown). After exposure to baicalein alone, the general aspect of the cells was not modified, however, a profusely flocculent extracellular material was seeing connecting yeast cells (Fig. 7A). Similar pattern was observed for cells exposed to baicalein in combination with fluconazole, although, the amount of extracellular was visible higher (Fig. 7B). SEM images also showed markedly reduced number of organisms due to baicalein.

On the other hand, the data obtained in this study showed that *C. tropicalis* underwent morphological alterations visible by SEM when treated with subinhibitory concentration of baicalein alone and in association with fluconazole.

Untreated cells (control) consisted of blastoconidia and pseudohyphae (not shown). For cells exposed to baicalein alone we observed the presence of elongated cells as well as a great capacity for producing pseudohyphae (Fig. 8A). Cells exposed to baicalein in combination with fluconazole showed an oval shape with profusely flocculent extracellular material connecting yeast cells (Fig. 8B). These data, suggest different inter-species response to baicalein alone as well as to in association with fluconazole.

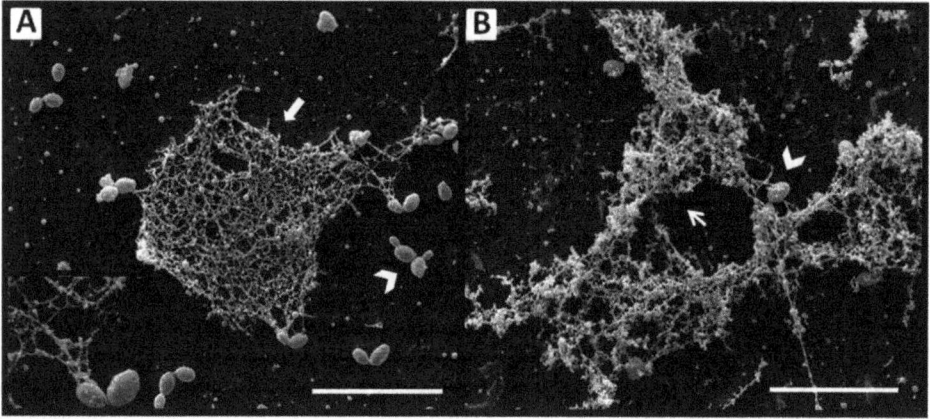

Fig. 7. *C. parapsilosis* treated with baicalein alone (MIC50) and in combination with fluconazole. (A) baicalein, (B) baicalein plus fluconazole. *C. parapsilosis* cells display a typical oval shape (heads arrow). Some of them are involved by an extracellular material constituted by irregular fibrils disposed as a network (large arrow). Note that fibrils appear like beads on a string (inset). In B, a flocculent extracellular material (thin arrow) is seeing connecting cells. Scale bar = 20µm. Inset =12,000 X

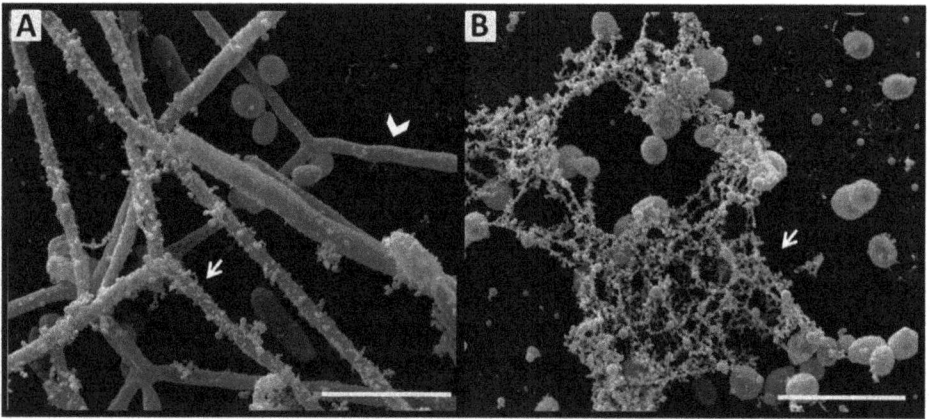

Fig. 8. *C. tropicalis* treated with baicalein alone (A) and in combination with fluconazole (B). Hyphae form is predominant in treatment with baicalein (head arrow), while in combination with antifungal blastoconidia prevails. Flocculent extracellular material (arrows) is seen surrounding hyphae and a prominent one is connecting blastoconidia. Scale bar = 20µm.

3. Conclusion

Yeast pathogenicity arises through complex interactions between the organism's virulence characteristics and the host's response. *Candida* species can exhibit several virulence factors such as adherence, biofilm formation and phenotypic switching that increase their persistence within the host as well as allow adapting to different anatomical sites in the human body. Therefore, the increase in the incidence and antifungal resistance of NCAC species, specifically, *C. parapsilosis* and *C. tropicalis*, and the unacceptably high morbidity and mortality associated with these species, make it essential to further enhance our knowledge on the virulence and resistance mechanisms associated with these species. An understanding of the virulence determinants of these species would provide insight into their pathogenic mechanisms. Our studies have shown that ultrastructural investigations at SEM of some of these virulence traits may help to elucidate general mechanisms of fungal virulence. Another approach that is in progress in our laboratory is the processing of yeast samples to analyse at transmission electron microscopy that may reveal inner ultrastructure at higher resolution.

4. Acknowledgments

The author gratefully acknowledges to all colleagues and students joined to the medical mycology group at State University of Londrina and Technological Federal University of Paraná in Londrina, Paraná-Brazil. Special thanks to the group members: Ana Flávia Leal Specian and Rosapa Serpa that contributed with results obtained during their master thesis work. The authors also thank Osvaldo Capello for technical support. Team´s members using original results made all the figures´ composition. This work was partially supported by Conselho Nacional de Desenvolvimento Científico e Tecnológico –CNPq-Brazil.

5. References

Anthony, G.; Saralaya, V.; Gopalkrishna Bhat, K; Shalini Shenoy, M. & Shivananda, P.G. (2009). Effect of phenotypic switching on expression of virulence factors by *Candida albicans* causing candidiasis in diabetic patients. *Revista Iberoamericana de Micologia*, Vol.26, No.3 (September 2009), pp. 202-205, ISSN 1130-1406.

Almirante, B.; Rodríguez, D. & Cuenca-Estrella, M. (2006). Epidemiology, risk factors, and prognosis of *Candida parapsilosis* bloodstream infections: case-control population-based surveillance study of patients in Barcelona, Spain, from 2002 to 2003. *Journal of Clinical Microbiology*, Vol.44, No.5 (May2006), pp. 1681-1685, ISSN 0095-1137.

Barnes, W.G.; Flesher, A.; Berger, A.E. & Arnold, D.J. (1971). Scanning electron microscopic studies of *Candida albicans*. *Journal of Bacteriology*, Vol.106, No.1 (April 1971), pp. 276-280, ISSN 0021-9193.

Bozzola, J.J. & Russel, L.D. 1999. *Electron microscopy. Principles and Techniques for Biologists*. (2nd ed.), Jones and Bartlett Publishers, ISBN 0-7637-0192-0, Sudbury, Massachusetts, USA.

Branchini, M.L.; Pfaller, M.A.; Rhine-Chalberg, J.; Frempong, T. & Isenberg, H.D. (1994). Genotypic variation and slime production among blood and catheter isolates of *Candida parapsilosis*. *Journal of Clinical Microbiology*, Vol.32, No.2 (February 1994), pp. 452-456, ISSN 0095-1137.

Brown, D.H.Jr.; Giusani, A.D.; Chen, X. & Kumamoto, C.A. (1999). Filamentous growth of *Candida albicans* in response to physical environmental cues and its regulation by the unique *CZF1* gene. *Molecular Microbiology*, Vol.34, No.4 (November 1999), pp. 651-662, ISSN 1365-2958.

Bruder-Nascimento, A.; Camargo, C.H.; Sugizaki, M.F.; Sadatsune, T.; Montelli, A.C.; Mondelli, A.L.; Bagagli, E. (2010). Species distribution and susceptibility profile of

Candida species in a Brazilian public tertiary hospital. *BCM Research Notes*, Vol.3, No.1 (January 2010), pp. 1-5, ISSN 1756-0500.

Cao, Y.Y.; Dai, B.D.; Wang, Y.; Huang, S.; Xu, Y.G.; Cão, Y.B.; Gao, P.H.; Zhu, Z.Y. & Jiang, Y.Y. (2008). In vitro activity of baicalein against *Candida albicans* biofilms. *International Journal of Antimicrobial Agents*, Vol.32, No.1 (July 2008), pp. 73–77, ISSN 0924-8579.

Colombo, A.L.; Nucci, M.; Park, B.J.; Nouer, A.S.; Arthington-Skaggs, B.; da Matta, D.A.; Warnock, D. & Morgan, J. (2006). Epidemiology of candidemia in Brazil: a nationwide sentinel surveillance of candidemia in eleven medical Centers. *Journal of Clinical Microbiology*, Vol.44: No.8 (August 2006), pp. 2816-23, ISSN 0095-1137.

De Bernardis, F.; Mondello, F.; San Millàn, R.; Ponón J. & Cassone, A. (1999). Biotyping and virulence properties of skin isolates of *Candida parapsilosis*. *Journal of Clinical Microbiology*, Vol.37, No.11 (November 1999), pp. 3481-3486, ISSN 0095-1137.

Douglas, J. (2003). *Candida* biofilms and their role in infection. *Trends in Microbiolgy*, Vol.11, No.1 (January 2003), pp.: 30–36, ISSN 0966-842X.

Duarte, M.C.T.; Figueira, G.M.; Sartoratto, A.; Rehder, V.L.G. & Delarmelina C. (2005). Anti-*Candida* activity of Brazilian medicinal plants. *Journal of Ethnopharmacology*, Vol.97, No.28 (February), pp. 305-311, ISSN 0378-8741.

Endo, E.H.; Cortez, D.A.G.; Ueda-Nakamura, T.; Nakamura, C.V. & Filho, B.P.D. (2010). Potent antifungal activity of extracts and pure compound isolated from pomegranate peels and synergism with fluconazole against *Candida albicans*. *Research in Microbiology*, Vol.161, No.7 (September 2010), pp. 534-540, ISSN 0923-2508.

Figueiredo, V.T.; Santos, D.A.; Resende, M.A. & Hamdan, J.S. (2007). Identification and in vitro antifungal susceptibility testing of 200 clinical isolates of *Candida* spp. responsible for fingernail infections. *Mycopathologia*, Vol.164, No.1 (June 2007), pp.27-33, ISSN 0301-486X.

França, E.J.G.; Andrade, C.G.T.J.; Furlaneto-Maia, L.; Serpa, R.; Oliveira, M.T.; Quesada, R.M.B. & Furlaneto, M.C. (2011). Ultrastructural architecture of colonies of different morphologies and biofilm produced by phenotypic switching of *Candida tropicalis*. *Micron*, Vol.42, No.7 (October 2011), pp. 726-732, ISSN 0968-4328.

França, E.J.G.; Furlaneto-Maia, L.; Quesada, R.M.B.; Favero, D.; Oliveira, M.T. & Furlaneto, M.C. (2010). Haemolytic and proteinase activities in clinical isolates of *Candida parapsilosis* and *Candida tropicalis* with reference to the isolation anatomic site. *Mycoses*, Vol.54, No.4 (July 2011), pp. e44-e51, ISSN 1439-0507.

Furlaneto, M.C.; Andrade, C.G.T.J.; Aragão, P.H.A.; França, E.J.G.; Moralez, A.T.P. & Ferreira, L.C.S. (2012). Scanning Electron Microscopy as tool for the analyses of colonies architecture of different morphologies produced by phenotypic switching of a human pathogenic yeast *Candida tropicalis*. *Journal of Physics. Conference Series*, ISSN 1742-6588, in press.

Hayat, M.A. 2000. *Principles and Techniques of Electron Microscopy. Biological Applications*. (4th ed.), Cambridge University Press, ISBN 0-521-63287-0, USA.

Haynes, K. (2001). Virulence in *Candida* species. *Trends in Microbiology*, Vol.9, No.12 (December 2001), pp. 591-596, ISSN 0966-842X.

Huang, S.; Cao, Y.Y.; Dai, B.D.; Sun, X.R.; Zhu, Z.Y.; Cao, Y.B.; Wang, Y.; Gao, P.H. & Jiang, Y.Y. (2008). In vitro synergism of fluconazole and baicalein against clinical isolates of *Candida albicans* resistant to fluconazole. *Biological Pharmaceutical Bulletim*, Vol.31, No.12 (December 2008), pp. 2234-2236, ISSN 1347-5215.

Iwazaki, R.S.; Endo, E.H.; Ueda-Nakamura, T.; Nakamura, C.V.; Garcia, L.B. & Filho, B.P.D. (2010). In vitro antifungal activity of the berberine and its synergism with fluconazole. *Antonie van Leeuwenhoek*, Vol.97, No.2 (November 2009), pp. 201-205, ISSN 0003-6072.

Kennedy, M.J.; Rogers, A.L.; Hanselman, L.R.; Soll, D.R. & Yancey, R.J. (1988). Variation in adhesion and cell surface hydrophobicity in *Candida albicans* white and opaque phenotypes. *Mycopathologia*, Vol.102, No.3 (June 1988), pp. 149-156, ISSN 0301-486X.

Jayatilake, J.A.M.S. (2011). A review of the ultrastructural features of superficial candidiasis. *Mycopathologia*, Vol.171, No.4 (October 2010), pp. 235-250, ISSN 0301-486X.

Jayatilake, J.A.; Samaranayake, Y.H.; Cheung, L.K. & Samaranayake, L.P. (2006). Quantitative evaluation of tissue invasion by wild type, hyphal and SAP mutants of *Candida albicans*, and non-albicans *Candida* species in reconstituted human oral epithelium. *Journal of Oral Pathology and Medicine*, Vol.35, No. 8 (August 2006), pp. 481–491, ISSN 1600-0714.

Jayatilake, J.A.M.; Samaranayake, Y.H. & Samaranayake, L.P. (2008). A comparative study of candidal invasion in rabbit tongue mucosal explants and reconstituted human oral epithelium. *Mycopathologia*, Vol.165, No.6 (March 2008), pp. 373–380, ISSN 0301-486X.

Jayatilake, J.A.M.; Tilakaratne, W.M. & Panagoda, G. J. (2009). Candidal onychomycosis: A Mini-Review. *Mycopathologia*, Vol.168, No.4 (May 2009), pp. 165–173, ISSN 0301-486X.

Krcmery, V. & Barnes, A. (2002). Non-*albicans Candida* spp. causing fungaemia pathogenicity and antifungal resistance. *Journal of Hospital Infection*, Vol.50, No.4 (April 2002), pp. 243-260, ISSN 0195-6701.

Laffey, S.F. & Butler, G. (2005). Phenotype switching affects biofilm formation by *Candida parapsilosis*. *Microbiology*, Vol.151, No.4 (January 2005), pp. 1073-1082, ISSN 0002-7739.

Lee, R.E. 1993. *Scanning Electron Microscopy and X-Ray Microanalysis*. PTR Prentice-Hall, ISBN 0-13-813759-5, New Jersey, USA.

Lima-Neto, R.G.; Beltrão, E.I.; Oliveira, P.C. & Neves, R.P. (2011). Adherence of *Candida albicans* and *Candida parapsilosis* to epithelial cells correlates with fungal cell surface carbohydrates. *Mycoses*, Vol.54, No.1 (January 2011), pp. 23-39, ISSN 0933-7407.

Lohse, M.B. & Johnson, A.D. (2009). White–opaque switching in *Candida albicans*. *Current Opininion in Microbiology*, Vol.12, No. 6 (December 2009), pp. 650-654, ISSN 1369-5274.

Martins, E.A.; Guerrer, L.V.; Cunha, K.C.; Soares, M.M.C.; & Almeida, M.T.G. (2007). Onychomycosis: clinical, epidemiological and mycological study in the municipality of São José do Rio Preto. *Revista da Sociedade Brasileira de Medicina Tropical*, Vol.40, No.5 (October 2007), pp. 596-598, ISSN 0037-8682.

Manzano-Gayosso, P.; Hernández-Hernández, F.; Méndez-Tovar, L.J.; Palacios-Morales, Y.; Córdova-Martínez, E.; Bazán-Mora, E. & Martinez-López, R. (2008). Onychomycosis incidence in type 2 diabetes mellitus patients. *Mycopathologia*, Vol.166, No.1 (March 2008), pp. 41-45, ISSN 0301-486X.

Nucci, C. & Colombo, A.L. (2007). Candidemia due to *Candida tropicalis*: clinical, epidemiologic, and microbiologic characteristics of 188 episodes occurring in tertiary care hospitals. *Diagnostic Microbiology and Infectious Diseases*, Vol.58, No.1 (May 2007), pp. 77-82, ISSN 0732-8893.

Oliveira, M.T.; Specian, A.F.; Andrade, C.G.T.J.; França, E.J.G.; Furlaneto-Maia, L. & Furlaneto, M.C. (2010). Interaction of *Candida parapsilosis* with human hair and nails surfaces revealed by scanning electron microscopy analysis. *Micron*, Vol.41, No.6 (August 2010), pp. 604-606, ISSN 0868-4328.

Oxman, D.A.; Chow, J.K.; Frendl, G.; Hadley, S.; Hershkovitz, S.; Ireland, P.; McDermott, L.A.; Tsai, K.; Marty, F.M.; Kontouiannis, D. & Golan, Y. (2010). Candidaemia associated with decreased in vitro fluconazole susceptibility: is *Candida* speciation predictive of the susceptibility pattern? *Journal of Antimicrobial Chemotherapy*, Vol.65, No.7 (April 2010), pp. 1460-1465, ISSN 0305-7453.

Panagoda, G.J.; Ellepola, A.N. & Samaranayake, L.P. (2001). Adhesion of *Candida parapsilosis* to epithelial and acrylic surfaces correlates with cell surface hydrophobicity. *Mycoses*, Vol.44, No.1-2 (March 2001), pp. 29–35, ISSN 1439-0507.

Pereira, G.H.; Mulles, P.R.; Szeszs, M.W.; Levin, A.S. & Melhem, M.S.C. (2010). Five-year evaluation of bloodstream yeast infections in a tertiary hospital: the predominance of non-*C. albicans Candida* species. *Medical Mycology*, Vol.48, No.6 (September 2010), pp. 839-842, ISSN 1369-3786.

Pfaller, M.A. & Diekema, D.J. (2004). Twelve years of fluconazole in clinical practice: global trends in species distribution and fluconazole susceptibility of bloodstream isolates of *Candida. Clinical Microbiology and Infection*, Vol.10, No.1 (January 2004), pp. 11–23.

Radford, D.R.; Challacombe, S.J. & Walter, J.D. (1994). A scanning electron microscopy investigation of the structure of colonies of different morphologies produced by phenotypic switching of *Candida albicans*. *Journal of Medical Microbiology*, Vol.40, No.6 (June, 1994), pp. 416-423, ISSN 1473-5644.

Radford, D.R.; Challacombe, S.J. & Walter, J.D. (1997). Scanning electron microscopy of the development of structured aerial mycelia and satellite colonies of phenotypically switched Candida albicans. *Journal of Medical Microbiology*, Vol.46, No. 4 (April, 1997), pp. 326-332, ISSN 1473-5644.

Segal, E. (2004). *Candida*, still number one – what do we know and where are we going from there? *Mycoses*, Vol.48, No.1 (April, 2005), pp. 3–11, ISSN 1439-0507.

Silva, S.; Hooper, S.J.; Henriques, M.; Oliveira, R.; Azeredo, J. & Williams, D.W. (2010). The role of secreted aspartyl proteinases in *Candida tropicalis* invasion and damage of oral mucosa. *Clinical Microbiology and Infection*, Vol.17, No. 2 (April 2010), pp. 264–272, ISSN 1469-0691.

Silva, S.; Negri, M.; Henriques, M.; Oliveira, R.; Williams, D.W. & Azeredo, J. (2011). Adherence and biofilm formation of non-*Candida albicans Candida* species. *Trends in Microbiology*, Vol.19, No.5 (May 2011), pp. 241-247, ISSN 0966-842X.

Slutsky, B.; Buffo, J. & Soll, D.R. (1985). High frequency switching of colony morphology in *Candida albicans*. *Science*, Vol.230, No. 4723 (November, 1985), pp. 666-669, ISSN 1535-9778.

Soll, D.R. (1992). High-frequency switching in *Candida albicans*. *Clinical Microbiology Reviews*, Vol.5, No.2 (April, 1992), pp. 183-203, ISSN 0893-8512.

Soll, D.R.; Staebell, M.; Langtimm, C.; Pfaller, M.; Hicks, J. & Gopala Rao, T.V. (1988). Multiple *Candida* strains in the course of a single systemic infection. *Journal of Clinical Microbiology*, Vol.26, No. 8 (August, 1988), pp. 1448-1459, ISSN 0095-1137.

Trofa, D.; Gácser, A. & Nosanchuk, J.D. (2008). *Candida parapsilosis*: an emerging fungal pathogen. *Clinical Microbiology Review*, Vol.21, No.4 (October 2008), pp. 606-625, ISSN 0893-8512.

van Asbeck, E.C.; Clemins, K.V. & Stevens, D.A. (2009) *Candida parapsilosis*: a review of its epidemiology, pathogenesis, clinical aspects, typing and antimicrobial susceptibility. *Critical Review in Microbiology*, Vol.35, No.4 (November, 2009), pp. 283–309, ISSN 1040-841X.

Yang, Y.L.; Ho, Y.A.; Cheng, H.H.; Ho, M. & Lo, H.J. (2004). Susceptibilies of *Candida* species to amphotericin B and fluconazole: the emergence of fluconazole resistance in *Candida tropicalis*. *Infection Control Hospital Epidemiology*, Vol.25, No.1 (January, 2004), pp. 60-64, ISSN 0899-823X.

Zucchi, P.C.; Davis, T.R. & Kumamoto, C.A. (2010). A *Candida albicans* cell wall-linked protein promotes invasive filamentation into semi-solid medium. *Molecular Microbiology*, Vol.76, No.3 (May 2010), pp. 733-748, ISSN 1365-2958.

Permissions

The contributors of this book come from diverse backgrounds, making this book a truly international effort. This book will bring forth new frontiers with its revolutionizing research information and detailed analysis of the nascent developments around the world.

We would like to thank Viacheslav Kazmiruk, for lending his expertise to make the book truly unique. He has played a crucial role in the development of this book. Without his invaluable contribution this book wouldn't have been possible. He has made vital efforts to compile up to date information on the varied aspects of this subject to make this book a valuable addition to the collection of many professionals and students.

This book was conceptualized with the vision of imparting up-to-date information and advanced data in this field. To ensure the same, a matchless editorial board was set up. Every individual on the board went through rigorous rounds of assessment to prove their worth. After which they invested a large part of their time researching and compiling the most relevant data for our readers. Conferences and sessions were held from time to time between the editorial board and the contributing authors to present the data in the most comprehensible form. The editorial team has worked tirelessly to provide valuable and valid information to help people across the globe.

Every chapter published in this book has been scrutinized by our experts. Their significance has been extensively debated. The topics covered herein carry significant findings which will fuel the growth of the discipline. They may even be implemented as practical applications or may be referred to as a beginning point for another development. Chapters in this book were first published by InTech; hereby published with permission under the Creative Commons Attribution License or equivalent.

The editorial board has been involved in producing this book since its inception. They have spent rigorous hours researching and exploring the diverse topics which have resulted in the successful publishing of this book. They have passed on their knowledge of decades through this book. To expedite this challenging task, the publisher supported the team at every step. A small team of assistant editors was also appointed to further simplify the editing procedure and attain best results for the readers.

Our editorial team has been hand-picked from every corner of the world. Their multi-ethnicity adds dynamic inputs to the discussions which result in innovative outcomes. These outcomes are then further discussed with the researchers and contributors who give their valuable feedback and opinion regarding the same. The feedback is then collaborated with the researches and they are edited in a comprehensive manner to aid the understanding of the subject.

Apart from the editorial board, the designing team has also invested a significant amount of their time in understanding the subject and creating the most relevant covers. They scrutinized every image to scout for the most suitable representation of the subject and create an appropriate cover for the book.

The publishing team has been involved in this book since its early stages. They were actively engaged in every process, be it collecting the data, connecting with the contributors or procuring relevant information. The team has been an ardent support to the editorial, designing and production team. Their endless efforts to recruit the best for this project, has resulted in the accomplishment of this book. They are a veteran in the field of academics and their pool of knowledge is as vast as their experience in printing. Their expertise and guidance has proved useful at every step. Their uncompromising quality standards have made this book an exceptional effort. Their encouragement from time to time has been an inspiration for everyone.

The publisher and the editorial board hope that this book will prove to be a valuable piece of knowledge for researchers, students, practitioners and scholars across the globe.

List of Contributors

Rahul Mehta
University of Central Arkansas, USA

Renaud Podor, Johann Ravaux and Henri-Pierre Brau
Institut de Chimie Séparative de Marcoule, UMR 5257 CEA-CNRS-UM2-ENSCM, Site de Marcoule, Bagnols sur Cèze cedex, France

Lahcen Khouchaf
Université Lille - Nord de France, Ecole des Mines de Douai, Douai, France

Zhongwei Chen and Yanqing Yang
State Key Laboratory of Solidification Processing, Shaanxi Materials Analysis & Research Center, Northwestern Polytechnical University, Xi'an, P.R. China

Huisheng Jiao
Oxford Instruments Shanghai Office, Shanghai, P.R. China

Timothy E. Kidd
Physics Department, University of Northern Iowa, Cedar Falls, IA, USA

Anna Rudawska
Lublin University of Technology, Poland

Jun Kawai, Yasukazu Nakaye and Susumu Imashuku
Department of Materials Science and Engineering, Kyoto University, Sakyo-ku, Kyoto, Japan

Guillermo San Martín and María Teresa Aguado
Departamento de Biología (Zoología), Facultad de Ciencias, Calle Darwin 2, Universidad Autónoma de Madrid, Canto Blanco, Madrid, Spain

Hong-Wei Xiao and Zhen-Jiang Gao
College of Engineering, China Agricultural University, Qinghua Donglu, Beijing, China

Bülent Gökçe
Ege University, School of Dentistry, Department of Prosthodontics, Turkey

Pinky Tripathi and Ajay Kumar Mittal
Banaras Hindu University, India

Takehiko Kenzaka and Katsuji Tani
Osaka Ohtani University, Faculty of Pharmacy, Japan

Cynthya Elizabeth González and Monika Inés Hamann
Centro de Ecología Aplicada del Litoral, Consejo Nacional de Investigaciones Científicas y Técnicas, Argentina

Cristina Salgado
Servicio de Microscopia Electrónica de Barrido, Universidad Nacional del Nordeste (UNNE), Corrientes, Argentina

Yasko Kodama
Instituto de Pesquisas Energéticas e Nucleares – IPEN–CNEN/SP, Brazil

Claudia Giovedi
Centro Tecnológico da Marinha em São Paulo – CTMSP, Brazil

Márcia Cristina Furlaneto, Emanuele Júlio Galvão de França and Alane Tatiana Pereira Moralez
Department of Microbiology, Brazil

Célia Guadalupe Tardeli de Jesus Andrade
Electronic Microscopy and Microanalysis Laboratory, State University of Londrina (UEL), Brazil

Luciana Furlaneto-Maia
Technological Federal University of Paraná (UTFPR), Londrina-PR, Brazil

www.ingramcontent.com/pod-product-compliance
Lightning Source LLC
Chambersburg PA
CBHW070731190326
41458CB00004B/1127